NANOTECHNOLOGY SCIENCE AND TECHNOLOGY

NANOFLUIDS

SYNTHESIS, PROPERTIES AND APPLICATIONS

NANOTECHNOLOGY SCIENCE AND TECHNOLOGY

Additional books in this series can be found on Nova's website under the Series tab.

Additional e-books in this series can be found on Nova's website under the e-book tab.

NANOTECHNOLOGY SCIENCE AND TECHNOLOGY

NANOFLUIDS

SYNTHESIS, PROPERTIES AND APPLICATIONS

S. M. SOHEL MURSHED, PH.D.
AND
C. A. NIETO DE CASTRO, PH.D.
EDITORS
University of Lisbon, Portugal

New York

Copyright © 2014 by Nova Science Publishers, Inc.

All rights reserved. No part of this book may be reproduced, stored in a retrieval system or transmitted in any form or by any means: electronic, electrostatic, magnetic, tape, mechanical photocopying, recording or otherwise without the written permission of the Publisher.

For permission to use material from this book please contact us:
Telephone 631-231-7269; Fax 631-231-8175
Web Site: http://www.novapublishers.com

NOTICE TO THE READER

The Publisher has taken reasonable care in the preparation of this book, but makes no expressed or implied warranty of any kind and assumes no responsibility for any errors or omissions. No liability is assumed for incidental or consequential damages in connection with or arising out of information contained in this book. The Publisher shall not be liable for any special, consequential, or exemplary damages resulting, in whole or in part, from the readers' use of, or reliance upon, this material. Any parts of this book based on government reports are so indicated and copyright is claimed for those parts to the extent applicable to compilations of such works.

Independent verification should be sought for any data, advice or recommendations contained in this book. In addition, no responsibility is assumed by the publisher for any injury and/or damage to persons or property arising from any methods, products, instructions, ideas or otherwise contained in this publication.

This publication is designed to provide accurate and authoritative information with regard to the subject matter covered herein. It is sold with the clear understanding that the Publisher is not engaged in rendering legal or any other professional services. If legal or any other expert assistance is required, the services of a competent person should be sought. FROM A DECLARATION OF PARTICIPANTS JOINTLY ADOPTED BY A COMMITTEE OF THE AMERICAN BAR ASSOCIATION AND A COMMITTEE OF PUBLISHERS.

Additional color graphics may be available in the e-book version of this book.

Library of Congress Cataloging-in-Publication Data

ISBN: 978-1-63321-677-8

Published by Nova Science Publishers, Inc. † New York

CONTENTS

Preface		**vii**
Chapter 1	Nanofluids Preparation Methodology *M. J. Lourenço and S. I. Vieira*	**1**
Chapter 2	Heat Transfer and Transport Properties of Nanofluids *Amyn S. Teja and Pramod Warrier*	**29**
Chapter 3	Thermophysical Properties and Heat Transfer Characteristics of Carbon Nanotubes Dispersed Nanofluids *S. M. Sohel Murshed and C. A. Nieto de Castro*	**53**
Chapter 4	Thermal Properties of Magnetic Nanofluids *P. D. Shima, Baldev Raj and John Philip*	**77**
Chapter 5	Viscosity of Nanofluids Containing Metal Oxide Nanoparticles *S. M. Sohel Murshed, F. J. V. Santos* *and C. A. Nieto de Castro*	**109**
Chapter 6	Capillary Wetting of Nanofluids *Milad Radiom, Chun Yang and Weng Kong Chan*	**133**
Chapter 7	Convective Heat Transfer of Nanofluids in Tubes *J. P. Meyer and C. C. Tang*	**155**
Chapter 8	Pool Boiling Heat Transfer of Nanofluids *Ehsan Ebrahimnia-Bajestan, Omid Mahian,* *Ahmet Selim Dalkilic and Somchai Wongwises*	**193**
Chapter 9	Nanofluids in Droplet-Based Microfluidics *S. M. Sohel Murshed and Nam-Trung Nguyen*	**215**
Chapter 10	Nanofluid-Based Optical Engineering: Fundamentals and Applications *R. A. Taylor and Y. L. Hewakuruppu*	**237**
Chapter 11	Progress and Challenges in Nanofluids Research *C. A. Nieto de Castro and S. M. Sohel Murshed*	**261**
Index		**279**

PREFACE

This book is the outcome of remarkable contributions from the renowned experts in the field of nanofluids and covers the entire spectrum of nanofluids starting from their preparation and various properties to potential applications. As an emerging field and having fascinating thermophysial and heat transfer properties as well as highly demanding numerous applications nanofluids have sparked immense interest worldwide and have been a subject of explosion of research in recent years which can be evidenced from the large number of nanofluids related research publications. Thus it is timely and of paramount importance to initiate and successfully complete the concept of a book edition on this rapidly growing field. Besides the state-of-art reviews and critical analysis on numbers of key areas of nanofluids this book presents comprehensive experimental and theoretical research efforts on the thermal conductivity, viscosity, convective heat transfer, capillary wetting, and transport properties of nanofluids. Study on application of nanofluids in microfluidic technology is presented and another new area of nanofluid-based optical engineering has also been explored. Featuring contributions from some of the leading researchers in the field this book is a unique reference source and invaluable guide to scientists, researchers, engineers, industrial people, graduate and postgraduate students and academicians across the science and engineering disciplines. It could also form the basis of a graduate-level course on this novel field of nanofluids.

Based on the thematic topics the organization of the book is presented chapter wise as follows:

Chapter 1 titled-nanofluids preparation methodology demonstrates that a careful nanomaterial selection and their characterization are crucial for the nanofluids preparation and stability. Actual existing suppliers must provide "nanomaterial identification" for achieving a special and focus functionality. A generic methodology for obtaining stable nanofluids using ultrasounds energy is presented. Some key points are the permanence of the identity of the nanoparticle without compromising the basic characteristic of the fluid involved. Aspects such as legislation, manipulation and toxicity are briefly referred.

A figure of merit analysis for electronics cooling is presented and the effect of nanoparticle addition on the figure of merit of base heat transfer fluids via their thermal conductivity and viscosity is critically reviewed in Chapter 2. It also describes a modified geometric mean model that takes into account the temperature dependence of the thermal conductivities of the individual phases as well as the size dependence of the thermal conductivity of the particles.

Chapter 3 critically reviews and addresses research progresses in thermophysical properties, heat transfer characteristics, and potential applications of carbon nanotubes dispersed nanofluids. It includes detailed discussions on preparation of these nanofluids, analysis of key thermophysical and heat transfer properties of this specific class of nanofluids and effect of various factors on these properties.

Chapter 4 summarizes the recent research on thermal properties of magnetic nanofluids. The effects of various factors such as volume fraction, magnetic field strength, nanoparticle size, temperature, base fluid material, aggregation and additives on the thermal conductivity of magnetic nanofluids are discussed in detail.

Recent development of viscosity of nanofluids containing metal oxide nanoparticles are critically reviewed in Chapter 5. New findings on rheological properties of two silicone oil-based oxide nanofluids are also reported.

Chapter 6 presents research advances and findings in the area of capillary wetting including spreading dynamics and dynamic contact angle of nanofluids.

A comprehensive experimental study on convective heat transfer of nanofluids in smooth tubes conducted in two different laboratories has been reported in Chapter 7. The convection heat transfer performance of nanofluids are compared.

Chapter 8 reviews studies and analyses recent progress on pool boiling heat transfer of nanofluids. Some representative results are also presented and analyzed.

Chapter 9 deals with the experimental investigations on applications of nanofluids in two droplet-based microfluidic devices. Heat-induced microfluidic T-junction and flow focusing devices are used and effects of nanoparticles, microchannel depths and flow rate on the droplet formation and size manipulation are studied and presented in this chapter.

Chapter 10 explores a new area of nanofluid-based optical engineering. Besides a state-of-the-art analysis on some key aspects of engineering nanoparticle-based optofluidic devices, selected developments in this area has presented in this chapter.

Chapter 11 summarizes the research progress in various areas of nanofluids and addresses the major challenges in this new field. The chapter ends with conclusions and outlook on research developments and challenges in this exciting field of nanofluids.

The editors are fully aware and it is known that some of the topics of the subject are controversial and the opinions expressed in the different chapters may demonstrate the beliefs and knowledge of the authors. The readers are advised to analyze this book critically regardless of the areas of their works. Nanofluids area is in a rapid evolution and the knowledge available is far from being mature.

Finally the editors would like to thank all the authors and co-authors for their time and efforts in preparing these high quality manuscripts for this book.

S. M. Sohel Murshed, Ph.D.
C. A. Nieto de Castro, Ph.D.

In: Nanofluids: Synthesis, Properties and Applications
Editors: S.M. Sohel Murshed, C.A. Nieto de Castro

ISBN: 978-1-63321-677-8
© 2014 Nova Science Publishers, Inc.

Chapter 1

NANOFLUIDS PREPARATION METHODOLOGY[*]

M. J. Lourenço[†] and S. I. Vieira
Centro de Ciências Moleculares e Materiais
Faculdade de Ciências, Universidade de Lisboa
Lisboa, Portugal

ABSTRACT

This chapter demonstrates that a careful nanomaterial selection and their characterization are crucial for the nanofluids preparation and stability. Actual existing suppliers must provide "nanomaterial identification" for achieving a special and focused functionality. A generic methodology for obtaining stable nanofluids, using ultrasounds energy is presented. A key point is the permanence of the identity of the nanoparticle without compromising the basic characteristic of the fluid involved. Aspects such as legislation, manipulation and toxicity are briefly referred.

1. INTRODUCTION

Stable suspensions of nanomaterials in fluids have many interesting properties and their distinct characteristics offer unprecedented potential for many applications. The nanomaterials are nanoscale (1 to 100 nm) entities (nanofibers, nanotubes, nanowires, nanorods, nanosheet or droplets). The common types of nanoparticles are: metals (gold, silver, copper, etc.), metal oxides (alumina, copper oxides, titanium oxides, etc.), carbon in various forms (diamond, graphite, carbon nanotubes, graphene). The main base fluids are water, ethylene glycol, ionic liquids and common refrigerants.

---

[*] Disclaimer: the identification of any commercial product or trade name does not imply endorsement or recommendation by the Centro Ciências Moleculares e Materiais, Lisboa, Portugal.

[†] E-mail: mjlourenco@fc.ul.pt.

The research conducted so far has shown that nanofluids have different properties from those of the base fluids, and if they are useful, they can replace, with advantages, fluids used in many engineering applications.

The enhanced experimental data for thermal conductivity, thermal diffusivity, viscosity and convective heat transfer coefficients compared with the base fluid "oil" or water already demonstrated potential applications in many fields. Thus, it is important the establishment of preparation methods, stability evaluation and create accepted procedures to improve the performance of nanofluids. The preparation and stability of a nanofluid are crucial steps in any experimental study with nanofluids, as confidence on experimental data starts by the confidence of stable samples.

Nanofluids are not simply mixtures of solids in liquids and its success depends on the method of preparation. The simpler nanofluids are bi-phasic systems with a solid phase dispersed in a liquid phase, forming systems that go from simple dispersions to gels and micellar solutions (different nanostructure organization), which require a fundamental property for any engineering or medical application – to be stable over a long period of time. However, today is still not possible to interpret the stability mechanisms of nanofluids, necessary for the wide range of applications foreseen (energy, biomedical and mechanical), a deemed requisite for its success.

There are other special and essential requirements, for example, the durability of the suspension, the prevention of agglomeration, and the constancy of the fluid chemistry.

Some review articles, refer to the progress of research in nanofluids [1-8], but most of the comments are directed to the experimental and theoretical studies of thermophysical properties or convective heat transfer of nanofluids. On the other hand most of the existing experimental data in the literature does not report the total required characterization of the nanomaterial (nanoparticle morphology) the temporal behavior of the suspension, existence or not of reactivity or physical interaction nanomaterial-fluid, how the integrity of the nanoparticle in the fluid base is kept (and if its kept), being very cautious or very reserved around the methods used in the preparation of the nanofluid and how stability was evaluated.

In this chapter we take a broad view for the preparation of nanofluids and alert for the need to establish a known methodology for their preparation and stability testing, accordingly to its intended use. Existing environmental concerns and the controversy that surrounds some nanomaterials must also be taken into account and develop a less controversial selection of nanomaterials that satisfy with the nanofluids the functional requirements, keeping in mind the environmental problems that might be associated. On the other hand it is our goal to leave an alert for the intelligent use of greener nanofluids, as the correct use of nanoparticles saves a lot of resources and produces smaller quantities of CO_2.

If it is possible to select the more functional nanofluids, i.e., those in which their properties increase the overall performance of equipment/systems, without environmental contamination effects, we can state that researchers and science areas involved are fit for the beginning of a new era of research: the nano-ecological and economic research. The term economic here only attaches to the resource savings that nanomaterials can ensure in the very near future.

In the production of nanomaterials used in the preparation of nanofluids, the suppliers / producers must select the most favorable procedures to obtain nanomaterials without contamination and with particle size distribution better defined. In nanosciences, some impurity contamination may damage the stability and durability of a nanofluid. We should reflect on some questions: Why use silver as antibacterial if small amounts of nanosilver have the same or a better effect? Is there a legislation on this subject? Are there toxicity data on silver nanofluids?

2. NANOFLUIDS PREPARATION METHODS

Generally we can define the existence of two methods of preparation/obtaining nanofluids: The one step and the two steps methods. The first consists in the combination of the process of synthesis and dispersion of nanoparticles in the base fluid in a single step, usually using evaporation of nanoparticles in vacuum conditions and subsequent condensation in the base fluid. These techniques reduce the effects of the agglomeration of nanoparticles (due to the high surface activity area, nanoparticles have the tendency to aggregate), but have a very high operational cost (a disadvantage). In addition they need to be developed in batches, i.e. normally in small quantities, making the commercial production of nanofluids non-rentable. Besides, the control of important parameters, such as size of nanoparticles, is limited. Another constraint is that the base fluid must have low vapor pressure, in order to avoid fluid evaporation and consequent particle agglomeration. In these cases it is very advantageous to use ionic liquids as base fluids, given the very low values of vapor pressures. The method of one step can be carried out through physical or chemical paths. If the nanofluid is prepared by a chemical path possible contamination due to secondary reagents used has to be foreseen and eliminated. For more detailed information on methods of preparation/obtaining nanofluids bibliographical analysis, for example [9-12] is recommended.

In two-step methods nanoparticles are synthesized in the form of ultra-thin dry powder and further dispersed in the base fluid through physical/mechanical processes. The main advantage of these methods is the variety of nanofluids that can be obtained, using many materials with different sizes and geometry. The cost is substantially less than in the one-step methods.

As already mentioned one of the ways to avoid the agglomeration of nanoparticles is the addition of dispersants (or surfactants) to the mixture base fluid/nanomaterial. These substances reduce the interfacial tension between the fluid and the nanoparticles, which keeps them scattered over more time. However the presence of surfactants, considered here as additives, alter the chemical and thermal properties of the mixture, being in most cases undesirable. The most common examples of surfactants are the CTAB (cetyl trymetil ammonium bromide), SDS (sodium dodecyl sulfate) and PVA (polyvinyl alcohol).

2.1. Two-Step Method

This method is the most widely used and more practical for nanofluids preparation. The nanomaterials are produce in the first place, in their most diverse geometric forms: nanofibers, nanotubes, nanorods, nanosheets or other, obtained generally as dry powders by chemical or physical methods. At this point the investigator shall reflect if the process of obtaining nanomaterials and/or other nanoparticles interferes in his study.

The researcher should study the techniques used today for the production of nanomaterials and select those which best satisfy his requirements (size and shape of nanoparticles, purity, etc.) The issues that must be dealt with are: What are the possible approaches to making nanomaterials? Which technologies can be used to produce nanostructures using a top-down approach? What is bottom-up method? Christina Raab et al. describe the most common nanomaterials production processes such as milling, gas phase and liquid phase technologies [13]. At this stage, the researcher must be aware of the nanofluid properties/parameters that are directly influenced by the nanomaterial identity.

Then the nanomaterial powder (which may have been subject to pre-grinding or ball-milling treatment) will be dispersed in fluid in the second stage of preparation, with the help of intense agitation, magnetic force, ultrasonic agitation or high pressure homogenization. This subject will be detailed below. The two-step method is more economical to produce large-scale nanofluids, because the synthesis techniques of nanopowders have already been scaled-up to industrial production levels.

As previously mentioned, one of the techniques used to avoid nanoparticle agglomeration is the addition of surfactants, which also increases the time stability (and sometimes the kinetic stability) of the nanofluid. However, the properties of these additives can cause problems and often change with the agitation processes/ultrasounds because they produce an uncontrolled increase in the temperature of the system.

Due to the difficulty in preparing stable nanofluids by two-step method, some authors prefer the various advanced techniques, including those based on one-step method.

2.2. One-Step Method

In this method the agglomeration of nanoparticles is minimized, the processes of transportation, storage, drying and dispersion of nanoparticles are avoided, so the stability of nanofluids is increased [14].

Zhu et al. obtained a non-agglomerated and stable suspended Cu nanofluids prepared by a novel one-step chemical method reducing $CuSO_4 \cdot 5H_2O$ with $NaH_2PO_2 \cdot H_2O$ in ethylene glycol under microwave irradiation. The authors reported the influences of $CuSO_4$ concentration, addition of NaH_2PO_2, and microwave irradiation on the reaction rate and expected that the method could be extended to synthesize other metallic nanofluids or nanoparticles [15].

To reduce the agglomeration of nanoparticles, Eastman et al. developed a one-step physical vapor condensation method to prepare Cu/ethylene glycol nanofluids [16].

Kumar et al. presented a novel in situ one-step method for the preparation of stable, non-agglomerated copper nanofluids by reducing copper sulphate pentahydrate with sodium hypophosphite as reducing agent in ethylene glycol as base fluid by means of conventional heating and with less time consumption and more economic [17].

Shenov et al. reported the synthesis of copper nanofluid by a one-step method involving the reduction of copper sulfate by glucose in the presence of sodium lauryl sulfate. The method relies on the simultaneous formation and dispersion of copper nanoparticles less than 50 nm in the base fluid. In this work sedimentation measurement reveals that the fluid is stable for a minimum period of six weeks, the method is simple, efficient, and low cost [18].

Lee et al. prepared an ethylene-glycol (EG) based nanofluid containing ZnO nanoparticles by a one-step physical method known as pulsed-wire evaporation (PWE). The thermal conductivity of the EG-based ZnO nanofluid at a higher concentration exhibited temperature-dependency due to the clustering and aggregation of nanoparticles to the total heat transfer in the fluid but all of the nanofluids showed Newtonian behavior [19]. The Submerged Arc Nanoparticle Synthesis System, vacuum-SANSS, is another method to prepare nanofluids using different dielectric liquids [20, 21]. The nanoparticles prepared exhibit needle-like, polygonal, square, and circular morphological shapes. The method avoids the undesired particle aggregation. One-step physical method cannot synthesize nanofluids in large scale, and the cost is also high, so the one-step chemical method is developing rapidly.

Ponmani et al. observed that between ZnO and CuO nanofluids in a biopolymer called xanthan gum (XG) aqueous solution CuO is more stable than the ZnO nanofluid. Both nanofluids were prepared using the two-step method in a 0.4 wt.% XG (as a dispersant) aqueous solution as a base with varying concentrations of nanoparticles (0.1, 0.3, and 0.5 wt.%). XG was observed to form a jelly-like structure around the nanoparticles, keeping them suspended in the solution. They expected to form the basis for the development of nanofluid-based technologies in which XG is the primary additive in upstream oil and gas industry. It is very promising because xanthan gum (XG) is used in the upstream oil and gas industry as a rheological modifier, drilling fluid additive, emulsion stabilizer, and fluid loss controlling agent [22].

An IoNanofluid [23] is referred by Zhang et al., synthesized non-spherical gold nanofluids in the presence of amino-functionalized ionic liquid by a one-step method in aqueous solution at room temperature. The results were observed by the usage of HRTEM and UV–vis spectroscopy [24].

Han et al. obtained indium nanoparticles of an average diameter of 30 nm by a one-step, economical nanoemulsion method to synthesize low-melting-point metallic nanoparticles. This nanoemulsion technique exploits the extremely high shear rates generated by the ultrasonic agitation and the relatively large viscosity of the continuous phase - polyalphaolefin (PAO), to rupture the molten metal down to diameter below 100 nm. The experimental results suggest that the nanoemulsion method is a viable route for mass production of low-melting nanoparticles such as metals, salts, and polymers. The nanoparticle size may be effectively regulated by changing the synthesis temperature in order to vary the viscosity of the continuous phase [25].

Silver nanoparticles are of great interest because of their bulk high thermal conductivity, 429 W/mK at 300 K and are therefore expected to have good heat transfer properties ideally suitable for thermal applications. Bernd Nowack, Harald F. Krug and Murray Height published, in 2011, a very interesting article about the history of nanosilver concept. They inform that nanosilver in the form of colloidal silver has been used for more than 100 years and has been registered as a biocidal material in the United States since 1954. The environmental standards are based on ionic silver in the form of small clusters or nanoparticles. The implications of this analysis for policy of nanosilver is that it would be a mistake for regulators to ignore the accumulated knowledge of our scientific and regulatory heritage in a bid to declare nanosilver materials as new chemicals, with unknown properties and automatically harmful simply on the basis of a change in nomenclature to the term "nano" [26]

Salehi et al. developed an in situ one-step chemical method for the preparation of stable, non-agglomerated silver nanofluids. Silver nanofluid is prepared by using silver nitrate as a source for silver nanoparticles, distilled water as a base fluid, and sodium borohydride and hydrazine as reducing agents by means of conventional heating using polyvinylpyrrolidone (PVP) as surfactant. The non-agglomerated and stably suspended silver nanofluids are obtained in a short time and the UV-visible spectra, XRD, and TEM image confirm the formation of crystalline silver nanoparticles dispersed in PVP. The authors says that the synthesized silver nanofluid may be used as an effective coolant in the industry in place of the conventional current fluids [27].

Nanofluids containing silver nanoparticles with a narrow-size distribution mineral oil-based were prepared and stable for about 1 month [28]. The microwave-assisted one-step method described in [29] originates stable ethanol-based nanofluids containing silver nanoparticles. The cationic surfactant octadecylamine (ODA) is also an efficient phase-transfer agent to synthesize silver colloids [30]. The phase transfer of the silver nanoparticles arises due to coupling of the silver nanoparticles with the ODA molecules present in organic phase via either coordination bond formation or weak covalent interaction.

2.3. Other Methods

In addition to the methods described we can mention references to new methods such as obtaining oxide nanoparticles in water under microwave irradiation through a novel precursor transformation method with the help of ultrasonic and microwave irradiation [31], phase-transfer method to obtain monodisperse noble metal colloids and Fe_2O_3 nanofluids [32, 33]. Feng et al. used the aqueous organic phase-transfer method for preparing gold, silver, and platinum nanoparticles on the basis of the decrease of the PVP's solubility in water with the temperature increase [34].

The preparation of nanofluids with adjustable microstructure is one of the vital issues. It is well known that the properties of nanofluids strongly depend on the structure and shape of nanomaterials. The recent research shows that nanofluids synthesized by chemical solution method have both higher conductivity enhancement and better stability than those produced by the other methods [33]. This method is distinguished from the others by its controllability. The nanofluid microstructure can be varied and manipulated by adjusting synthesis parameters such as temperature, acidity, ultrasonic and microwave irradiation, types and concentrations of reactants and additives, and the order in which the additives are added to the solution.

3. NANOFLUIDS STABILITY

After selection of the most suitable method for the preparation of nanofluid it is imperative to use a methodology to evaluate the degree of stability as a function of time. For the measurement of the properties of the nanofluid its stability is a key factor. On the other hand in terms of industrial/laboratory production the lack of nanofluid stability could compromise your storage for long periods or impose very special and expensive methods like permanent agitation.

Research on the stability of nanofluid is critical because it influences their properties being urgent to study and analyze the factors that influence dispersion stability of nanofluids. This section refers to the methods of assessment of stability and some processes to improve stability mechanisms. The result of the agglomeration of nanoparticles is manifested not only by precipitation and blockage of microchannels, but also in the reduction of thermal conductivity of nanofluids.

3.1. The Stability Evaluation Methods for Nanofluids

3.1.1. Spectral Absorbency Analysis

Spectral absorbency analysis is an efficient way to evaluate the stability of nanofluids. There is a linear relationship between the absorbency intensity and the concentration of nanoparticles in fluid. It was believed that the stability of nanofluids was strongly affected by the characteristics of the suspended particles and the base fluid such as particle morphology. Several studies have been conducted with the measurement of this property MWCNT, Al_2O_3, CuO, FePt, etc [35, 36, 37, 38]. The sedimentation kinetics could also be determined by examining the absorbency of particle in solution [39]. If the nanomaterials dispersed in fluids have characteristic absorption bands in the wavelength 190–1100 nm, it is an easy and reliable method to evaluate the stability of nanofluids using UV-vis spectral analysis. The variation of supernatant particle concentration of nanofluids with sediment time can be obtained by the measurement of absorption of nanofluids, because there is a linear relation between the supernatant nanoparticle concentration and the absorbance of suspended particles. The outstanding advantage comparing to other methods is that UV-vis spectral analysis can present the quantitative concentration of nanofluids.

3.1.2. Electrokinetic Potential Analysis

Many of the important properties of colloidal systems are determined directly or indirectly by the electrical charge (or potential) on the particles. Adsorption of ions and dipolar molecules is determined by this charge and potential distribution. The potential distribution itself determines the interaction energy between the particles, and this is responsible for the stability of particles towards coagulation and for many aspects of the flow behavior of the colloidal suspension. The *shear plane* (slipping plane) is an imaginary surface separating the thin layer of liquid bound to the solid surface and showing elastic behavior from the rest of liquid showing normal viscous behavior. The electric potential at the shear plane is called the electrokinetic potential or zeta potential. IUPAC defined zeta potential as the electric potential difference between the fixed charges on the immobile support and the diffuse charge in the solution. The responsible for the electrokinetic phenomena is the potential drop across the mobile part of the double layer. If one of the phases is a polar liquid, like water, its (dipolar) molecules will tend to be oriented in a particular direction at the interface and this will generate a potential difference. This potential is positive if the potential increases from the bulk of the liquid phase towards the interface. Zeta potential can be defined also the potential difference between the dispersion medium and the stationary layer of fluid attached to the dispersed particle. This potential indicates the degree of repulsion between adjacent, similarly charged particles in a dispersion. The significance of zeta potential is that its value can be related to the stability of colloidal dispersions. For molecules and particles that are small enough, a high zeta potential will confer stability, i.e., the solution or dispersion will resist aggregation. When the potential is low, attraction exceeds repulsion and the dispersion will break and flocculate. So, colloids with high zeta potential (negative or positive) are electrically stabilized while colloids with low zeta potentials tend to coagulate or flocculate. In general, a value of 25mV (positive or negative) can be taken as the arbitrary value that separates low-charged surfaces from highly charged surfaces. The colloids with zeta potential from 40 to 60 mV are believed to be very stable, and those with more than 60 mV have excellent stability. Also the velocity of coagulation of different particles depends on the zeta potentials of both kinds of particles. Therefore, the zeta potential is an important parameter characterizing colloidal dispersion and nanofluids. It is also possible to correlate the zeta potential with the sedimentation behavior of colloidal systems and with the flotation behavior of mineral ores [40, 41].

Zhu et al. [39] measured the zeta potential of Al_2O_3-H_2O nanofluids under different pH values and different SDBS concentration. Zeta potential measurements were employed to study the absorption mechanisms of the surfactants on the MCWNT surfaces with the help of Fourier transformation infrared spectra. Kim et al. prepared Au nanofluids with an outstanding stability even after 1 month although no dispersants were observed [42]. The stability is due to a large negative zeta potential of Au nanoparticles in water. The influence of pH and sodium dodecylbenzene sulfonate (SDBS) on the stability of two water-based nanofluids was studied [43], and zeta potential analysis was an important technique to evaluate the stability.

3.1.3. Sedimentation and Centrifugation Methods

The simplest method to evaluate the stability of nanofluids is sedimentation method [44, 45]. The nanofluids are considered to be stable when the concentration or particle size of supernatant particles keeps constant. Sedimentation photograph of nanofluids in test tubes taken by a camera is also a usual method for observing the stability of nanofluids. The variation of concentration or particle size of supernatant particle with sediment time can be obtained by special apparatus [5]. The suspension fraction of nanoparticles at a certain time could be calculated. The disadvantage for the sedimentation method is the long period for observation. Therefore, centrifugation method is developed to evaluate the stability of nanofluids. Singh et al. applied the centrifugation method to observe the stability of silver nanofluids prepared by the microwave synthesis in ethanol by reduction of $AgNO_3$ with PVP as stabilizing agent [29]. It has been found that the obtained nanofluids are stable for more than 1 month in the stationary state and more than 10 h under centrifugation at 3,000 rpm without sedimentation. Excellent stability of the obtained nanofluid is due to the protective role of PVP, as it retards the growth and agglomeration of nanoparticles by steric effect.

3.2. Improvement of Nanofluids Stability

3.2.1. Surfactants

Surfactants used in nanofluids are also called dispersants that can markedly affect the surface characteristics of a system in small quantity. Adding dispersants in the two-phase systems is an easy and economic method to enhance the stability of nanofluids. Dispersants consists of a hydrophobic tail portion, usually a long-chain hydrocarbon, and a hydrophilic polar head group. Dispersants are employed to increase the contact of two materials, sometimes known as wettability. In a two-phase system, a dispersant tends to locate at the interface of the two phases, where it introduces a degree of continuity between the nanoparticles and fluids. According to the composition of the head, surfactants are divided into four classes: nonionic surfactants without charge groups in its head, anionic surfactants with negatively charged head groups, cationic surfactants with positively charged head groups, and amphoteric surfactants with zwitterionic head groups (charge depends on pH). Here the key issue is to select suitable dispersants. In general, when the base fluid of nanofluids is polar solvent, we should select water-soluble surfactants; otherwise, we will select oil soluble ones. For nonionic surfactants, we can evaluate the solubility through the term hydrophilic/lipophilic balance (HLB) value. The lower the HLB number, the more oil soluble the surfactants, and the higher the HLB number, the more water-soluble the surfactant is. The HLB value can be obtained easily by many handbooks. The surfactants might cause several problems [46]. For example, the addition of surfactants may contaminate the heat transfer media or may produce foams when heating. Furthermore, surfactant molecules attaching on the surfaces of nanoparticles can enlarge the thermal resistance between the nanoparticles and the base fluid, which may limit the enhancement of the effective thermal conductivity.

3.2.2. Surface Modification Techniques: Surfactant-Free Method

The surfactant-free technique is a promising approach to achieve long-term stability of nanofluid with functionalized nanoparticles. The chemical modification to functionalize the surface of CNT is a common method to enhance the stability of CNT in solvents. Here, we recommend a review about the surface modification of CNT [47]. Results from the infrared spectrum and zeta potential measurements showed that the hydroxyl groups had been introduced onto the treated CNT surfaces [48]. Plasma treatment was used to modify the surface characteristics of diamond nanoparticles [49]. Yang and Liu presented a work on the synthesis of functionalized silica (SiO_2) nanoparticles [50]. Hwang et al. introduced hydrophilic functional groups on the surface of the nanotubes by mechanochemical reaction with no contamination to medium, good fluidity, low viscosity, high stability, and high thermal conductivity, would have potential applications as coolants in advanced thermal systems. [38]. A stable dispersion of titania nanoparticles in an organic solvent of diethylene glycol dimethylether (diglyme) was successfully prepared using a ball milling process [51]. In order to enhance dispersion stability of the solution, surface modification of dispersed titania particles was carried out during the centrifugal bead mill process. Zinc oxide nanoparticles could be modified by polymethacrylic acid (PMAA) in aqueous system PMAA enhanced the dispersibility of nano-ZnO particles in water. The modification did not alter the crystalline structure of the ZnO nanoparticles [52].

3.2.3. Stability Mechanisms in Nanofluids

Stability means that the particles do not aggregate at a significant rate. The rate of aggregation is in general determined by the frequency of collisions and the probability of cohesion during collision. Particles in dispersion may adhere together and form aggregates of increasing size which may settle out due to gravity. Derjaguin, Verway, Landau, and Overbeek (DVLO) developed a theory which dealt with colloidal stability [53, 54]. DLVO theory suggests that the stability of a particle in solution is determined by the sum of Van der Waals attractive and electrical double layer repulsive forces that exist between particles as they approach each other due to the Brownian motion they are undergoing. If the attractive force is larger than the repulsive force, the two particles will collide, and the suspension is not stable. If the particles have a sufficient high repulsion, the suspensions will exist in stable state. For stable nanofluids or colloids, the repulsive forces between particles must be dominant. According to the types of repulsion, the fundamental mechanisms that affect colloidal stability are divided into two kinds, one is steric repulsion, and another is electrostatic (charge) repulsion [55]. Kamiya et al. studied the effect of polymer dispersant structure on electrosteric interaction and dense alumina suspension behavior [56]. An optimum hydrophilic to hydrophobic group ratio was obtained from the maximum repulsive force and minimum viscosity.

4. ULTRASONIC ENERGY METHODOLOGY

Acoustic methods are very efficient, precise and fast, and much less sensitive to contamination compared to traditional techniques used for mixing/dispersing particles in a liquid medium, since there isn't such a higher risk of bringing more residues into a fresh sample.

Sonication or the use of ultrasonic energy can be applied by two methods: a probe or an ultrasonic bath. Ultrasonic waves are then generated directly or indirectly, respectively. The longitudinal waves propagate in a fluid and the energy is transported across space for long distances. The sound wave can also induce shear and thermal waves which propagate only in the close vicinity of phase boundaries. The progression of a sound wave through a liquid causes the molecules to oscillate in alternating cycles of compression and stretching or rarefaction. During the rarefaction cycle, the stretching force is so strong that the molecules are pulled apart beyond the critical distance separating them, which leads to the creation of voids or microbubbles between molecules, referred to as cavities. In the compression cycle that follows, the cavities will be forced to contract, leading to the generation of a large amount of energy. The shock waves produced on the total collapse of the cavities are thought to be the cause of the erosion of the particles in the vicinity of the cavities.

When a tip is immersed in a liquid suspension, coming into direct contact with the sample, the physical barriers are reduced and the waves are disseminated directly between particles, liquid and the probe. When the dispersion is put in a bath it doesn't experience immediately the acoustic energy. The waves travel first through the bath liquid and the wall of the sample container and only then reach the suspension. Direct sonication is recommended for dispersing powders since the energy output received by the materials is higher than the one sensed when the sample is in an ultrasonic bath, which typically operates at lower energy levels than those attained with probes. Sonication with baths is suggested for re-suspensions of samples that were previously sonicated directly or when higher energy levels can cause unintentional alterations or damages to the particles.

Increasing the pressure applied to the suspension, leads to an increase in both the cavitation border and in the intensity of the cavity collapse. The agglomerates come apart and the particles disperse more efficiently in the liquid. The external pressure is applied with a piston and the agglomerates are separate into smaller particles by a combination of strong and regular impacts between particles themselves and the interior wall of the vessel, cavities and sheer force of the liquid. This results in a more homogeneous suspension.

Due to its nature, where the level of contamination is reduced, this method gathers even more support when dealing with the toxicity and waste disposal of the nanomaterials. These considerations are of the upmost importance for evaluating environmental, health and safety risks. Nanomaterials and nanofluids have been subjected to greater efforts in identifying their toxicological effects but reproducibility is still a challenge. Different reasons are pointed, such as the lack of standard protocols where an effective and homogeneous dispersion is reached in biological and environmental matrixes [57,58]. National Institute of Standards and Technology (NIST) as well as the Center for the Environmental Implications of NanoTechnology (CEINT) have been responsible to close this gap by creating protocols with generic and relevant guidelines that allow to reproduce consistent dispersions with nanomaterials based in ultrasonic disruption.

As a result of its physical nature, the noise that is produced when such technique as ultrasonic disruption is used may lead to hearing loss, which has no cure, when occupational exposure exceeds 120 dB (discomfort threshold). Therefore workers should be under safe conditions by programs that practice noise monitoring, hearing protection, hearing tests and training. The personal protective equipment (hearing, respirators, goggles and clothes) should be used at all time. An acoustic box is desirable in procedures of direct sonication. When in close proximities to ultrasonic sources, the worker should use ear muffs, ear plugs or canal caps.

For a more profound evaluation of the considerations described in this chapter about ultrasonic disruption methods, the reader is advised to consult the reference [58]. For more depth about relevant reports, conditions and critical parameters related to ultrasonic disruption it is recommended to check [59]. For more general considerations about this method, references [60-64].

Sonication is a specific method for dispersion and involves a series of complex physico-chemical interactions that can result in efficient suspensions or chemical reactions between the different components, but if done incorrectly, can end in larger agglomerates. Consequently, some parameters must be well defined for a given system and by certain equipment, by evaluating the particle size and suspension situation.

4.1. Temperature

Temperature is a parameter that must be carefully controlled during sonication. During the cycles of compression and stretching, extreme heating will take place at the site where cavities are formed. This will result in bulk heating of the suspension over time, leading to loss of liquid or degradation of the components (fluid phase and nanomaterial). The overheating can be avoided with cooling baths (ice water, ice-salt), vessels with a high thermal conductivity in order to guaranty a fast release of the heat created during sonication (aluminium, stainless steel, glass, polyethylene). The suspension volume should be checked before and after sonication.

4.2. Sonication Time, Frequency and Operation Mode

The amount of energy (E) input in the system will depend on the applied power (P) and the amount of time (t) of the sonication. So, the same suspension can have altered results if different ultrasonic treatments are applied.

$$E = P \, x \, t$$

Also, it is known that as the frequency increases, the cavitation effect tends to be lower. At these frequencies, the cycles of compression and stretching are very short, leading to a reduction in the extent of cavitation and its effects such as the shock wave pressure. Then the power or amplitude of irradiation will have to be increased considerably to maintain cavitation. Lower frequencies in the range of 20-30 kHz are normally selected for sonochemical applications.

The ultrasonic disruptors can either work in pulse or continuous mode. In the first, the ultrasonic input is alternated with static intervals, which allows a better control of the temperature since the risk of overheated sites in the suspension is avoided.

4.3. Sample Volume and Concentration

The same sample with different volume or particle concentration will result in a different dispersion if the same time and sonication power are employed. At constant volume, higher concentrations signify an increase in particle-particle collision frequency which in principle can enhance the dispersion level. On the other side, depending on the chemical nature of the particles, these collisions can promote the aggregates formation and coalescence. At constant particle concentration, the sample volume is often measured as energy density ($WsmL^{-1}$) that is the quantity of energy supplied by unit of volume. In practice, this means that if the same power output is used for the same concentration, a lower volume of suspension will suffer a higher disruptive effect, which can lead to a faster increase in temperature and cooling steps are more advisable.

4.4. Probe and Vessel

Ultrasonic energy is produced in a medium by using transducers which are devices that convert electrical or mechanical energy into sound energy. The chemical effects of ultrasound are in the frequency range of 20-100 kHz. By using any transducer design, it is possible to introduce power ultrasound into the reacting system by means of a probe. The amount of power in the suspension depends on the equipment shape and immersion.

The vessel geometry is also important in the way the sound waves distribute inside the medium. It is advisable to use vessels with smaller diameters and longer heights in order to increase the liquid-probe surface area and the dissipation of heat. The bottom shape should also be taken into account since the energy distribution varies with flat, conical or round flasks.

The minimum distance between the probe tip and the vessel bottom should never be less than 1cm, to avoid the erosion of the vessel material and consequent contamination of the sample. By its operation characteristics, the probes are subjected to erosion. Consequently, it is essential to proceed with a frequent tip maintenance to avoid the release of microscopic tip residues that can contaminate a suspension and to guaranty an adequate power output. This erosion can be recognized by the matting appearance of the tip, opposed to the shiny one, very characteristic in new probes. A gentle abrasion with a grit carbide paper or emery cloth can prolong good working operations for the equipment.

4.5. Aerosoling and Foaming

When the tip of the probe is not sufficiently immersed in the suspension, it can cause the formation and release of aerosols, which can cause unwanted side effects. It is a phenomenon easily recognizable by the appearance of a fine spray in the probe vicinity (which lowers the probe performance in the energy input to the sample) or by the fluctuation in the noise during operation. By using the pulsed technique, allowing the foam to dissipate during the static intervals, or by immersing the probe deeply into the suspension, aerosoling and foaming can be prevented.

4.6. Solvent Characteristics

In addition to the important factors previously considered which influence sonication of a suspension, a few others also need to be considered since they affect the amount of energy given to disperse the particles in the liquid. Among these are surface tension, viscosity, density, velocity of the sound wave, chemical composition and solubility. Solvents with higher viscosity require higher amplitudes (or power) for cavitation to occur.

Gases that are extremely soluble may dissolve before cavitation is initiated or the bubble may increase its volume, causing it to float to the surface. Hence there is considerable amount of ultrasonic energy that may be attenuated and not used for dispersion.

4.7. The Calorimetric Method

This method is suggested as a guideline for standardization for sonochemistry studies in order to best assess the amount of acoustic energy involved in a sonication of a suspension and the cavitation effect felt by it. It is fast, simple and inexpensive, based on the measurement of the temperature increase in a liquid over time. It allows determining the delivered power which is the recommended value to use in a sonication and let different laboratories with different sonication equipment's to reproduce similar results. It is fully described in Taurozzi et al. [58].

4.8. Optimizing Sonication

The main objective of an optimization process in a sonochemistry study to obtain a homogeneous and stable dispersion is to get the best result with the minimum energy involved, to minimize unwanted side effects. In many cases, this task is done by trial and error. In the first approach, the focal point should be to determine if a stable dispersion is possible, using a high power output, low sample volume, high concentration of the solid component that can be diluted later on and prolonged sonication time. This should lead to an efficient and systematic decision-making approach, which can be better assessed by measuring the medium size distribution particle. Generally the reduction of the particle size allows the user to achieve a better dispersion. Once the right parameters of time, concentration and volume are selected, the power output of the equipment can be varied and the determined the best value of energy to supply to the sample.

4.9. Final Remarks and Authors Observations

The authors are aware of the difficulties associated with the selection of the sonochemical equipment. In this way the suppliers Sonics, Hielscher, Heidolph, Omni-inc, Bandelin, Terra Universal, Midsci, Bibby Scientific Web Shop can be found on the respective websites and we suggest a direct discussion with each to ensure the specificity of the nanofluid work.

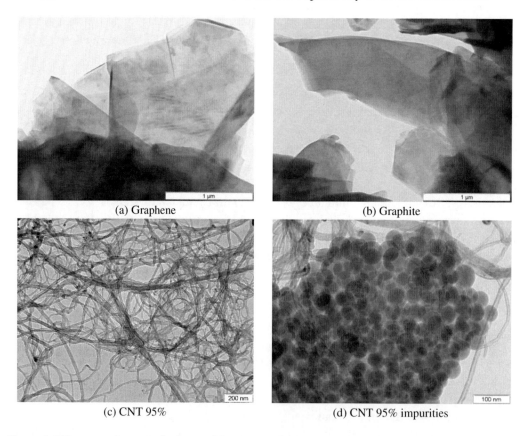

Figure 1. Microscopy images of commercial nanomaterials.

In a study conducted between 2008 and 2012 the authors obtained microscopy images of some commercial samples of Graphene SSNano, Graphite SSNano and CNT Bayer 95% in order to justify the discrepancies founded in the laboratory experiments compared with the literature results.

In Figure 1 we can observe the morphology associated with graphene (a) and graphite (b) that the supplier (SSNano) did not have at the purchase date. The Bayer CNT 95% demonstrates how unknown impurities may influence the experimental results as they are quite visible. The image (c) appears to resemble MWCNT without impurities compared to others in literature. However, a closer scan of the sample image (d) highlights the presence of impurities which can then justify the anomalies found in the experimental results. We stress that the dimension of "spheres" in image (d) are also nano! Note that in nanomaterials a small amount of impurities, not chemical identified, can significantly modify the conclusion and expected data of the research.

The figure 2 aims to elucidate the researchers to follow a methodology oriented and with well-defined goals. It should be noted that the price of nanomaterials and fluids involved here can difficult the decision because the more expensive products also have high degrees of purity as it happens with all chemicals.

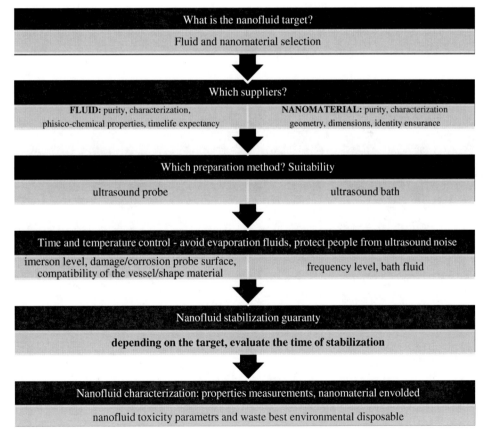

Figure 2. Elucidative diagram to orient researchers to decide about the best methodology for which nanofluid target.

5. A ROAD TO SUCCESSFUL NANOFLUID: THE NANO-ECOLOGICAL AND ECONOMIC RESEARCH

Nanomaterials are nowadays a common expression for the scientific community but also for the public. Their usefulness for industry is unmistakable. The application field is vast and the continuous development in science and technology makes their global market estimated at 11 million tonnes at a market value of €20 billion and the direct employment in this sector is estimated at 300 000 to 400 000 in Europe [65]. These values combine materials such as aluminium oxide, bismuth, calcium, carbon, carbon nanotube, carnauba wax, cellulose nanofibers, ceramics, cerium oxide, clays, cobalt, copper, fullerene, gallium, gold, graphene, graphite, iron, lead, liposome, magnesium, nano micelles, nanocellulose, nanoceramic, nanoclay, organics, platinum, polymer, quantum dots, retinol, selenium, silicon, silicon dioxide, silver, titanate, titanium, titanium dioxide, tungsten disulfide, zeolite and zinc oxide. However the market is still dominated by carbon black and synthetic amorphous silica. Its extent is still growing fast every year, especially because many commercial products incorporating nanomaterials are being developed with increasing swiftness in fields of information and communications technology; aerospace, aviation and automotive; military and defence; civil engineering; electronics and semiconductors; energy sources; biological, medical and life sciences; food, agricultural and environmental sciences; cosmetics; textile; material science and others (figure 3). Nonetheless this increase is mainly due to new applications of existing nanomaterials and not to new ones.

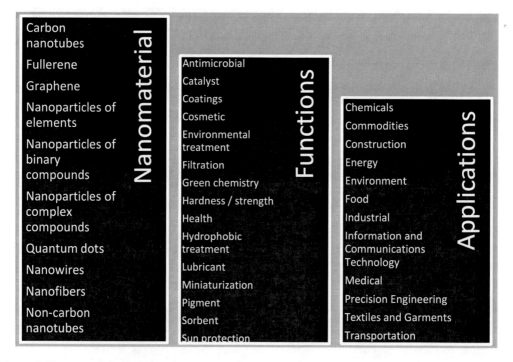

Figure 3. Nanomaterials, their functions and applications.

The R&D has been essential in the great surge of new spin-out companies, largely from universities, which has helped to the market growth. The great difficulties the new players face are related to commercialization risks. The nanomaterial's radical nature (potential for delivering dramatically better product performance or lower production costs, or both [66]) and generic nature (may yield benefits for a wide range of sectors of the economy and/or society [67]) makes their initial introduction through its mass production and adoption more capital intensive and may require executive and management innovation. Cases where the firm's already existing capabilities and complementary assets can be applied make the ideal setting. The development of new high-tech products is typically a high-cost, long-timeframe and high-risk activity. To achieve commercial success in the field of nanomaterials, barriers such as high levels of both market and technological uncertainties, large scale and long term investment must be overcome with a well thought, well planned infrastructure.

The universities spin-out companies fail in accessing the right strategy or identifying the rapidly emerging trends in the market, thus building unviable and unsustainable routes. Also, the vast number of possible applications of nanomaterials or the process to tailor-made them and alter their properties, thus opening even more promising areas, puts new ventures in difficulties, specifically in prioritising development objectives. Combining life cycle and value chain in a single framework for approaching nanomaterial risk research and achieving business growth must be one of the main focuses of new ventures, with more systematic approach to testing ideas and designing processes. The market is dominated by large established firms in the chemical, advanced materials, micro-electronics and pharmaceutical sectors, with more product and technology patents [68]. Since these ventures see the commercialization of radical innovative products facing higher market uncertainty in industry value chain and unknown potential impact of the technology, the big firms prefer to delegate the cutting edge work in nanomaterials to small companies [69], employing their bigger resources into the acquisition of universities spin-out companies, either helping them or when they have already established that have overcome market and technological uncertainty and proved commercial value [70].

There has always been a key challenge in engineering to successfully apply nanomaterials into more domains since many of their physical and chemical properties are not well known, but simulations and predictions have been able to orient industry into their more promising functions, such as anticaking agent, antimicrobial protection, catalyst, coatings, cosmetic, environmental treatment, filtration, green chemistry, hardness and strength, health applications, hydrophobic treatment, lubricant, miniaturization, pigment, sorbent, sun protection. In spite of the success reached in the market with these new products, stricter regulations require manufacturers to increase the information in labels and datasheets to provide critical data about how to safely handle and use different nanomaterials or products containing nanomaterials.

Fundamental research has been done on the synthesis of nanomaterials but the difficulty to characterize them, particularly in thermophysical, mechanical and toxicological properties, still leaves most Material Safety Data Sheets only with information regarding the appearance, specific gravity and bulk density, melting point, solubility in common solvents, pH and chemical composition. Remarkable features of nanomaterials are related to their durability, thermal and mechanical properties and other functions such as low density, permeability, colour, hydrophobicity, etc., so their accurate characterization is fundamental in order to enhance or optimize a product performance. While they remain poorly understood or completely unknown, they can never be properly valuable. Concerns from an environment, health and safety effect perspective also impact in their enormous economic potential.

Reactions in humans, principally in occupational settings where nanomaterials are manufactured and high exposures via inhalation are a tangible condition, a large attention and knowledge in their toxicological nature and handling care must be put into action. There may be new effects produced by nanomaterials that are not understood at this time and unexpected repercussions must be predicted.

The European Community regulation on chemicals and their safe use — REACH (Registration, Evaluation, Authorisation and Restriction of CHemical substances) — [71] aims to improve the protection of human health and the environment from the risks that can be posed by chemicals, while enhancing the competitiveness of the EU chemicals industry. REACH registration, in force since 2007, is mandatory for any chemical produced in the quantity over one tonne per year. Although it is being criticised for its ever-increasing cost and the number of animals employed in testing [72], REACH undoubtedly provides a meaningful, and necessary, framework to raise human and environmental safety. Currently, multi-walled carbon nanotubes, graphite, calcium carbonate are registered in REACH.

In nanotechnology commercialization, including nanomaterials, some applications require high levels of quality and purity (occasionally 99.99%). In the manufacture of semiconductors, sensors, microelectronics parts, contamination-free and defect-free materials are essential for the accurate operation of these devices. This is also true for, biological applications such as cosmetics and medicine, advanced ceramics or even in civil engineering, to improve mechanical properties [73]. Information on the sources and manufacturing processes of nanomaterials is important, since different methods can give different yields and generate products with different characteristics, such as size, shape, composition, function (charge, surface energy, functionalization; magnetic, electrical or optical properties), contaminants, surface chemistry, including the surface coating material and the dispersion, either in solid or liquid phase, etc. All these characteristics can contribute to diverse ecological and toxicological properties, purity, product variability, performance, and use.

Large databases where anyone can find much information about nanomaterials, suppliers, functions, locations, or the products currently commercialized that employ nanomaterials, can be found at Project on Emerging Nanotechnologies CPI [74] and Nanotechnology Databases from Nanowerk [75]. The biggest manufacturers can be found in the USA, followed by Europe and East Asia. Silver was the most sold nanomaterial in 2013, followed by titanium based products, carbon based products, silicon/silica, zinc based products and gold [74].

Table 1. Nanomaterial suppliers (purity ≥ 99%, average particle size ≤ 100nm)

Supplier	APS (nm)	Purity	Country
Silver			
Applied Nanotech Holdings	45	99	USA
EPRUI Nanoparticles & Microspheres	20-80	99	China
Grafen Chemical Industries	100	99.95	Turkey
	20-30	99.95	
		99.95 Coated with PVP	
	15	99.9 Coated with polymer	
IoLiTec	35	99.5	Germany
	50-60	99.9	
	20	99 Coated silver powder	

Supplier	APS (nm)	Purity	Country
Silver			
Kemix	20	99.95	Australia
	90	99.9	
MKnano	90	99.9	Canada
	15	99.99 Coated with polymer (10% Ag, 90% Polymer)	
	15	99.99 Coated with polymer (25% Ag, 75% Polymer)	
	40	99.9 Coated w/~0.2wt% oleic acid	
Nano Powder R&D Center	30-50	99.9 Coated w/~0.3% PVP	China
	35	99.5	
	30-50	99.5 Coated with 0.2 wt% PVP	
		99.9 Coated with 0.2 wt% oleic acid	
	20	99.9 Coated w/~0.3% PVP	
	80	99.9 Coated w/~0.3% PVP	
NTbase	80	99.99	South Korea
Payamavaran Nanotechnology Fardanegar	70	99.9	Iran
Shanghai HuZheng Nano Technology	15	99.99	China
Sisco Research Laboratories	90	99.9	India
SkySpring Nanomaterials	20-30	99.95	USA
	50-60	99.9	
	100	99.95	
	15	99.99 Coated with polymer (25% Ag, 75% Polymer)	
	20-30	99.95 Coated w/~0.2 wt% PVP	
	30-50	99.95 Coated w/~0.2wt% oleic acid	
Stanford Advanced Materials	50	99.9	USA
Sun Innovation's Nanomaterial Store	35	99.8	USA
Titanium and Titanium Dioxide			
EPRUI Nanoparticles & Microspheres	60	Ti, 99.9	China
IoLiTec	60-80	Ti, 99.9	Germany
Grafen Chemical Industries	40-60	Ti, 99.9	Turkey
Kemix	50	Ti, 99.9	Australia
	50	TiO_2, Rutile, 99 Hydrophilic (with SiO_2 coating)	
NaBond	60-100	Ti, 99	China
Nanostructured & Amorphous Materials	30-50	Ti, 99	USA
Payamavaran Nanotechnology Fardanegar	10-15	TiO_2, 99	Iran
SkySpring Nanomaterials	40-60	Ti, 99.9 Partially passivated	USA
	60-80	Ti, 99.9 Partially passivated	

Table 1. (Continued)

Supplier	APS (nm)	Purity	Country
Carbon Nanotubes			
Stanford Advanced Materials	40	Ti, 99.9	USA
Supplier	APS (nm)	Purity	Country
Catalytic Materials LLC	OD: 8	99, MWCNT	USA
HeJi	OD: <8	99, MWCNT	China
	OD: 8-15		
	OD: 10-20		
	OD: 20-30		
	OD: 30-50		
	OD: >50		
Intelligent Materials Pvt.	OD: 10-20 Length: 3-8 μm	99, MWCNT	India
Kemix	OD: 30-50 Length: 20 μm	99.9, MWCNT (graphitized)	Australia
	OD: 10-20 Length: 20 μm	99.9, MWCNT (graphitized, -COOH functionalized) -COOH content: 1.0wt%	
	OD: 30-50 Length: 20 μm	99.9, MWCNT (graphitized, -OH functionalized) -OH content: 1.53wt%	
MKnano	OD: 1-2 Length: 0.3- 5μm	99, SWCNT	Canada
	OD: 8-15 Length: 50 μm	99.9, MWCNT (graphitized)	
	OD: 10-20 Length: 20 μm		
	OD: 20-30 Length: 30 μm		
	OD: 30-50 Length: 20 μm		
	OD: >50 Length: 30 μm		
	OD: 8-15 Length: 50 μm	99.9, MWCNT (graphitized, -COOH functionalized) -COOH content: 0.25-1.28wt%	
	OD: 10-20 Length: 20 μm		
	OD: 20-30 Length: 30 μm		
	OD: 30-50 Length: 20 μm		
	OD: >50 Length: 30 μm		
	OD: 8-15 Length: 50 μm	99.9, MWCNT (graphitized, -OH functionalized) -OH content: 0.36-1.85wt%	
	OD: 10-20 Length: 20 μm		
	OD: 20-30 Length: 30 μm		

Supplier	APS (nm)	Purity	Country
Carbon Nanotubes			
(cont.) MKnano	OD: 30-50 Length: 20 μm	99.9, MWCNT (graphitized, -OH functionalized) -OH content: 0.36-1.85wt%	Canada
	OD: >50 Length: 30 μm		
NanoIntegris	OD: 1.2-1.7 Length: 0.3-5 μm	99, SWCNT, Ultrapure	Canada
		99, SWCNT, Semiconducting	
		99, SWCNT, Metallic	
Nanostructured & Amorphous Materials	OD: 8-15 ID: 3-5 Length: 50 μm	99.9, MWCNT (graphitized, high-temperature treated)	USA
	OD: 10-20 ID: 5-10 Length: 10-30 μm		
	OD: 20-30 ID: 5-10 Length: 10-30 μm		
	OD: 30-50 ID: 5-12 Length: 10-20 μm		
	OD: 50-80 ID: 5-15 Length: 10-20 μm		
	OD: 8-15 ID: 3-5 Length: 50 μm	99.9, MWCNT (graphitized, -COOH functionalized) -COOH content: 0.24-1.34wt%	
	OD: 10-20 ID: 5-10 Length: 10-30 μm		
	OD: 20-30 ID: 5-10 Length: 10-30 μm		
	OD: 30-50 ID: 5-12 Length: 10-20 μm		
	OD: 50-80 ID: 5-15 Length: 10-20 μm		
	OD: 8-15 ID: 3-5 Length: 50 μm	99.9, MWCNT (graphitized, -OH functionalized) -OH content: 0.34-1.94wt%	
	OD: 10-20 ID: 5-10 Length: 10-30 μm		
	OD: 20-30 ID: 5-10 Length: 10-30 μm		

Table 1. (Continued)

Supplier	APS (nm)	Purity	Country
Carbon Nanotubes			
(cont.) Nanostructured & Amorphous Materials	OD: 30-50 ID: 5-12 Length: 10-20 μm OD: 50-80 ID: 5-15 Length: 10-20 μm	99.9, MWCNT (graphitized, -OH functionalized) -OH content: 0.34-1.94wt%	USA
Sisco Research Laboratories	OD: 30-50 Length: 20 μm	99, MWCNT (graphitized)	India
Fullerene (C_{60} and C_{70})			
American Dye Source	-	C_{60} 99.5, 99.9 and 99.95 C_{70}, 99	Canada
EMFUTUR Technologies	-	C_{60}, 99 C_{70}, 99	Spain
Grafen Chemical Industries	0.7 -	C_{60}, 99.5 C_{60}, 99.5, purified C_{60}, 99.9, ultra-pure, vacuum oven dried	Turkey
IoLiTec	-	C_{60} 99, 99.5, 99.9 and 99.95 C_{70}, 99.5	Germany
Kemix	-	C_{60}, 99.5	Australia
Materials and Electrochemical Research	0.7	C_{60}, 99 C_{60}, 99.9, sublimed 20-30% ^{13}C enriched C_{60}, 99 C_{70}, 99, sublimed	USA
Materials Technologies Research	-	C_{60} 99, 99.9 and 99.95 C_{60}, 99.95, sublimed C_{70} 99 and 99.5 C_{70}, 99 and 99.5, sublimed	USA
Nano-C	-	C_{60}, 99.5 C_{60}, 99.95, sublimed	USA
NeoTechProduct Research & Production Company	-	C_{60}, 99.5 C_{70}, 99	Russia
PlasmaChem	-	C_{60}, 99 C_{70}, 99	Germany
SES Research	-	C_{60} 99.5, 99.9 and 99.95 C_{70}, 99	USA

Supplier	APS (nm)	Purity	Country
Graphene			
Sisco Research Laboratories	-	C_{60}, 99.5	India
		C_{70}, 99	
TCI Chemicals	-	C_{60}, 99 and 99.5	Japan
Term USA	-	C_{60} 99, 99.5 and 99.98	USA/Russia
		C_{70} 99 and 99.5	
Anderlab	Thickness: 3-6 X-Y Thickness: 10μm	99, 3-6 layers	India
Graphene Supermarket	Thickness: 8 Size: ~ 550	99.9, 20-30 monolayers	USA
IoLiTec	Thickness: 0.6-3.8	99, 1-10 layers	Germany
	Thickness: ~ 2	99.5 Platelet nanopowder	
	Thickness: 6-8	99.5 Platelet nanopowder	
	Thickness: 11-15	99.5 Platelet nanopowder	
SkySpring Nanomaterials	Thickness: 6-8 Size: 15μm	99.5 Platelet nanopowder	USA
	Thickness: 11-15 Size: 15μm	99.5 Platelet nanopowder	
Timesnano	Thickness: 0.55-3.74 Size: 0.5-3μm	99, 1-10 layers	China
XG Sciences	Thickness: 2 Size: < 2μm	99.5	USA
	Thickness: 6-8 Size: 5, 15 or 25μm	99.5	
	Thickness: 15 Size: 5, 15 or 25μm	99.5	
Silicon and Silica			
Grafen Chemical Industries	50	Si, 99	Turkey
IoLiTec	50-100	Si, 99.99	Germany
Kemix	40	Si, 99 Crystalline	Australia
Meliorum Technologies	5 or 30	Si, 99.9, deliverable with no surface treatment if specified by user	USA
MKnano	15	SiO_2, 99.5 Hydrophilic,	Canada
		SiO_2, 99.5 Hydrophobic, Coated w/ silane coupling agent	
		SiO_2, 99.5 Lipophilic, Coated w/ lauric acid	

Table 1. (Continued)

Supplier	APS (nm)	Purity	Country
Payamavaran Nanotechnology Fardanegar	10-15	SiO_2, 99.9	Iran
Sisco Research Laboratories	40	Si, 99 Crystalline	India
SkySpring Nanomaterials	100	Si, 99	USA
	15-20	SiO_2, 99.5 Porous	
	20	SiO_2, 99.5 Non-porous	
	10-30	SiO_2, 99.5 Coated w/ silane coupling agent	
	10-20	SiO_2, 99.8 Modified with amino group	
	10-20	SiO_2, 99.8 Modified with epoxy group	
Silicon and Silica			
(cont.) SkySpring Nanomaterials	10-20	SiO_2, 99.8 Modified with double bond	USA
	10-20	SiO_2, 99.8 Modified with double layer	
	5-15	SiO_2, 99.8 Modified with single layer organic chain	
	10-20	SiO_2, 99.8 Modified with single layer organic chain	
Stanford Advanced Materials	30	Si, 99.9 Polycrystalline	USA
Zinc and Zinc oxide			
Applied Nanotech Holdings	45	Zn, 99	USA
	30	ZnO, 99	
EPRUI Nanoparticles & Microspheres	60-80	Zn, 99.99	China
	15	ZnO, 99.9	
	30	ZnO, 99.6	
	50	ZnO, 99.5	
Grafen Chemical Industries	40-60	Zn, 99.9	Turkey
	10-30	ZnO, 99.8	
Inframat Advanced Materials	30	ZnO, 99.7	USA
IoLiTec	35	Zn, 99.9	Germany
Kemix	50	Zn, 99.9	Australia
	50	ZnO, 99.99	
MKnano	20	ZnO, 99.9	Canada
	20	ZnO, 99.9	
NaBond	35	Zn, 99.9	China
	20	ZnO, 99	

Supplier	APS (nm)	Purity	Country
Nano Pars Spadana	6-12	ZnO, 99.8	Iran
Nanophase	20	ZnO, 99	USA
	40	ZnO, 99	
	60	ZnO, 99	
Nano Powder R&D Center	35	Zn, 99.9	China
	30	ZnO, 99.5	
Nanostructured & Amorphous Materials	80	Zn, 99.5	USA
	100	Zn, 99.9	
	20	ZnO, 99.5	
NTbase	100	Zn, 99.99	South Korea
Payamavaran Nanotechnology Fardanegar	10-15	ZnO, 99.9	Iran
Sisco Research Laboratories	50	Zn, 99.9	India
	30	ZnO, 99.9	
SkySpring Nanomaterials	40-60	Zn, 99.9	USA
	10-30	ZnO, 99.8	
	10-30	ZnO, 99 treated with stearic acid	
	10-30	ZnO, 99 treated with silicone oil	
Zinc and Zinc oxide			
Stanford Advanced Materials	20	Zn, 99.9	USA
	20	ZnO, 99.9	
Sukgyung AT	10-20	ZnO, 99.9	South Korea
	20-40	ZnO, 99.9	
Sun Innovation's Nanomaterial Store	35	Zn, 99.8	USA
Gold			
Goodfellow	15	99.95	England
Grafen Chemical Industries	50-100	99.99	Turkey
	15	99.99 Coated with polymer	
IoLiTec	50-100	99.99	Germany
Kemix	90	99	Australia
MKnano	90	99.99	Canada
NaBond	10-50	99.95	China
Nanostructured & Amorphous Materials	10-20	99.95	USA
	50-100	99.99	
Sisco Research Laboratories	90	99	India
SkySpring Nanomaterials	50-100	99.99	USA

PVP: Polyvinylpyrrolidone.

MWCNT: Multi Walled Carbon Nanotubes.

SWCNT: Single Walled Carbon Nanotubes.

The following table (1) gives a list of many suppliers of nanomaterials. They were chosen taking into account their phase (powder), grain size below 100nm, a purity higher than 99%, and a combination between the most sold and studied by our research group at Centre of Molecular Sciences and Materials (silver, titanium and titanium dioxide, carbon nanotubes, fullerene, graphene, silicon, silica, zinc, zinc oxide and gold).

ACKNOWLEDGMENTS

This research was partially supported by Fundação para a Ciência e a Tecnologia-FCT, Portugal, through projects Pest-OE/QUI/UI0536/2011 and PTDC/EQU-FTT/104614/2008 attributed to Centro de Ciências Moleculares e Materiais (CCMM).

REFERENCES

[1]   Sidik, N. A. C.; Mohammed, H. A.; Alawi, O. A.; Samion, S. *Int. Comm. Heat Mass Transfer* 2014, 54, 115–125.

[2]   Hussein, A. M.; Sharma, K. V.; Bakar, R. A; Kadirgama, K. *Ren. Sust. En. Rev.* 2014, 29, 734–743.

[3]   Özerinç, S.; Kakaç, S.; Yazıcıoğlu, A. G. *Microfluid. Nanofluid.* 2010, 8,145–170.

[4]   Kakac,S.; Pramuanjaroenkij, A. *Int. J. Heat Mass Transfer* 2009, 52, 3187–3196.

[5]   Li, Y.; Zhou, J.; Tung, S.; Schneider, E.; Xi, S. *Powder Technol.* 2009, 196, 89–101.

[6]   Wang X. Q.; Mujumdar, A. S. *Brazil. J. Chem. Eng.* 2008, 25, 613–630.

[7]   Trisaksri, V.; Wongwises, S. *Ren. Sust. En. Rev.*2007, 11, 512–523.

[8]   Wang X. Q.; Mujumdar, A. S. *Int. J. Therm. Sci.* 2007, 46, 1–19.

[9]   Haddad, Z.; Abid, C.; Oztop, H. F.; Mataoui, A. *Int. J. Therm. Sci.* 2014,76, 168-189.

[10] Kharissova, O. V.; Kharisov, B. I.; Ortiz, E. G. de C. *RSC Adv.*2013, 3, 24812-24852.

[11] Hassan, A. A.; Sandre, O.; Cabuil, V. *Angew. Chem.* 2010, 49, 6268–6286.

[12] Richter, K.; Birkner, A.; Mudring, A. V. *Angew. Chem.*2010, 49, 2431–2435.

[13] Raab, C.; Simkó, M.; Fiedeler, U.; Nentwich, M.; Gazsó, A. *Institute of Technology Assessment of the Austrian Academy of Sciences*, N 006, February 2011.

[14] Li, Y.; Zhou, J.; Tung, S.; Schneider, E.; Xi, S. *Powder Technol.*2009, 196, 89–101.

[15] Zhu, H.; Lin, Y.; Yin, Y. S. *J. Colloid. Interface Sci.* 2004, 277, 100–103.

[16] Eastman, J. A.; Choi, S. U. S.; Li, S.; Yu, W.; Thompson, L. J. *Appl. Phys. Lett.* 2001, 78, 718–720.

[17] Kumar, S. A.; Meenakshi, K. S.; Narashimhan, B. R. V.; Srikanthb, S.; Arthanareeswaran, G. *Mater. Chem. Phys.*2009, 113, 57–62.

[18] Shenoy, U. S.; Shetty, A. N. *Syn. React. Inorg. Metal-Orga. Nano-Metal Chem.* 2013, 43, 343–348.

[19] Lee, G. J.; Kim, C. K.; Lee, M. K.; Rhee, C. K.; Kim, S.; Kim, C. *Thermochim. Acta* 2012, 542, 24– 27.

[20] Lo, C. H.; Tsung, T. T.; Chen, L. C. *J. Crys. Growth* 2005, 277, 636–642.

[21] Lo, C. H.; Tsung, T. T.; Chen, L. C.; Su, C. H.; Lin, H. M. *J. Nanopart. Res.* 2005, 7, 313–320.

[22] Ponmani, S.; William, J. K. M.; Samuel, R.; Nagarajan, R.; Sangwai, J. S. *Colloid. Surf. A: Physicochem. Eng. Aspects* 2014, 443, 37–43.

[23] Nieto de Castro, C. A.; Lourenço, M. J. V.; Ribeiro, A. P. C.; Langa, E.; Vieira, S. I. C.; Goodrich, P.; Hardacre, C. *J. Chem. Eng. Data,* 2010, 55, 653-661.

[24] Zhang, H.; Cui, H.; Yao, S.; Zhang, K.; Tao, H.; Meng, H. *Nanoscal. Res. Lett.* 2012, 7, 583.

[25] Han, Z. H.; Yang, B.; Qi, Y.; Cumings, J. *Ultrasonics* 2011, 51, 485-488.

[26] Nowack, B.; Krug, H. F.; Height, M. *Environ. Sci. Technol.* 2011, 45, 1177–1183.

[27] Salehi, J. M.; Heyhat, M. M.; Rajabpour, A. *Appl. Phys. Lett.* 2013, 102, 231907.

[28] Onnemann, H. B.; Botha, S. S.; Bladergroen, B.; Linkov, V. M. *Appl. Organometal. Chem.* 2005, 19, 768–773.

[29] Singh, A. K.; Raykar, V. S. *Colloid Poly. Sci.* 2008, 286, 1667–1673.

[30] Kumar, A.; Joshi, H.; Pasricha, R.; Mandale, A. B.; Sastry, M. *J. Colloid Interface Sci.* 2003, 264, 396–401.

[31] Zhu, H. T.; Zhang, C. Y.; Tang, Y. M.; Wang, J. X. *J. Phys. Chem. C* 2007, 111, 1646–1650.

[32] Chen Y.; Wang, X. *Mater. Lett.* 2008, 62, 2215–2218.

[33] Yu, W.; Xie, H.; Chen, L.; Li, Y. *Colloid. Surfaces A* 2010, 355, 109–113.

[34] Feng, X.; Ma, H.; Huang S. et al., *J. Phys. Chem. B,* 2006, 110, 12311–12317.

[35] Chen, L.; Xie, H. *Thermochim. Acta* 2010, 506, 62–66.

[36] Huang, J.; Wang, X.; Long, Q.; Wen, X.; Zhou, Y.; Li, L. *In Proceedings of Symp. Photon. Optoelectron.* (SOPO2009), Wuhan, China, August 14-16, 2009.

[37] Farahmandjou, M.; Sebt, S. A.; Parhizgar, S. S.; Aberomand, P.; Akhavan, M. *Chin. Phys. Lett.* 2009, 26, 027501.

[38] Hwang, Y.; Lee, J. K.; Lee C. H. et al., *Thermochim. Acta* 2007, 455, 70–74.

[39] Zhu, D.; Li, X.; Wang, N.; Wang, X.; Gao, J.; Li, H. *Curr. Appl. Phys.* 2009, 9, 131–139.

[40] Evans, F. D.; Wennerstrçm, H. *Adv. Interfacial Eng.,* Wiley-VCH, Weinheim, 1999.

[41] Hunter, R. J. *Zeta Potential in Colloid Science-Principles and Applications*, Academic Press Limited, 1988.

[42] Kim, H. J.; Bang, I. C.; Onoe, J. *Optics Lasers Eng.* 2009, 47, 532–538.

[43] Wang, X. J.; Li, X.; Yang, S. *Energy Fuel.* 2009, 23, 2684–2689.

[44] Wei, X.; Wang, L. *Particuology* 2010, 8, 262–271.

[45] Li, X.; Zhu, D.; Wang, X. *J. Colloid Interface Sci.* 2007, 310, 456–463.

[46] Chen, L.; Xie, H.; Li, Y.; Yu, W. *Thermochimica Acta* 2008, 477, 21–24.

[47] Wepasnick, K. A.; Smith, B. A; Bitter, J. L.; Fairbrother, D. H. *Anal. Bioanaly. Chem.* 2010, 396, 1003–1014.

[48] Chen L.; Xie, H. *Thermochim. Acta* 2010, 497, 67–71.

[49] Yu, Q.; Kim, Y. J.; Ma, H. *Appl. Phys. Lett.* 2008, 92, 103111.

[50] Yang, X.; Liu, Z. H. *Nanoscal. Res. Lett.* 2010, 5, 1324–1328.

[51] Joni, I. M.; Purwanto, A.; Iskandar, F.; Okuyama, K. *Ind. Eng. Chem. Res.* 2009, 48, 6916–6922.

[52] Tang, E.; Cheng, G.; Ma, X.; Pang, X.; Zhao, Q. *Appl. Surf. Sci.* 2006, 252, 5227–5232.

[53] Missana, T.; Adell, A. *J. Colloid Interface Sci.* 2000, 230, 150–156.

[54] Popa, I.; Gillies, G.; Papastavrou, G.; Borkovec, M. *J. Phys. Chem. B* 2010, 114, 3170–3177.

[55] Yu W.; Xie, H.; *J. Nanomat.* 2012, (17p).

[56] Kamiya, H.; Fukuda, Y.; Suzuki, Y.; Tsukada, M.; Kakui, T.; Naito, M. *J. Am. Cer. Soc.* 1999, 82, 3407–3412.

[57] Roebben, G.; Ramirez-Garcia, S.; Hackley, V. A. et al. *J. Nanopart. Res.* 2011, 13, 2675-2687.

[58] Taurozzi, J. S.; Hackley, V. A.; Wiesner, M. R. *Nanotoxicology* 2011, 5, 711-729.

[59] Taurozzi, J. S.; Hackley, V. A.; Wiesner, M. R. *Special Publication 1200-1*, National Institute of Standards and Technology, Gaithersburg, MD, USA, June 2012

[60] Berlan, J.; Mason, T. J. *Ultrasonics* 1992, 30, 203-212.

[61] Suslick, K. S.; *Ultrasound: Its Chemical, Physical and Biological Effects*, VCH Publishers, 1988.

[62] Mason, T. J. *Sonochemistry: Theory, applications and uses of ultrasound in chemistry*, Ellis Horwood, 1989.

[63] Mason, T. J.; Peters, D. *Practical sonochemistry: Power ultrasound uses and applications*, Woodhead Publishing, 2003.

[64] Hielscher, T. Ultrasonic production of nano-sized dispersions and emulsions, In *Dans European Nano Systems Workshop*, Paris, France 2005.

[65] Nanotechnology (2014). European Commission Website. Retrieved [14-04-2014], from http://ec.europa.eu/nanotechnology/index_en.html

[66] Utterback, J. *Mastering the Dynamics of Innovation: How Companies Can Seize Opportunities in the Face of Technological Change*, HBS Press, Boston, 1994.

[67] Keenan, M. *J. Forecasting* 2003, 22, 129–149.

[68] Avenel, E.; Favier, A.; Ma, S.; Mangematin, V.; Rieu, C. *Research Policy* 2007, 36, 864–870.

[69] Graff, G. The nanomaterials market is starting to climb the growth curve. *Purchasing Magazine*, August 28, Service, R. F., 2000.

[70] Maine, E.; Lubik, S.; Garnsey, E. *Technovation* 2012, 32, 179–192.

[71] Enterprise and Industry (2014) REACH—Registration, Evaluation, Authorisation and Restriction of Chemicals. Retrieved [22-04-2014], from http://ec.europa.eu/enterprise/sectors/chemicals/reach/index_en.htm

[72] Hartung, T.; Rovida, C. *Nature* 2009, 460, 1080–1081.

[73] Olar, R. *Nanomaterials and nanotechnologies for civil engineering*, Gheorghe Asachi Technical University, 2011

[74] Project on Emerging Nanotechnologies (2014). Consumer Products Inventory. Retrieved [10-04-2014], from http://www.nanotechproject.org/cpi

[75] Nanowerk (2014). Nanotechnology Databases. Retrieved [10-04-2014], from http://www.nanowerk.com/nanotechnology_databases.php

In: Nanofluids: Synthesis, Properties and Applications ISBN: 978-1-63321-677-8
Editors: S.M. Sohel Murshed, C.A. Nieto de Castro © 2014 Nova Science Publishers, Inc.

Chapter 2

HEAT TRANSFER AND TRANSPORT PROPERTIES OF NANOFLUIDS

Amyn S. Teja[1] and Pramod Warrier[2]*
[1]School of Chemical and Biomolecular Engineering,
Georgia Institute of Technology, Atlanta, GA, US
[2]REC Silicon Inc., Moses Lake, WA, US

ABSTRACT

The addition of nanoparticles has been suggested as a means to enhance the thermal conductivity of existing heat transfer fluids. This is of considerable interest in microelectronics systems because thermal management is a key bottleneck in their advancement. In this chapter, a figure of merit (FOM) analysis for electronics cooling is presented, and the effect of nanoparticle addition on the FOM of heat transfer fluids via their thermal conductivity and viscosity is critically reviewed. A modified geometric mean model is described that takes into account the temperature dependence of the thermal conductivities of the individual phases, as well as the size dependence of the thermal conductivity of the particles. Based on experimental data on the rheological properties of a pentadecane-bentonite nanofluid and an FOM analysis, it is surmised that the addition of particles to fluorocarbon coolants will not improve their performance in electronics cooling.

1. INTRODUCTION

Performance enhancement while reducing system size, known as process intensification in the chemical industry, is a universal characteristic of technological development. In the microelectronics industry, it is exemplified by Moore's law which states that the number of transistors in an integrated chip doubles every 18-24 months [1]. Similar intensification is also observed in nuclear, automobile, and aircraft systems, often accompanied by processes

---

*Corresponding author: amyn.teja@chbe.gatech.edu

that generate heat that must be dissipated from smaller and smaller areas. As a result, there is increasing demand for improving thermal management technologies to match the rate at which heat removal demands are increasing.

Thermal management is widely regarded as the key bottleneck in the further development of microelectronic systems [2, 3]. Such systems currently generate heat fluxes that exceed 100 W cm^{-2}, and future electronic devices are likely to generate heat fluxes of over 1 kW cm^{-2}. The latter is of the same order of magnitude as the heat flux on the surface of the sun, albeit over a much smaller area and at much lower temperatures. This makes heat dissipation even more challenging in microelectronic systems because of small temperature gradients [3]. Heat fluxes of over 1 kW cm^{-2} are likely to require forced convection of liquids and/or phase change heat transfer for cooling electronic devices of the future. However, the advancement of liquid cooling systems has been rather slow because of the low thermal conductivities of typical heat transfer fluids (the highest thermal conductivity being that of water 0.6 W m^{-1} K^{-1}). This is an even bigger issue in direct immersion cooling systems where liquid comes in direct physical contact with electronic systems. Electrical and chemical compatibility of liquid and electronic components restricts the use of coolants to dielectric fluids such as fluoroinerts (FC-72, FC-86, FC-77) and Novec fluids (HFE-7100, HFE-7200) for such applications. The performance of direct immersion cooling is severely limited by the low thermal conductivity of fluorocarbon coolants which is an order of magnitude lower than that of water.

Nanofluids (or fluids containing dispersed nanoparticles) have attracted a lot of attention in recent years due to their superior thermal conductivity compared to that of the base liquid in which nanoparticles are dispersed. Work by Choi [4] and Eastman et al. [5] has shown that an ethylene glycol-based nanofluid can exhibit a 40 % enhancement in the thermal conductivity over that of pure ethylene glycol when a small amount (0.3 %) of Cu nanoparticles are added. An even greater (150 %) enhancement in the thermal conductivity was observed by Choi et al. [6] upon the addition of 1 % (v/v) carbon nanotubes to synthetic oil. These enhancements in the thermal conductivity are much higher than those predicted for dispersions of micro- and macro-scale particles [7] which suggests that the addition of nanoparticles to common heat transfer fluids may be promising in the development of thermal management technologies.

This chapter reviews the thermal conductivity and viscosity of nanofluids and examines the applicability of nanofluids for electronics cooling based on a figure of merit (FOM) analysis. Since thermal conductivity and viscosity have the most significant effect on FOM, the effect of nanoparticle addition on the thermal conductivity and viscosity of heat transfer fluids is critically reviewed. Fluorocarbon-based nanofluids are then evaluated in terms of their applicability for electronic cooling.

2. FIGURE OF MERIT ANALYSIS

The efficacy of coolants cannot be evaluated by consideration of their thermophysical properties in isolation. Instead, all fluid thermophysical property-dependent terms in existing heat transfer correlations must be grouped to define FOMs [8-15] that allow comparisons to be made of the heat transfer characteristics of different fluids.

In electronics cooling, very large heat fluxes must be dissipated over small temperature gradients. Therefore, the heat transfer regime of interest is likely to be the phase change or boiling heat transfer regime. The FOMs listed below provide reasonable estimates of heat transfer performance in the boiling regime, as discussed by several authors [16, 17].

Lazarek and Black FOM for flow boiling [9]:

$$\text{FOM-LB} = \frac{10^3 \times k}{\eta^{0.857} H_{vap}^{0.714}} \tag{1}$$

Tran FOM for flow boiling [10]:

$$\text{FOM-T} = \frac{10^6 \times k}{\left(\eta H_{vap}\right)^{0.62}} \left(\frac{\rho_v}{\rho_l}\right)^{0.607} \tag{2}$$

Rohsenow FOM for pool boiling [8]:

$$\text{FOM-R} = \frac{10^3}{\left(\dfrac{H_{vap}}{C_p}\right)\left(\dfrac{C_p\eta}{k}\right)^{1.7}\left(\dfrac{1}{\eta H_{vap}}\sqrt{\dfrac{\sigma}{g(\rho_l - \rho_v)}}\right)^{0.33}} \tag{3}$$

In Eqs. (1-3), k is the thermal conductivity in W m^{-1} K^{-1}, η is the dynamic viscosity in kg m^{-1} s^{-1}, H_{vap} is the enthalpy of vaporization at the normal boiling point in J kg^{-1}, C_p is the specific heat capacity in J kg^{-1} K^{-1}, σ is the surface tension in N m^{-1}, g is the acceleration due to gravity in m s^{-2}, and ρ_l and ρ_v are liquid and vapor densities in kg m^{-3}. FOM-LB, FOM-T and FOM-R values at 298 K for HFE 7200 are shown in Table 1. For electronics cooling, it is desired to select nanoparticles that yield nanofluids that have higher FOM values than those listed in Table 1.

Table 1. FOMs for HFE 7200

Figure of Merit	HFE 7200
Lazarek and Black (FOM-LB)	9.21
Tran (FOM-T)	260.87
Rohsenow (FOM-R)	7.20

It should be noted that FOM values generally increase with increasing thermal conductivity and decrease with increasing viscosity (Eq. 1-3). Since the addition of solid particles to a liquid is expected to lead to an increase in both the effective thermal conductivity and the effective viscosity, its effect on FOM values is not straightforward. In developing nanofluid coolants, the detrimental effect of an increase in viscosity must be balanced by the beneficial effect of an increase in the thermal conductivity. Below, we examine the thermal conductivity and viscosity behavior of nanofluids.

3. THERMAL CONDUCTIVITY OF NANOFLUIDS

Heterogeneous mixtures of micro- and macro-scale particles in liquids have been studied for many years [7, 18] and it is well known that the effective thermal conductivity of particle dispersions depends on the thermal conductivity of the two phases that comprise the dispersion, in proportion to the volume fraction of each phase. However, as mentioned previously, dispersions of nanoparticles are known to exhibit unusually high thermal conductivity enhancements that are much higher than those predicted by the classical Maxwell model [7].

Whereas most studies of nanofluid thermal conductivity report large enhancement of the thermal conductivity over that of the base fluid, a few studies have also produced thermal conductivity decreases. There are also seemingly conflicting results with respect to the effect of particle size and temperature. This has resulted in different mechanisms being proposed to explain some of the conflicting results. Proposed mechanisms include Brownian motion, ordering of the liquid at the particle surface, and nanoparticle clustering. However, models based on these mechanisms have generally proved ineffective in predicting the thermal conductivity of nanofluids. In the following section, we briefly review the effect of different variables on the thermal conductivity and highlight some of the shortcomings of nanofluid models. It will also be shown that the thermal conductivity can be predicted with a simple volume fraction weighted geometric mean model.

3.1. Thermal Conductivity of Liquids and Solids

As nanofluids are heterogeneous mixtures, the first step in understanding their thermal conductivity behavior is to examine the behavior of the individual phases.

3.1.1 Thermal Conductivity of Liquids

The thermal conductivity of liquids is at least one order of magnitude lower than that of solids. It ranges from about 0.06 W m^{-1} K^{-1} for fluorocarbon liquids to about 0.6 W m^{-1} K^{-1} for water at ambient conditions. Furthermore, the thermal conductivity of nonpolar liquids is generally lower than that of polar liquids, and decreases monotonically with increasing temperature because of thermal expansion of the liquid. [19] By contrast, associating liquids such as water and ethylene glycol display a maximum in their thermal conductivity vs. temperature behavior [20, 21]. This is due to changes in their hydrogen bonding network with temperature. At low temperatures, some of the energy being transferred is stored in hydrogen bonds as they form a network, leading to a lower thermal conductivity. As the temperature increases, less energy is captured by the hydrogen bonding network, leading to increased thermal conductivity. This phenomenon competes with the typical decrease in thermal conductivity with temperature due to thermal expansion, and results in a maximum in the thermal conductivity vs. temperature behavior of the fluid.

Since nanofluids generally contain only a small amount of solid nanoparticles, the temperature dependence of nanofluid thermal conductivity is expected to follow that of the base fluid. That is, nanofluids consisting of nonpolar base fluids are expected to exhibit a monotonous decrease in thermal conductivity with increasing temperature. Similarly, nanofluids consisting of hydrogen-bonding base fluids are expected to display a maximum in their thermal conductivity vs. temperature behavior.

3.1.1 Thermal Conductivity of Solids

3.1.1.1. Thermal Conductivity of Insulator and Semiconductor Solids

In insulating and semiconducting solid lattices, heat is conducted predominantly by lattice vibrational waves or phonons [22]. These phonons travel at the speed of sound when the atoms in the crystal lattice oscillate harmonically. However, anharmonicity is often observed [23], and leads to phonon scattering or a change in the direction of the phonon wave. Momentum is conserved in elastic phonon scattering, but not in inelastic scattering. Inelastic scattering creates resistance to thermal transport and leads to a decrease in the thermal conductivity. Scattering can result from collisions of phonons with each other (Umklapp scattering), or defects in the crystal structure such as impurities and grain boundaries. Thermal conductivities of insulators and semiconductors [24] therefore generally increase and then decrease with temperature due to these competing phenomena.

The effect of particle size on the solid thermal conductivity has received considerable attention. A number of studies [25-27] have concluded that the thermal conductivity of submicron semiconductor thin films decrease with decreasing thickness of the film. Indeed, Liu and Asheghi [27] have reported that the out-of-plane thermal conductivity of a 20 nm-thick silicon film is nearly an order of magnitude smaller than the thermal conductivity of bulk silicon. Nanowires, which are confined in two dimensions, should exhibit a lower thermal conductivity than nanofilms, which are only confined in one dimension. Li et al. [28] have validated this hypothesis using measurements of axial heat conduction in silicon nanowires. They reported that the axial thermal conductivity of a 22 nm (diameter) silicon nanowire was approximately 6 W m^{-1} K^{-1}, while the out-of-plane thermal conductivity of a 20-nm-thick silicon film was approximately 22 W m^{-1} K^{-1}.

Semiconductor nanoparticles should exhibit an even lower thermal conductivity than nanowires or nanofilms, because nanoparticles are confined in three dimensions. Fang et al. [29] used molecular dynamic simulations to estimate the thermal conductivity of silicon nanoparticles and reported that the thermal conductivity of particles smaller than 8 nm was about 2 W m^{-1} K^{-1}. In contrast, the thermal conductivity of bulk silicon is 237 W m^{-1} K^{-1}. The nanoparticle results have yet to be confirmed experimentally, although it seems clear that the thermal conductivity of semiconductor or insulator particles must decrease with particle size when the particle size approaches the mean free path of phonons in the solid. The studies described above imply that the contribution of the particle thermal conductivity to the effective thermal conductivity of a solid dispersion should also decrease as the size of the dispersed particles approaches the phonon mean free path.

Size dependence of the thermal conductivity of semiconductors and insulators can be calculated using a phenomenological model proposed by Liang and Li [30]. Their model takes into account the intrinsic size effect on phonon velocity and mean free path, as well as surface scattering. According to Liang and Li, the thermal conductivity of a nanostructure is given by:

$$\frac{k_p(L)}{k_b} = p \exp\left(-\frac{l_0}{L}\right)\left[\exp\left(\frac{1-A}{L/L_0 - 1}\right)\right]^{3/2} \quad (4)$$

where $k_p(L)$ is the thermal conductivity of the nanostructure of characteristic size L, k_b is the thermal conductivity of the bulk material, l_0 is the phonon mean free path at room temperature, and L_0 is the critical size when almost all atoms of the crystal are located on its surface. L_0 may be obtained from:

$$L_0 = 2(3-n)\sigma \quad (5)$$

where σ is the atomic or molecular diameter, and $n = 0$, 1, and 2 for nanoparticles, nanowires, and thin films, respectively. Parameter A depends on the bulk vibrational entropy of melting S_v and is given by:

$$A = 1 + (2/3) S_v / R \quad (6)$$

where R is the universal gas constant. For III-V and II-VI solids, $S_v = (H_m/T_m - R)$ where H_m is the enthalpy of melting and T_m is the bulk melting temperature. For molecular crystals [31], $S_v \approx S_m = H_m/T_m$. The adjustable parameter p ($0 < p \leq 1$) in Eq. (4) provides a measure of surface roughness.

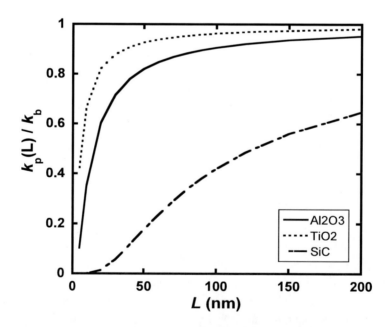

Figure 1. Size effects on the thermal conductivity of alumina, titania and silicon carbide.

Figure 1 illustrates the effect of size on the thermal conductivity of several solid materials using Eqs. (4-6). The Liang and Li model has been incorporated in the geometric mean model to account for particle size effects in nanofluid thermal conductivity [32], as discussed further below.

3.1.1.2. Thermal Conductivity of Metals

In metallic solids, thermal energy is transferred predominantly by the movement of free electrons. We may treat these free electrons qualitatively as a classical gas (the so-called electron gas) and apply the kinetic theory expression for the thermal conductivity k_b as given by [24]:

$$k_b = \frac{1}{3} \rho C_{v,e} v_F \lambda_{e,b} \tag{7}$$

where ρ is the mass of electrons per unit volume, $C_{v,e}$ is the volumetric specific heat of electrons, v_F is the Fermi velocity and $\lambda_{e,b}$ is the mean free path of electrons in the bulk material. Upon substituting for ρ and $C_{v,e}$, Eq. (7) can becomes:

$$k_b = \frac{k_B^2 \pi^2 n_e T \lambda_{e,b}}{3 m_e v_F} \tag{8}$$

where k_B is the Boltzmann constant, T is the temperature and n_e and m_e are the number of free electrons per atom and the mass of an electron, respectively. Eq. (8) can be used to calculate the mean free path of electrons in the solid $\lambda_{e,b}$ if the bulk thermal conductivity and Fermi energy are known.

As particle size (L) becomes of the same order as the electron mean free path, boundary or interface scattering will lead to a decrease in particle thermal conductivity. When $L \ll \lambda_{e,b}$, the thermal conductivity of the particle $k_p(L)$ can be expressed as [24]:

$$\frac{k_p(L)}{k_b} = \frac{\lambda_{e,P}}{\lambda_{e,b}} = \frac{1}{Kn} \tag{9}$$

where $Kn = \lambda_{e,b} / L$ is the Knudsen number. When L is of the same order as $\lambda_{e,b}$, the effective mean free path of the electron in the particle can be calculated using the following relation:

$$\frac{1}{\lambda_{e,P}} = \frac{1}{\lambda_{e,b}} + \frac{1}{L} \tag{10}$$

This leads to the following relationship for the thermal conductivity of the particle [24]:

$$\frac{k_p(L)}{k_b} = \frac{\lambda_{e,P}}{\lambda_{e,b}} = \frac{1}{1+Kn} \tag{11}$$

Eqs. (8-11) relate the thermal conductivity of metallic nanoparticles to their characteristic size, which is illustrated in Figure 2 for copper nanoparticles. The thermal conductivity of copper nanoparticles shown in Figure 2 was calculated using Eq. (9) when $Kn > 5$, and Eq. (11) when $Kn < 1$. In the intermediate region ($1 < Kn < 5$), the thermal conductivity was obtained by interpolation. Although no data are available to validate these calculations, the measurements of Nath and Chopra [33] for the thermal conductivity of thin films of copper (also plotted in Figure 2) clearly show a decrease in the thermal conductivity as the thickness of the film decreases. It is expected that metallic nanoparticles will exhibit similar trends with size.

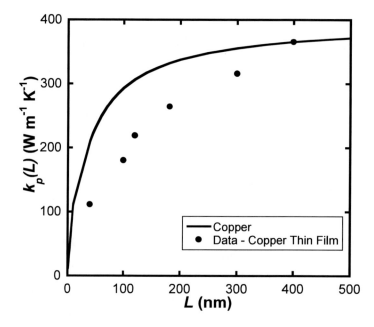

Figure 2. Size effects on thermal conductivity of metals. Data on Cu thin films by Nath and Chopra [33].

The studies described above imply that the contribution of the particle thermal conductivity to the effective thermal conductivity of a solid dispersion should also decrease as the size of the dispersed particles approaches the phonon or electron mean free path. The following section reviews models related to the thermal conductivity of dispersions of solid particles in liquids in light of the above observations.

3.2. Thermal Conductivity Models

3.2.1. Classical Heterogeneous Models

Maxwell [7] derived the following relationship for the thermal conductivity of dilute suspensions of spherical particles:

$$\frac{k_{eff}}{k_l} = 1 + \frac{3(\alpha-1)\phi}{(\alpha+2)-(\alpha-1)\phi} \tag{12}$$

where k_{eff} is the effective thermal conductivity of the dispersion, α is the ratio (k_p/k_l) of thermal conductivity of the particle to that of the fluid, and ϕ is the particle volume fraction. The model is applicable to uniform dispersions of spherical particles when there are no particle interactions. Maxwell's model has been extended by Rayleigh [34], Hamilton and Crosser [18], Jeffrey [35], Progelhof et al. [36], Landauer [37] and others. Maxwell-type models generally imply an effective thermal conductivity that increases with the volume fraction of particles, is dependent mostly on the thermal conductivity of the base liquid, and is 0-15 % greater than that of the base liquid when $\phi < 0.05$. Turian et al. [38] demonstrated that models such as those proposed by Maxwell [7], Jeffrey [35] and Progelhof et al. [36], are able to fit experimental data for dilute suspensions within 2 % when $0.4 < \alpha < 2.4$. Agreement with experiment becomes less satisfactory as α and/or ϕ increases. Note that increasing ϕ often leads to particle aggregation and this cannot be accounted for by Maxwell-type models.

Krischer [39] considered an array of elements of specific resistance distributed in a matrix, and obtained:

$$k_{eff} = \left[\frac{1-f}{(1-\phi)k_l+\phi k_P} + f\left(\frac{1-\phi}{k_l}+\frac{\phi}{k_P}\right) \right]^{-1} \tag{13}$$

where f is an empirical factor equivalent to the fraction of parallel resistances in a rectangular array of elements. This model can be used to determine upper and lower bounds for the thermal conductivity of a heterogeneous system, with $f = 0$ signifying that all particles are arranged in series (creating a high thermal conductivity pathway) and $f = 1$ signifying that all particles are arranged in parallel (creating a low thermal conductivity arrangement). The actual value of f must be obtained by experiment. Tsao [40] developed a similar model by considering different geometries of the discrete phase, whereas Hashin and Shtrikman [41] derived more restrictive bounds using the variational theorem. Their bounds may be expressed as follows:

$$1+\frac{3\phi(\alpha-1)}{\alpha+2-\phi(\alpha-1)} \leq \frac{k_{eff}}{k_l} \leq \alpha\left[1-\frac{3(\alpha-1)(1-\varphi)}{3\alpha-\phi(\alpha-1)}\right] \tag{14}$$

In Eq. (14), the lower bound is the Maxwell limit, whereas the upper bound implies an effective thermal conductivity that is higher than the Maxwell limit when particle aggregation is significant. Turian et al. [38] noted that the upper and lower bounds of the Krischer model are equal to the volume-fraction-weighted arithmetic and harmonic means of the thermal conductivities of the two phases. The geometric mean of the two thermal conductivities falls between these bounds and also falls within the more restrictive Hashin - Shtrikman bounds when $\alpha > 5$. Turian et al. therefore used the volume fraction weighted geometric mean of the thermal conductivities of the individual phases defined by:

$$\frac{k_{eff}}{k_l} = \alpha^{\phi} \tag{15}$$

to calculate thermal conductivities of their suspensions. They found that, in the range $3.5 < \alpha < 70$, the Maxwell model was able to fit data within 14.3 % whereas Eq. (15) could predict data within 5.7 % of experimental values. When $70 < \alpha < 200$, the average deviation was 26.3 % for the Maxwell equation and 9.9 % for Eq. (15). Turian et al. [38] concluded that Eq. (15) provides good estimates of the effective thermal conductivity for particle suspensions when $\alpha > 3.5$. It should be added here that particle sizes in their suspensions were relatively large (of the order of microns).

Similar "mixture models" for the effective thermal conductivity of composites have also been proposed [42, 43]. These models may be summarized as follows:

$$\left(k_{eff}\right)^n = \phi\left(k_p\right)^n + \left(1 - \phi\right)\left(k_l\right)^n \qquad \text{-1 < n < 1} \tag{16}$$

When $n = 1$, Eq. (16) reduces to the arithmetic mean of the thermal conductivities of the two materials, appropriate for conduction in materials arranged in parallel. Similarly, when $n = -1$, Eq. (16) reduces to the harmonic mean of the two thermal conductivities, suitable for conduction in materials arranged in series. Finally, for n approaching zero, Eq. (16) reduces to the geometric mean of the thermal conductivities of the two materials. As discussed previously, Turian et al. [38] have shown that the geometric mean works well for describing the thermal conductivity of heterogeneous suspensions of micron-sized particles. Their conclusions agree with those of Prasher et al. [44] who suggested further that particle aggregation would result in thermal conductivity enhancements that are greater than those predicted by the Maxwell equation. None of these models, however, account for any particle size dependence of the thermal conductivity.

3.2.2. Nanofluid Thermal Conductivity Models

Several mechanistic models for heat transport in nanofluids have been proposed to account for thermal conductivities that exceed values predicted by the Maxwell equation. These models are discussed below.

Yu and Choi [45] proposed a contribution to the thermal conductivity from an ordered liquid layer at the solid-liquid interface. This ordered layer is assumed to have a higher thermal conductivity than the bulk liquid, leading to an effective thermal conductivity given by:

$$\frac{k_{eff}}{k_l} = \frac{k_{pe} + 2k_l + 2\left(1+\beta_l\right)^3\left(k_{pe}-k_l\right)\phi}{k_{pe} + 2k_l - \left(1+\beta_l\right)^3\left(k_{pe}-k_l\right)\phi} \tag{17}$$

where β_l is the ratio of the ordered liquid layer thickness to the nanoparticle radius, and k_{pe} is the effective thermal conductivity of the particle defined by:

$$k_{pe} = \frac{\left[2\left(1-\gamma\right)+\left(1+\beta_l\right)^3\left(1+2\gamma\right)\right]\gamma}{\left(1+\beta_l\right)^3\left(1+2\gamma\right)-\left(1-\gamma\right)}k_p \tag{18}$$

Here, γ is the ratio of the thermal conductivity of the ordered liquid layer to that of the solid particle. Similar models based on an effective particle size that includes the surrounding ordered liquid layer have been proposed by others [46-51]. These models imply an inverse relationship of the effective thermal conductivity with particle size, and generally treat the thickness and thermal conductivity of the ordered liquid layer as adjustable parameters. However, the ordered layer thickness obtained by fitting data for nanofluids was found to be between 1 nm and 3 nm [48], and the thermal conductivity of the ordered layer was reported [45, 49] to be about 5-10 times the thermal conductivity of the base fluid. In contrast, Li et al. [52] and Evans et al. [53] used MD simulations to estimate an ordered layer thickness of about 0.5 nm, and an ordered layer thermal conductivity of crystalline water to be about three times that of liquid water. Neither of these values is in agreement with that obtained by fitting data. Moreover, use of values from MD simulations leads to enhancements that are about the same as those predicted by the Maxwell equation, except when particle diameters are less than 5 nm. A number of models attribute the enhanced thermal conductivity of nanofluids to a local microscale convective effect created by Brownian motion of particles. [54-58] For instance, Xuan et al. [57] included the microconvective effect of the dynamic particles in the Maxwell equation to obtain:

$$\frac{k_{eff}}{k_l} = \frac{\alpha+2+2\phi\left(\alpha-1\right)}{\alpha+2-\phi\left(\alpha-1\right)} + \frac{18\phi H A k T}{\pi^2 \rho d^6 k_l}\tau \tag{19}$$

where H is the overall heat transfer coefficient between the particle and the fluid, A is the corresponding heat transfer area, and τ is a comprehensive relaxation time constant. The heat transfer area is proportional to the square of the diameter, leading to an effective thermal conductivity that is inversely proportional to the fourth power of the particle diameter. Such strong inverse dependence on particle size has not been demonstrated experimentally. In addition, Eq. (19) reduces to the Maxwell equation with increasing particle size, and cannot therefore account for thermal conductivities that are greater than the Maxwell limit.

A model that incorporates an interfacial thermal resistance into the Maxwell equation has been published by Nan et al. [59]. The interfacial thermal resistance is related to the different rates of transport in the two phases and leads to a temperature discontinuity at the solid-liquid interface. The interfacial resistance is sometimes referred to as the Kapitza resistance [60] and includes the effects of phonon scattering at the interface, as well as other phenomena that create resistance to heat transport such as poor contact between the phases. For spheres, the Nan et al. [59] model can be written as follows:

$$\frac{k_{eff}}{k_l} = \frac{\alpha(1+2\chi) + 2 + 2\,\phi\left[\alpha(1-\chi)-1\right]}{\alpha(1+2\chi) + 2 - \phi\left[\alpha(1-\chi)-1\right]}$$

(20)

where $\chi = R_B\,k_l\,/\,d$ and R_B is the interfacial thermal resistance. Eq. (20) reduces to the Maxwell equation when $\chi \ll 1$ and is thus incapable of representing enhancements that are greater than those predicted by the Maxwell equation. However, it correctly predicts a decrease in effective thermal conductivity with particle size.

Another type of model takes account of particle aggregation in nanofluids and was proposed by Prasher et al. [44]. Their model assumes that particle aggregates form conductive pathways in the fluid resulting in enhancements that are greater than those predicted by the Maxwell model. However, the model requires information on aggregate size, as well as the fraction of particles forming conductive pathways. These quantities are seldom available, although the hypothesis that particle clustering enhances conduction in suspensions is supported by numerical simulations and molecular dynamics studies [61-63].

It was observed by Beck et al. [64-66] that predictions from the geometric mean model (Eq. 15) agree well with experimental values for a variety of nanofluids when the particle size is large enough to ignore size effects on thermal conductivity. The geometric mean model was modified by Warrier et al. [32, 67] to incorporate size effects as follows:

$$\frac{k_{eff}(L,T,\varphi)}{k_l(T)} = \left[\frac{k_p(L,T)}{k_l(T)}\right]^{\varphi}$$

(21)

where $k_p(L,T)$ is the size dependent thermal conductivity of the nanoparticle given by Eqs. (4, 9, and 11). The modified geometric mean model was verified with literature data for various particle − fluid systems including nanofluids containing metallic, semiconducting, and insulator particles. Figure 3 compares the predictions of modified geometric mean model with experimental data for alumina nanofluids [66].

An advantage of this model is that the adjustable parameter p (see Eq. 4) depends only on the particle and is independent of the base fluid. Therefore, once the value of p is established, the model can be used to "predict" the thermal conductivity of nanofluids based on any fluid.

Variation in the thermal conductivity of solids with particle size was used by Teja et al. [68] to design nanofluids that exhibit a lower effective thermal conductivity compared with that of the base liquid. Nanofluid models that do not incorporate size effects on intrinsic thermal conductivity of particles cannot predict this behavior. The modified geometric mean model correctly predicted the decrease in thermal conductivity of the fluid upon addition of nanoparticles using only one value of the adjustable parameter p (see Figure 4).

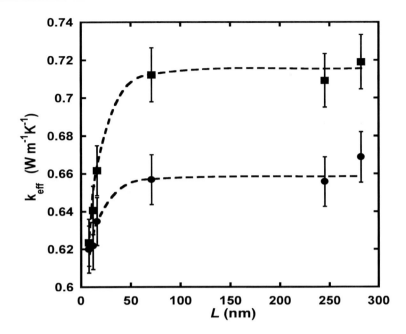

Figure 3. Thermal conductivity of aqueous nanofluids containing 2 (●) and 4 (■) % (v/v) alumina particles at room temperature. Points represent experimental values of Beck et al. [66]. Dashed lines represent calculations using Eqs. (4) and (21).

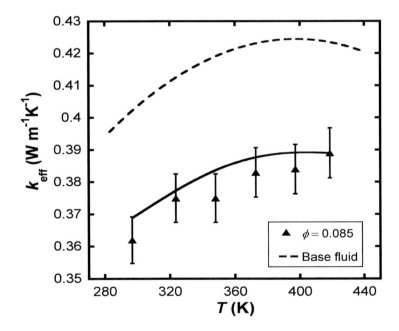

Figure 4. Thermal conductivity of 2 nm titania nanoparticles dispersed in ethylene glycol + water mixture. The data points are from Teja et al. [68]. Solid lines are predictions using the modified geometric mean model with $p = 0.41$.

Based on the above discussion, the thermal conductivity behavior of nanofluids can be summarized as follows:

- The addition of solid particles to liquids generally leads to a change in the effective thermal conductivity of nanofluids, in proportion to the amount (volume fraction) of particles added, and to the thermal conductivity of the solid particles.
- Thermal conductivity enhancements in dilute nanofluids ($\phi < 0.05$) are generally less than 30 %, although a few studies have reported much larger enhancements. It should be noted that an International Nanofluid Property Benchmark Exercise (INPBE) carried out in 2009 [69] also concluded that nanofluids do not exhibit any anomalous enhancement in thermal conductivity.
- The temperature dependence of the effective thermal conductivity of nanofluids conforms closely to that of the base fluid.
- The effective thermal conductivity decreases with decreasing size of dispersed particles, when particles are very small.
- Particle aggregation has a significant effect on the thermal conductivity of nanofluids, although the magnitude of this effect must be determined by experiment.

4. VISCOSITY OF NANOFLUIDS

Most studies of nanofluids have focused on their thermal properties and shown that the addition of nanoparticles to a liquid results in an enhancement in the thermal conductivity. Very few studies have investigated the rheological properties of nanofluids, although it is known that an increase in viscosity is detrimental to the heat transfer performance of the base fluid. In the following sections, we discuss the rheological properties of nanofluids using data for a pentadecane nanofluid with dispersed bentonite nanoparticles.

4.1. Data Discussion

Published work on nanofluids has shown that the viscosity of heterogeneous mixtures increases with particle volume fraction. Most studies [70-75] have also concluded that nanofluids exhibit non-Newtonian shear thinning behavior, even when the base fluid is Newtonian. A few studies [76, 77], however, did note that nanofluid behavior is Newtonian at low particle concentrations. The temperature dependence of nanofluid viscosity was found to be the same as that of the base fluid, with viscosity decreasing as the temperature increased [70, 74, 78]. Also, as noted by Chang et al. [79], Kang et al. [80], Zhao et al. [81] and others, the viscosity of nanofluids increases with decreasing particle size. In addition, Zhao et al. [81] observed that for particle sizes below ~ 20 nm, the ratio of aggregate size to particle diameter increased significantly and resulted in a nonlinear increase in the viscosity with decreasing particle size. In general, all these studies agree qualitatively with regard to the general behavior of nanofluid viscosity as a function of temperature, particle volume fraction, and particle size. Any quantitative differences can be attributed to different techniques being employed to prepare and stabilize the nanofluids investigated. However, none of the published studies have investigated all the variables for a single nanofluid. This makes it difficult to draw general conclusions regarding the viscosity behavior of nanofluid coolants. This is particularly true for fluorocarbon-based nanofluids of interest in direct immersion cooling of electronics. The high volatility and low viscosity of fluorocarbons does not permit extensive evaluation of their rheological properties. Therefore, in the following sections, we discuss nanofluid viscosity using a fluid with very low volatility and relatively high viscosity. This nanofluid was prepared in our laboratory by dispersing bentonite nanoparticles in pentadecane.

4.1.1. Rheological Characteristics of Bentonite-Pentadecane Nanofluids

The rheological characteristics of pentadecane nanofluids containing dispersed bentonite particles of average diameter 15.7 nm and porosity of 12.2 % were investigated using an Anton Paar MCR 300 rheometer.

4.1.1.1. Effect of Temperature

Figure 5 illustrates the viscosity-temperature behavior of the base fluid (pentadecane) and of pentadecane nanofluids containing 1 % and 4 % (v/v) bentonite nanoparticles. Viscosities of the nanofluids were measured as a function of temperature at a constant 10 s^{-1} shear rate. Note that the viscosity-temperature behavior of the two nanofluids follows that of the base fluid, as expected.

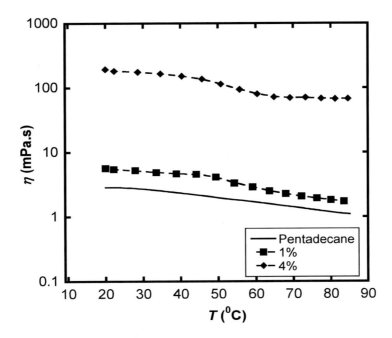

Figure 5. Effect of temperature on the viscosity of a pentadecane nanofluid containing dispersed bentonite nanoparticles.

4.1.1.2. Effect of Shear Rate

Figure 6 illustrates the viscosity-shear rate behavior of the pentadecane nanofluid containing 1 % and 4 % bentonite nanoparticles at 20 and 100 °C. The nanofluids exhibit shear thinning behavior, as shown by the decrease in viscosity with shear rate.

4.1.1.3. Effect of Particle Concentration

The effect of particle concentration on the viscosity of the two nanofluids was investigated at 20 and 50 °C. The results are shown in Figure 7. At low concentrations ($\phi < 0.03$), the viscosity appears to be increasing linearly with particle volume fraction. However, the increase is nonlinear at higher volume fractions.

We conclude that the viscosity of nanofluids changes by several orders of magnitude with temperature, particle concentration, and shear rate (as illustrated in Figures 5-7). Such changes are much larger than those observed in the case of the thermal conductivity. The viscosity of nanofluids apparently decreases with shear rate, follows the temperature behavior of the base fluid, and increases in a non-linear fashion with the volume fraction of particles. In the following section, we review viscosity models and comment on their applicability to nanofluids.

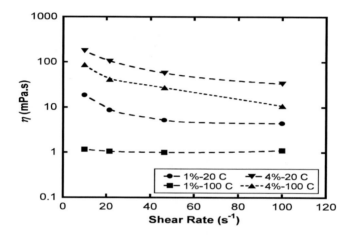

Figure 6. Shear rate vs. viscosity for bentonite nanofluid at 20 and 100 °C.

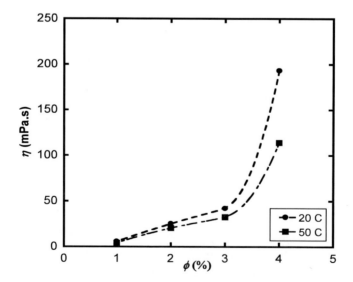

Figure 7. Effect of particle concentration on viscosity of bentonite nanofluid.

4.2. Viscosity Models

Einstein [82] investigated the viscosity of colloidal dispersions and developed the following equation for the viscosity of very dilute dispersions of hard spheres:

$$\eta = \eta_0 \left(1 + [\eta]\phi\right) \tag{22}$$

where η is the viscosity of the dispersion, η_0 is the viscosity of the base fluid, ϕ is the volume fraction and $[\eta]$ is the intrinsic viscosity of the suspension. For hard spheres, $[\eta] = 2.5$, which works well for dilute suspensions ($\phi < 0.01$). Batchelor [83] extended the model to concentrated suspensions by introducing the Huggins constant to account for interparticle interactions.

A semi-empirical relationship for the viscosity of suspensions covering the full range of particle volume fraction was obtained by Krieger and Dougherty [84, 85], who showed that:

$$\eta_r = \left(1 - \frac{\phi_a}{\phi_m}\right)^{-[\eta]\phi_m} \tag{23}$$

where $\phi_a = \phi / \phi_{ma}$. In Eq. (23), ϕ_{ma} is the packing fraction of aggregates in the suspension and ϕ_m is the maximum particle packing fraction. For power law fluids of consistency index D,

$$\phi_a = \phi \left(\frac{a_a}{a}\right)^{3-D} \tag{24}$$

where a_a/a is the ratio of the effective radius of the aggregate to that of the primary particle. ϕ_m generally varies from 0.495 to 0.54 under quiescent conditions, and is approximately 0.605 at high shear rates [86]. The Krieger and Dougherty model is widely used to estimate the viscosity of suspensions, as it covers the entire range of particle concentrations. However, aggregation alone cannot describe the rheological properties of nanofluids [87]. This has led many researchers to develop nanofluid viscosity models by drawing parallels with the nanofluid thermal conductivity. For example, Avsec and Oblak [88] modified the Taylor series expansion in Einstein's model by incorporating liquid layering to obtain:

$$\eta_r = 1 + 2.5\phi_e + \left(2.5\phi_e\right)^2 + \left(2.5\phi_e\right)^3 + \left(2.5\phi_e\right)^4 + \ldots \tag{25}$$

where ϕ_e is the effective volume fraction of the particles expressed as:

$$\phi_e = \phi \left(1 + \frac{h}{r}\right)^3 \tag{26}$$

In Eq. (26), h is the thickness of the nanolayer and r is the particle radius.

A Brownian motion based model was developed by Masoumi et al. [89] to obtain the following expression for viscosity:

$$\eta = \eta_0 + \frac{\rho_p V_B d_p^2}{72C\delta} \qquad (27)$$

where, ρ_p is the particle density, δ is the average interparticle separation, d_p is the particle diameter, V_B is the Brownian velocity, and C is a correction factor given by:

$$C = \frac{\left(c_1 d_p + c_2\right)\phi + \left(c_3 d_p + c_4\right)}{\eta_0} \qquad (28)$$

c_1 - c_4 are adjustable parameters subject to the constraint $\phi \ \square < (c_1 d_p + c_2)/(c_3 d_p + c_4)$.

The models discussed above only describe the behavior of nanofluid viscosity with temperature and particle concentration and do not include the effect of shear rate. As nanofluids generally display shear thinning behavior [70-75], their rheological behavior can be modeled using a power law expression:

$$\eta = K\gamma^{n-1} \qquad (29)$$

where K is the consistency index and n is the flow behavior index. The power $n = 1$ for Newtonian fluids, >1 for shear thickening fluids, and < 1 for shear thinning fluids. Very few nanofluid models are able to describe the rheological behavior of nanofluids as a function of temperature, particle concentration, and shear rate. For this reason, nanofluid viscosities are often correlated using the power law model with different values of K and n for different volume fractions of particles.

In the following section, we evaluate fluorocarbon-based nanofluids for electronics cooling, based on the above observations.

5. NANOFLUIDS FOR ELECTRONICS COOLING

The heat transfer characteristics of nanofluids have been investigated by a number of authors [90, 91] who have reported enhancements in boiling and convective heat transfer. However, these studies were performed with conventional flat plate or wire heaters and did not specifically address microelectronics cooling. Moreover, the increase in critical heat flux obtained with nanofluids was attributed to a change in surface wettability induced by deposition of nanoparticles on the heater surface. [92] Deposition of nanoparticles in electronics systems is highly undesirable as it can lead to localized hot spots.

The application of nanofluids for electronics cooling was investigated by Prasher et al. [93] and Escher et al. [94]. Using the Nusselt number and pressure drop for fully developed flow inside a tube, Prasher et al. [93] concluded that the addition of nanoparticles to conventional microelectronics coolants would not be beneficial if the viscosity enhancement is 4 times greater than the thermal conductivity enhancement. A more restrictive limit was obtained by Escher et al. [94] using a coefficient-of-performance analysis for a microchannel heat sink. Escher et al. reported severe deterioration in the heat sink performance when the increase in viscosity was greater than the increase in thermal conductivity.

To evaluate the effectiveness of nanofluids in electronics cooling, we have performed an FOM analysis based on a 30 % increase in thermal conductivity and an order of magnitude increase in viscosity. We found that the FOM-LB, FOM-T, and FOM-R values would decrease by 82 %, 69 %, and 93 %, respectively, in this case.

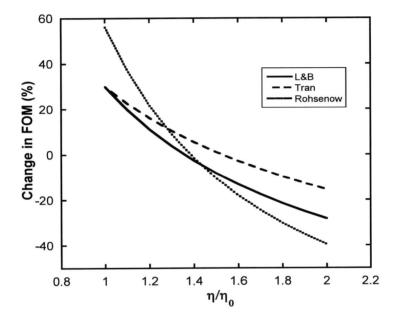

Figure 8. Change in FOM with increasing relative viscosity.

A more detailed analysis is shown in Figure 8, where changes in FOM values (over those of pure HFE 7200) are plotted against the relative viscosity at a constant relative thermal conductivity k_{eff}/k_f of 1.3 (*i.e.* for a 30 % enhancement in the thermal conductivity). We can see that when the enhancement in thermal conductivity is 30 %, then the enhancement in viscosity must be less than 36 %, 53%, and 39 % to produce no changes in FOM-LB, FOM-T, and FOM-R, respectively. Enhancements in viscosity that are greater than these values will have a detrimental effect on heat transfer. Therefore, in designing nanofluids for electronics cooling, it is desirable to find particle characteristics that will produce viscosity enhancements that are no greater than thermal conductivity enhancements. In addition, significant attention must be devoted to the stability of the dispersion in order to avoid any particle deposition.

CONCLUSION

The heat transfer and transport properties of nanofluids have been reviewed, with an emphasis on the thermal conductivity and viscosity of nanofluids. The addition of nanoparticles to fluorocarbon coolants was considered on the basis of a figure-of-merit analysis for phase change and boiling heat transfer. It was concluded that the addition of particles to fluorocarbon coolants such as HFE 7200 would only be beneficial in electronics cooling if the increase in viscosity is less than the increase in thermal conductivity. However, a review of the literature on nanofluids indicates that this is unlikely to be the case in practice. Therefore, the addition of nanoparticles to dielectric coolants should not be considered further for electronics cooling applications.

REFERENCES

[1] Moore, G. E. *Electronics*. 1965, 38(8), 114-117.
[2] Incropera, F. P. *Liquid Cooling of Electronic Devices by Single-Phase Convection*; John Wiley & Sons: New York, U.S.A., 1999.
[3] Tummala, R. R. and Swaminathan, M. *Introduction to System-On-Package (SOP): Miniaturization of the Entire System*, McGraw-Hill: New York, U.S.A., 2008.
[4] Choi, S. U. S. In: Siginer, D. A. and Wang, H. P.; Ed; *Developments and Applications of Non-Newtonian Flows*, ASME-FED: New York, U.S.A., 1995, 231, pp. 99-105.
[5] Eastman, J. A.; Choi, S. U. S.; Li, S.; Yu, W.; and Thompson, L. J. *Applied Physics Letters*. 2001, 78, 718-720.
[6] Choi, S. U. S.; Zhang, Z. G.; Yu, W.; Lockwood, F. E.; Grulke, E. A. *Applied Physics Letters*. 2001, 79, 2252-2254.
[7] Maxwell, J. C., *A Treatise on Electricity and Magnetism*, University Press: Oxford, U.K., 3rd edition, Vol. II., 1892.
[8] Rohsenow, W. M. A. *Method of Correlating Heat Transfer Data for Surface Boiling of Liquids*, Division of Industrial Cooperation, Massachusetts Institute of Technology, Massachusetts, U.S.A., 1951, Technical Report No. 5.
[9] Lazarek, G. M. and Black, S. H. *International Journal of Heat and Mass Transfer*. 1982, 25(7), 945-960.
[10] Tran, T. N.; Wambsganss, M. W.; France, D. M., *International Journal of Multiphase Flow*. 1996, 22(3), 485-498.
[11] Chen, T. and Garimella, S. V. *IEEE Transactions in Components and Packaging Technology*. 2007, 30, 24-31.
[12] Klimenko, V. V. *International Journal of Heat and Mass Transfer*. 1990, 33(10), 2073-2088.
[13] Liu, D.; Garimella, S. V. *Journal of Heat Transfer*. 2007, 129, 1321-1332.
[14] Riehl, R. R.; Seleghim, P.; Ochterbeck, J. M. *Sixth Intersociety Conference on Thermal and Thermomechanical Phenomena in Electronic Systems*. 1998, 409-416.
[15] Dhir, V. K. *Annual Reviews of Fluid Mechanics*. 1998, 30, 365-401.
[16] Warrier, P.; Sathyanarayana, A.; Patil, D. V.; France, S.; Joshi, Y.; Teja, A. S. *International Journal of Heat and Mass Transfer*, 2012, 55, 3379–3385.
[17] Warrier, P; Sathyanarayana, A.; Bazdar, S.; Joshi, Y.; Teja, A. S. *Industrial and Engineering Chemistry Research*. 2012, 51, 10517-10523.
[18] Hamilton, R. L. and Crosser, O. K. *Industrial & Engineering Chemistry Fundamentals*. 1962, 1, 187-191.

[19] Bird, R. B.; Stewart, W. E.; Lightfoot, E. N. *Transport Phenomena*. John Wiley & Sons, Inc: New York, U.S.A., 2002.

[20] Meyer, C. A.; Ed.; *ASME Steam Tables: Thermodynamic and Transport Properties of Steam*. American Society of Mechanical Engineers: New York, U.S.A., 1993.

[21] Diguilio, R. and Teja, A. S., *Journal of Chemical and Engineering Data*. 1990, 35, 117-121.

[22] Tien, C. -L.; Majumdar, A.; Gerner, F. M.; Ed; *Microscale Energy Transport*. Taylor & Francis: Washington D.C., 1998.

[23] Tien, C. -L.; Ed.; *Annual Review of Heat Transfer*. Begell House, Inc.: New York, U.S.A., Vol. 7, 1996.

[24] Zhang, Z. M. *Nano/Microscale Heat Transfer*. McGraw Hill Professional: Nanoscience and Nanotechnology Series, 2007.

[25] Ju, Y. S. *Applied Physics Letters*. 2005, 87(15), 153106.

[26] Behkam, B.; Yang, Y. Z.; Asheghi, M. *International Journal of Heat and Mass Transfer*. 2005, 48, 2023-2031.

[27] Liu, W. and Asheghi, M., *Applied Physics Letters*. 2004, 84, 3819-3821.

[28] Li, D. Y.; Wu, Y. Y.; Kim, P.; Shi, L.; Yang, P. D.; Majumdar, A. *Applied Physics Letters*. 2003, 83, 2934-2936.

[29] Fang, K. C.; Weng, C. I.; Ju, S. P. *Nanotechnology*. 2006, 17, 3909-3914.

[30] Liang, L. H. and Li, B. *Physical Review B*. 2006, 73, 153303.

[31] Zhang, Z.; Zhao, M.; Jiang, Q. *Semiconductor Science and Technology*. 2001, 16, L33-L35.

[32] Warrier, P.; Yuan, Y.; Beck, M. P.; Teja, A. S. *AIChE Journal*. 2010, 56(12), 3243-3256.

[33] Nath, P. and Chopra, K. L. *Thin Solid Films*. 1974, 20(1), 53-62.

[34] Rayleigh, L., *Philosophical Magazine*. 1892, 34, 481-502.

[35] Jeffrey, D. J. *Proceedings of the Royal Society (London) A*. 1973, 335, 355-367.

[36] Progelhof, R. C.; Throne, J. L.; Ruetsch, R. R. *Polymer Engineering & Science*. 1976, 16, 615-625.

[37] Landauer, R. *Journal of Applied Physics*. 1952, 23, 779-784.

[38] Turian, R. M.; Sung, D. J.; Hsu, F. L. *Fuel*. 1991, 70, 1157-1172.

[39] Krischer, O. *Die Wissenschaftlichen Grundlagen der Trocknungstechnik (The Scientific Fundamentals of Drying Technology)*, Springer-Verlag: Berlin, Germany, 1963.

[40] Tsao, G. T. N. *Industrial & Engineering Chemistry*. 1961, 53, 395-397.

[41] Hashin, Z. and Shtrikman, S. *Journal of Applied Physics*. 1962, 33, 3125.

[42] Nielsen, L. E. *Predicting the Properties of Mixtures: Mixing Rules in Science and Technology*, Mercer Dekker: New York, U.S.A., 1978.

[43] Nan, C. W. *Progress in Material Science*. 1993, 37, 1-117.

[44] Prasher, R.; Evans, W.; Meakin, P.; Fish, J.; Phelan, P.; Keblinski, P. *Applied Physics Letters*. 2006, 89, 143119.

[45] Yu, W. and Choi, S. U. S., *Journal of Nanoparticle Research*. 2003, 5, 167-171.

[46] Yu, W. and Choi, S. U. S., *Journal of Nanoparticle Research*. 2004, 6, 355-361.

[47] Xie, H. Q.; Fujii, M.; Zhang, X. *International Journal of Heat and Mass Transfer*. 2005, 48, 2926-2932.

[48] Xue, Q. Z. *Physica B*. 2005, 368, 302-307.

[49] Leong, K. C.; Yang, C.; Murshed, S. M. S. *Journal of Nanoparticle Research*. 2006, 8, 245-254.

[50] Feng, Y. J.; Yu, B. M.; Xu, P.; Zou, M. Q. *Journal of Physics D: Applied Physics*. 2007, 40, 3164-3171.

[51] Lee, D. *Langmuir*. 2007, 23, 6011-6018.

[52] Li, L.; Zhang, Y. W.; Ma, H. B.; Yang, M. *Physics Letters A*. 2008, 372, 4541-4544.

[53] Evans, W.; Fish, J.; Keblinski, P. *Journal of Chemical Physics*. 2007, 126, 154504.

[54] Koo, J. and Kleinstreuer, C. *Journal of Nanoparticle Research*. 2004, 6, 577-588.

[55] Prasher, R.; Bhattacharya, P.; Phelan, P. E. *Physical Review Letters*. 2005, 94, 025901.

[56] Ren, Y.; Xie, H.; Cai, A. *Journal of Physics D: Applied Physics*. 2005, 38, 3958-3961.

[57] Xuan, Y. M.; Li, Q.; Zhang, X.; Fujii, M. *Journal of Applied Physics*. 2006, 100, 043507.

[58] Prakash, M. and Giannelis, E. P., *Journal of Computer Aided Materials Design*. 2007, 14, 109-117.

[59] Nan, C. W.; Birringer, R.; Clarke, D. R; Gleiter, H. *Journal of Applied Physics*. 1997, 81, 6692-6699.

[60] Swartz, E. T. and Pohl, R. O. *Reviews of Modern Physics*. 1989, 61, 605-668.

[61] Prasher, R.; Evans, W.; Meakin, P.; Fish, J.; Phelan, P.; Keblinski, P. *Applied Physics Letters*. 2006, 89, 143119.

[62] Kumar, S. and Murthy, J. Y. *Numerical Heat Transfer Part B*. 2005, 47, 555-572.

[63] Gao, L. and Zhou, X. F. *Physics Letters A*. 2006, 348, 355-360.

[64] Beck, M. P.; Yuan, Y.; Warrier, P.; Teja, A. S. *Journal of Applied Physics*. 2010, 107, 066101.

[65] Beck, M. P.; Yuan,Y.; Warrier, P.; Teja, A. S. *Journal of Nanoparticle Research*. 2010, 12, 1469-1477.

[66] Beck, M. P.; Yuan,Y.; Warrier, P.; Teja, A. S. *Journal of Nanoparticle Research*. 2009, 11, 1129-1136.

[67] Warrier, P. and Teja, A. S. *Nanoscale Research Letters*. 2011, 6, 247.

[68] Teja, A. S.; Beck, M. P.; Yuan, Y.; Warrier, P. *Journal of Applied Physics*. 2010, 107, 114319.

[69] Buongiorno, J. et al., *Journal of Applied Physics*. 2009, 106, 094312.

[70] Teipel, U. and Förter-Barth, U. *Propellants, Explosives, Pyrotechnics*. 2001, 26(6), 268-272.

[71] Tseng, W. J. and Wu, C.H., *Acta Materialia*. 2002, 50(15), 3757-3766.

[72] Kulkarni, D. P.; Das, D. K.; Chukwu, G. A. *Journal of Nanoscience and Nanotechnology*. 2006, 6(4), 1150-1154.

[73] Lu, K., *Powder Technology*. 2007, 177(3), 154-161.

[74] Chen, H. and Ding, Y. *Journal of Nanoparticle Research*. 2009, 11(6), 1513-1520.

[75] Phuoc, T. X. and Massoudi, M. *International Journal of Thermal Sciences*. 2009, 48(7), 1294-1301.

[76] Xie, H.; Chen, L.; Wu, Q. *High Temperatures-High Pressures*. 2008, 37, 127–135.

[77] Yu, W.; Xie, H. Q.; Chen, L. F.; Li, Y. *Thermochimica Acta*. 2009, 491(1-2), 92–96.

[78] Duangthongsuk, W. and Wongwises, S. *Experimental Thermal and Fluid Science*. 2009, 33(4), 706-714.

[79] Chang, H.; Jwo, C. S.; Lo, C. H.; Tsung, T. T.; Kao, M. J.; Lin, H. M. *Reviews on Advanced Material Science*. 2005, 10(2), 128-132.

[80] Kang, H. U.; Kim, S. H.; Oh, J. M. *Experimental Heat Transfer*. 2006, 19, 181-191.

[81] Zhao, J. F.; Luo, Z. Y.; Ni, M. J.; Cen, K. F. *Chinese Physics Letters*. 2009, 26(6), 066202.

[82] Einstein A., *Annals of Physics*. 1906, 19, 289.

[83] Batchelor G. K. *Journal of Fluid Mechanics*. 1977, 83, 97-117.

[84] Krieger I. M. and Dougherty T. J. *Transactions of the Society of Rheology*. 1959, 3, 137-152.

[85] Wang L. Q.; Ed; *Advances in Transport Phenomena*. Springer-Verlag: Berlin, Germany, Vol. 1, 2009.

[86] Chen, H.; Ding, Y.; Tan, C. *New Journal of Physics*. 2007, 9, 367.

[87] Nwosu, P. N.; Meyer, J. P.; Mohsen, S., *In Proceedings of the ASME 2012 3rd Micro/Nanoscale Heat & Mass Transfer International Conference*, Atlanta, Georgia, U.S.A., MNHMT2012-75314, 2012.

[88] Avsec J. and Oblak M., *International Journal of Heat and Mass Transfer*. 2007, 50, 4331-4341.

[89] Masoumi N.; Sohrabi N.; Behzadmehr A. *Journal of Physics D: Applied Physics*. 2009, 42(1-6), 055501.

[90] Murshed, S. M. S.; Nieto De Castro, C. A.; Lourenco, M. J. V.; Lopes, M. L. M.; Santos, F. J. V. *Renewable and Sustainable Energy Reviews*. 2011, 15, 2342–2354.

[91] Das, S. K.; Choi, S. U. S.; Yu, W.; Pradeep, T. *Nanofluids: Science and Technology*. Wiley-Interscience: New Jersey, U.S.A., 2008.

[92] Kim, S. J.; Bang. I. C.; Buongiorno, J.; Hu, L. W. *Applied Physics Letters*. 2006, 89, 153107.

[93] Prasher, R.; Song, D.; Wang, J.; Phelan, P. *Applied Physics Letters*. 2008, 89, 133108.

[94] Escher, W.; Brunschwiler, T.; Shalkevich, T.; Burgi, T.; Michel, B.; Poulikakos, D. *Journal of Heat Transfer*. 2011, 133, 051401.

[95] Murshed, S. M. S., *Heat Transfer Engineering*, 2012, 33(8), 722-731.

In: Nanofluids: Synthesis, Properties and Applications
Editors: S.M. Sohel Murshed, C.A. Nieto de Castro

ISBN: 978-1-63321-677-8
© 2014 Nova Science Publishers, Inc.

Chapter 3

THERMOPHYSICAL PROPERTIES AND HEAT TRANSFER CHARACTERISTICS OF CARBON NANOTUBES DISPERSED NANOFLUIDS

S. M. Sohel Murshed[*] *and C. A. Nieto de Castro*
Centro de Ciências Moleculares e Materiais
Faculdade de Ciências, Universidade de Lisboa
Lisboa, Portugal

ABSTRACT

Research progress in thermophysical properties and convective heat transfer characteristics of carbon nanotubes (CNT)-laden nanofluids has been reviewed and addressed in this chapter. Besides briefing the preparation of these nanofluids, available studies on thermal conductivity, viscosity, specific heat and convective heat transfer of this specific class of nanofluids are discussed in detail. Effects of different parameters such as carbon nanotube concentration and temperature on thermal conductivity, viscosity, specific heat and convective heat transfer coefficient are also demonstrated. Despite inconsistencies among the available data, substantial increases in these thermal features of CNT-nanofluids compared to their base fluids are undisputed. Nevertheless, reported data are still limited and scattered to clearly understand the underlying mechanisms for the observed anomalous enhancements of thermal properties of these nanofluids. In addition to research works on specific heat and thermal diffusivity, available theoretical models and heat transfer mechanisms of this particular type of nanofluids are presented and discussed. Review reveals that CNT-nanofluids exhibit superior thermal properties compared to their base fluids and these properties further increase with increasing concentration of CNT as well as temperature. With the fascinating properties carbon nanotubes dispersed nanofluids show great potential as advanced heat transfer fluids in many important applications.

---

[*] E-mail: smmurshed@fc.ul.pt

1. INTRODUCTION

Over the last several decades scientists and engineers have attempted to develop fluids, which can offer better heat transfer performance compared to conventional fluids. However, there was no success until Steve Choi of Argonne National Laboratory of USA coined the concept of "nanofluids" in 1995 [1]. Nanofluids (NF) are a new class of engineered heat transfer fluids which is defined by the suspensions of nanomaterials of any shape such as spherical particle, rod, tubes, and flakes in conventional heat transfer fluids. It is also noted that such suspensions of nanoparticles (i.e. nanofluids) were used in a boiling heat transfer study as early as 1984 by Yang and Maa [2] and another experimental study on the thermal conductivity and viscosity of several types of nanoparticles-suspensions was performed by Masuda and co-workers [3] in 1993. Furthermore, Arnold Grimm [4], a German researcher also won a German patent on the enhanced thermal conductivity of suspensions of nano- and micro-sized particles in 1993. Nevertheless, with anomalously high thermophysical properties and heat transfer characteristics nanofluids showed great promises as advanced heat transfer fluids and can meet the cooling and heating challenges facing numerous high-tech industries and thermal management systems.

Carbon nanotubes (CNT) are often known as wonder nanomaterials which have very large aspect ratio and a very broad range of unique thermal, mechanical, chemical, optical and electronic properties. The remarkable and unique properties of carbon tubes have placed them right among the hottest topics in multidisciplinary fields, particularly in materials sciences. Thus research on carbon nanotubes has become a hot topic in multidisciplinary fields. Although innovation of CNT is attributed to a Japanese scientist, Iijima [5] in 1991, Endo and co-workers [6] first reported TEM images of CNT in 1976. Iijima [5] demonstrated synthesizing of needle-like nano-sized (diameter ranging from 4 to 30 nanometers) carbon tubes using arc-discharge evaporation technique. Later in 1993 the growth process of single wall CNT was reported by two research groups, one by Iijima and Ichihashi [7] and the other by Bethune and co-workers [8]. Since this revolutionary discovery in early 1990 carbon nanotubes have attracted immense interest from both the researcher and the industrial communities due to their fascinating properties and potential applications in numerous fields such as aerospace, automotive, electronic, optical, and energy conversion [9-10]. These nanotubes can behave like metals or semiconductors and can conduct better electricity and heat compared to copper and diamond, respectively. Thus they can be used in nanoelectronics like diodes and transistors and in supercapacitors as electromechanical actuators and sensors, in lithium-ion batteries, as well as fillers in composite materials such as polymer-based composites [10-12]. The remarkable thermal property of carbon nanotube is their ultra-high thermal conductivity which is order magnitude higher compared to those of the metallic or oxide nanomaterials commonly used in nanofluids as heat transfer enhancer. Thus like carbon nanotubes, nanofluids have attracted great interest from researchers worldwide because of their anomalously high thermophysical properties and potentials applications in numerous important fields such as microelectronics, MEMS, microfluidics, transportation, manufacturing, instrumentation, medical, and HVAC systems [13-23]. Even after almost two decades of innovation of these new fluids (i.e., nanofluids)[1] and having extensive research performed thereafter, mechanisms for anomalous thermal features particularly thermal conductivity of nanofluids are still inconclusive [13-16,23]. Researchers are still facing enormous challenges to uncover the true mechanisms behind such anomalously thermal properties of nanofluids.

With such ultrahigh thermal conductivity of carbon nanotubes, their nanofluids exhibit much higher thermophysical properties and heat transfer features such as thermal conductivity, viscosity, specific heat and convective as well boiling heat transfer performance as compared to their base fluids as well as nanofluids containing other types of nanomaterials [13-17, 24-34]. With very large aspect ratio carbon nanotubes also exhibit excellent dispersion behavior in most of the commonly used solvents. In order to fully utilize these superior thermal characteristics, it is very important to prepare CNT-nanofluids that show long term stability and well-dispersion of CNT. Literature results on the thermal features of CNT-nanofluids reveal their great potentials as advanced heat transfer fluids. There is therefore growing research interest on various heat transfer features of this particular type of nanofluid. This can be evidenced from the reported growth of annual publications on this nanofluid in the last decade as shown in Figure 1. According to Web of Science searched results, out of total 5831 publications on nanofluids only 455 carbon nanotubes-related publications have appeared until March 19, 2014. Note that publications include all types of journals and conference articles, patent, news, letter and others. This publications record of CNT-nanofluids reveals that despite large increase in thermophysical properties and recent growing interest, research activities on CNT-nanofluids are not promising as compared to other nanofluids. The main reason could be the comparatively high price of CNT. Nevertheless, research efforts on the CNT-nanofluid are scattered and findings from different research groups lack consistencies. Given the prospect and potentials of CNT-nanofluids, it is imperative and timely to provide an informative review of available research findings on this special type of nanofluids.

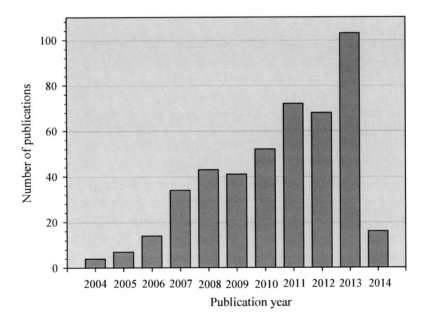

Figure 1. Annual publications on CNT-nanofluids (publications searched by topic "nanofluids" and refined by "carbon nanotubes" in Web of Science on March 19, 2014).

This chapter addresses the recent research on various thermal features of CNT-nanofluids. Besides discussing the sample preparation and stability, the available studies on conduction, convection and viscosity of these nanofluids are thoroughly reviewed. Effects of various parameters such as concentration of CNT and fluid temperature on these thermal features are reported and analyzed. The specific heat capacity of this nanofluid is also discussed.

2. PREPARATION OF CNT -NANOFLUIDS

Stable nanofluids which are important for the optimization of their thermal properties can be ensured by proper preparation and dispersion procedures. Techniques for good dispersion and production of uniform-sized nanoparticles in liquids or directly producing stable nanofluids are crucial. A nanofluid does not mean a simple mixture of liquid and nanoparticles. Proper dispersion and stabilization of the CNT in base fluids (BF) are essential in order to prepare CNT-nanofluids. Homogenization of dispersed nanomaterials is commonly performed using ultrasonication which is very effective in breaking the larger clusters into smaller clusters. There are two types of techniques commonly used to prepare nanofluids and they are the one-step method and the two-step method. The two-step method, which is the dispersion of purchased or produced dry CNT in base fluids, is widely used to prepare any nanofluids. The one-step method is directly synthesizing CNT in base fluids by applying various chemical or physical methods such as chemical vapor deposition. Although the one-step method gives better dispersion and stability of nanofluids, it is less popular mainly due to the complexity in directly synthesizing suspensions of CNT. Nonetheless, most of the studies used the two-step method to prepare CNT-nanofluids. As mentioned previously, with very large aspect ratio carbon nanotubes usually show excellent dispersion behavior in most of the commonly used heat transfer fluids like water (W), ethylene glycol (EG), olefin oil (OO), engine oil (EO), decene (DE), silicon oil (SO) and poly (α-olefin) oil (PAO). Thus, researchers used various types of host fluids to prepare CNT-nanofluids. While most studies used multi-walled carbon nanotubes (MWCNT) dispersed in water or ethylene glycol, very few studies employed single-walled carbon nanotubes (SWCNT) and other types of base fluids. As proper dispersion of CNT and long term stability of their nanofluids are of utmost importance for the optimum properties, most of the researchers added surfactants as well as functionalized or surface treated CNT in order to obtain better stability of sample nanofluids. Sometimes functionalization of CNT is found to be more effective for stability of the nanofluids. For example, Nasiri et al. [35] reported that the functionalized CNT-nanofluids showed better stability and higher thermal conductivity compared to surfactant added CNT-nanofluids. The surfactants commonly used in CNT-nanofluids include sodium dodecyl sulfate (SDS), cetyltrimethyl ammonium bromide (CTAB), hexamethyldisiloxane (HMDS), sodium dodecyl benzene sulfonate (SDBS), polyvinyl pyrrolidone (PVP), chitosan, and gum arabic (GA). In addition, regardless of addition of surfactant and functionalization of CNT, sonication is always used for breaking the agglomerations of nanoparticles including CNT dispersed in base fluids. Sonication time can be varied from several minutes to hours depending on the CNT concentration, base fluids, power or frequency setting of the sonicator. Studies showed that through the state of dispersion sonication time also influence the thermal conductivity of any nanofluids [13, 14, 17]. A summary of preparation of various CNT-nanofluids and CNT treatment or surfactant addition for better dispersion is provided in Table 1.

Table 1. CNT-nanofluids and their stabilization methods from the literature

CNT/Base fluids	Addition of surfactant/ Surface treatment	Researchers
MWCNT/EG	12-3-12,2Br^{-1} addition	Xie and Chen [17]
MWCNT/OO	None	Choi et al. [24]
MWCNT/W	Surface treatment	Xie et al. [25]
MWCNT/W/EG	Surface treatment	Chen et al. [26]
MWCNT/W	SDS addition	Assael et al. [27]
CNT/W	GA addition	Ding et al. [28]
MWCNT/W	SDS addition	Hwang et al. [29]
MWCNT/EO	NHS addition	Liu et al. [30]
SWCNT/EG	None	Amrollahi et al. [31]
MWCNT/W	SDS addition and functionalization	Nasiri et al. [35]
MWCNT/EG/W	None	Liu et al. [36]
MWCNT/SO	HMDS/Functionalization	Chen and Xie [37]
MWCNT/W/EG	Functionalization	Aravind et al. [38]
MWCNT/W	PVP addition	Park et al. [39]
MWCNT/W	SDS addition	Kathiravan et al. [40]
MWCNT/EG/ W	SDS, CTAB, and Triton X-100 addition	Assael et al. [41]
MWCNT/W+EG	Chitosan addition	Teng and Yu [42]
CNT/R113	None	Jiang et al. [43]
MWCNT/W	Plasma-treatment of MWCNT	Kim et al. [44]

3. THERMAL CONDUCTIVITY OF CNT-NANOFLUIDS

3.1. Studies on Thermal Conductivity

As mentioned before, CNT-nanofluids exhibit higher thermal conductivity as compared to their base fluids and they further increase with increasing concentration and temperature. In addition to different types of CNT-nanofluids, researchers employed different techniques to measure the thermal conductivity of their nanofluids. Figure 2 demonstrates that regardless of some inconsistencies in literature data, the significant enhancement of thermal conductivity with CNT concentration is obvious. Interestingly, the enhanced thermal conductivity increases even more pronouncedly with the temperature as shown in Figure 3. Such increase in thermal conductivity with temperature makes these nanofluids more attractive for their applications at elevated temperatures. A summary of room temperature thermal conductivity

(TC) results on CNT-nanofluids from literature is presented in Table 2 which clearly agrees with the aforementioned statement on the enhanced thermal conductivity of these nanofluids.

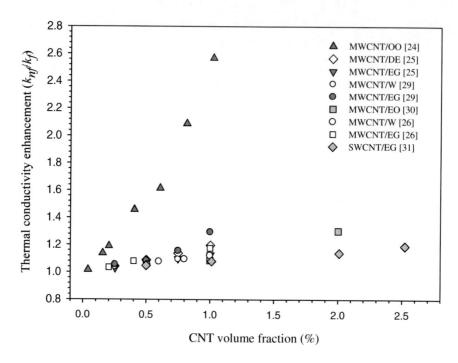

Figure 2. Effect of CNT concentration on the enhancement of thermal conductivity of nanofluids.

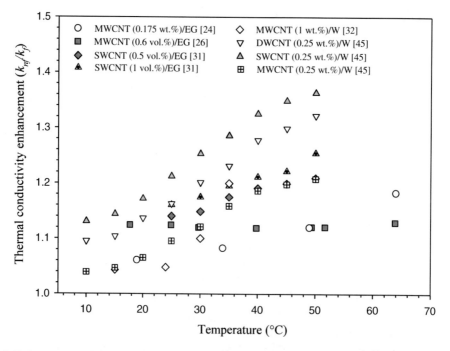

Figure 3. Enhancement of thermal conductivity of CNT-nanofluids as a function of temperature.

In addition to summarizing available results (Table 2), some representative experimental investigations on thermal conductivity of CNT-nanofluids are elaborated here.

Table 2. Summary of literature studies and results on thermal conductivity (TC) of CNT-nanofluids

Researcher group	Nanofluids (+surfactant)	Measuring method	Main findings (Maximum increase in TC)
Choi et al. [24]	MWCNT/Oil	THWM	TC increased 160% at 1 vol. % of CNT.
Xie et al. [25]	MWCNT/W	THWM	At 1 vol. % of CNT, TC increased 7%.
Chen et al. [26]	MWCNT/W	THWM	TC increased 12% at 1 vol. % of CNT.
Chen et al. [26]	MWCNT/EG	THWM	TC increased 17.5% at 1 vol. % of CNT.
Assael et al. [27]	MWCNT/W (+SDS)	THWM	At 0.6 vol. % of CNT, TC increased 38%.
Ding et al. [28]	CNT/W (+GA)	KD2 Pro	TC increased 25% at 0.5 wt. % of CNT.
Hwang et al. [29]	MWCNT/W (+SDS)	THWM	TC increased 13.3% at 1 vol. % of CNT.
Liu et al. [30,36]	MWCNT/EO(+NHS)	THWM	At 2 vol. % of CNT, TC increased 30%.
Amrollahi et al. [31]	SWCNT/EG	Steady method	At 2.5 vol. % of CNT, TC increased 20%.
Liu et al. [36]	MWCNT/EG	THWM	TC increased 12.4% At 1 vol. % of CNT.
Chen and Xie [37]	MWCNT/SO (+HMDS)	TSHWM	TC increased 19% at 1 vol. % of CNT.
Aravind et al. [38]	MWCNT/W/EG	Lambda Instruments (THWM)	At 0.03 vol. % of CNT, TC of water and EG increased up to 33% and 40%, respectively.
Assael et al. [41]	MWCNT/EG (+SDS)	THWM	At 0.6 vol. % of CNT, TC increased 21%.
Assael et al. [41]	MWCNT/W (+TritonX-100)	THWM	At 0.6 vol. % of CNT, TC increased 13%.
Jiang et al. [43]	CNT/R113	Thermal analyzer	At 1 vol. % of CNT, TC increased 104%.
Kim et al. [44]	MWCNT/W	THWM	TC increased 25% at 0.01 vol. % CNT.
Harish et al. [46]	SWCNT/EG	THWM	TC increased 14.8% at 0.2 vol. % CNT.
Hwang et al. [47]	MWCNT/W (+SDS)	THWM	TC increased 7% at 1 vol. % of CNT.
Nanda et al. [48]	SWCNT/EG/PAO	Transient planar source	At 1.1 vol. % of CNT, TC of EG and PAO increased 35% and 12%, respectively.
Assael et al. [41, 49]	MWCNT/W (+CTAB)	THWM	At 0.6 vol. % CNT, TC increased 34%.
Han et al. [50]	Hybrid sphere/CNT /PAO	3ω-wire (THWM)	TC increased 21% at 0.2 vol. % of hybrid sphere/CNT particles.
Glory et al. [51]	MWCNT/W (+GA)	Steady state	At 3 wt. % of CNT, TC increased 64%.
Jha and Rama-prabhua [52]	Ag-MWCNT/W	KD2 Pro	At 0.03 CNT vol. %, TC increased 37.3%.

Yang et al. [53] studied the effect of MWCNT and dispersant concentrations as well as dispersing energy on thermal conductivity of CNT/poly (α-olefin) oil (PAO6)-based nanofluids. Polyisobutene succinimide (PIBSI) dispersant was also used in their study. The thermal conductivity of these nanofluids correlate with each other and vary with the size of large scale agglomerates of nanoparticles as nanofluids with large scale agglomerates have high thermal conductivities. However, the thermal conductivity was found to increase considerably with increasing the concentration of MWCNT in this oil. For instance, thermal conductivity of PA06 oil increased more than three times due to addition of 0.35 vol. % of CNT in this base fluid. On the other side, PIBSI surfactant concentration showed a mixed effect on the thermal conductivity of nanofluids. The enhanced thermal conductivity of nanofluids decreased significantly with increasing surfactant concentration from 0 to 3 wt. % and then started increasing with further increasing the loading of surfactant. The effect of dispersion energy (i.e., sonication power) on thermal conductivity was found to be significant as the thermal conductivity was decreasing almost exponentially with increasing specific dispersing energy.

Following their previous work [27], Assael et al. [49] measured the thermal conductivity of aqueous MWCNT and DWCNT (double-walled CNT) nanofluids and a maximum 34% enhancement of thermal conductivity was observed at 0.6 vol.% loading of MWCNT in CTAB surfactant added water.

Using a 3ω hotwire method Han et al. [50] measured the thermal conductivity of a new type of nanofluid containing hybrid sphere/CNT nanoparticle in poly-alpha-olefin (PAO) oil over a temperature range from 10 to 90 °C. At room temperature the effective thermal conductivity of this nanofluid increased 21% for 0.2% loading of their hybrid sphere/CNT. They also found that the enhanced thermal conductivity increased almost linearly with increasing temperature.

Jana et al. [54] experimentally determined the thermal conductivity of one CNT/w nanofluid and two hybrid CNT nanofluids which are Au and Cu nanoparticles mixed with CNT in water. Their results showed a nonlinear dependence of thermal conductivity of these nanofluids with volume fraction and a maximum increase in thermal conductivity was 34% at CNT volumetric loading of 0.8%. However, no obvious enhancement of thermal conductivity of these nanofluids was found due to addition of CNT to Au and Cu nanoparticles.

Using a thermal constants analyzer (i.e., transient plane source method) Jiang et al. [43] measured the thermal conductivity of nanofluids containing CNT of various aspect ratios and concentrations in a refrigerant (i.e., R113). They found that the maximum enhancement of thermal conductivity of refrigerant was 104% at 1 vol. % of CNT. Results of their study also revealed that the aspect ratio and diameter of CNT significantly influence the thermal conductivity of CNT-nanofluids.

Chen et al. [26] measured the thermal conductivity of treated multiwalled carbon nanotubes (TCNT) dispersed in ethylene glycol using a short hot-wire method (TSHWM) and they reported up to 17.5 % enhancement of thermal conductivity of this nanofluid at 0.01volume fraction of TCNT. In order for better dispersion, their nanotubes were treated by using mechanochemical reaction method. They however did not observe any obvious effects of temperature on thermal conductivity enhancement. Same group later used silicone oil (SO) as base fluid and reported thermal conductivity and viscosity of SO-based nanofluid containing the same TCNT [37]. They found substantial increase in thermal conductivity of this nanofluid which further increased with increasing TCNT loading as well as temperature.

Employing a THWM-based Lambda instrument the thermal conductivity of nanofluids having oxidized MWCNT dispersed in two base fluids (DI water and EG) was measured by Aravind et al. [38]. At a volume fraction of 0.03 the maximum thermal conductivity enhancements for MWCNT/DI water and MWCNT/EG were found to be 33% and 40%, respectively. The enhanced thermal conductivity of these nanofluids was further increased with increasing temperature.

In another study, Amrollahi et al. [31] investigated the effects of temperature, volume fraction, and vibration time on the thermo-physical properties of a SWCNT/EG nanofluid. For 2.5 vol. % concentration of SWCNT the thermal conductivity of ethylene glycol was observed to increase up to 20%. They also reported a strong influence of temperature on the enhancement of the thermal conductivity of this nanofluid.

Nasiri et al. [45] studied the effect of CNT structures on thermal conductivity and stability of nanofluids. They used several different structures, single-wall, double-wall (DWCNT), few-wall (FWCNT) and two different multiwall carbon nanotubes in water. Their results showed that both the stability and thermal conductivity of nanofluids decrease with increasing number of walls of the carbon nanotubes. However, all these CNT nanofluids exhibited enhanced thermal conductivity which further increases with increasing temperature. Same group previously reported another work [35], where they studied the effect of dispersion method on thermal conductivity and stability of nanofluids containing five different CNT structures. Their [35] results showed that the effective thermal conductivity of these nanofluids increases with temperature over a temperature range of 10 to 50°C and the best stability and thermal conductivity were associated with the functionalized CNT dispersed nanofluids.

Thermal conductivity, viscosity, and stability of MWCNT-nanofluids were measured by Phuoc et al. [55]. The chitosan dispersant of different concentrations was used for better stability of nanofluids. The thermal conductivity of water increased up to 13% for 3 wt. % loading of MWCNT and a popular classical model under-predicted the enhanced thermal conductivity of their nanofluids. Surprisingly they also demonstrated that MWCNT can be used either to enhance or reduce the viscosity of base fluid.

The effect of temperature on the thermal conductivity of SWCNT/water nanofluids was investigated by Harish et al. [56]. Their results showed that the effective thermal conductivity of their nanofluids increased considerably with increasing SWCNT concentration as well as temperature. A maximum thermal conductivity increase of 16% was obtained at a temperature of 333 K and at a 0.3% volumetric loading of SWCNT.

Singh et al. [57] performed an experimental investigation on the thermal conductivity of ethylene glycol-based CNT nanofluids. In their study, the chemically treated CNT was dispersed in EG to prepare stable nanofluids. A linear increase in thermal conductivity of ethylene glycol was observed with increasing the loading of CNT. The maximum enhancement of thermal conductivity of this nanofluid was 72% at 0.4 wt. % of CNT in EG and at a temperature of 33°C.

Many other studies also reported significant enhancement of thermal conductivity of CNT-nanofluids with increasing loading of CNT and temperature [29, 44, 47, 48, 52, 58-60].

Very recently Gu et al. [61] measured thermal conductivity of three different nanofluids including CNT-nanofluids and showed that nanofluids containing higher aspect ratio fillers exhibit larger thermal conductivity enhancement. When the volume fraction of CNT is 0.2%, the nanofluids showed only 3.7 % increase in thermal conductivity. This enhancement was much smaller as compared to Ag-nanofluids also used in their study. Their results demonstrated that the shape of the particle has a substantial effect on the effective thermal conductivity of suspension.

Despite inconsistent data and controversy regarding the heat transfer mechanisms of nanofluids, the reported thermal conductivity enhancements justify their applications as advanced heat transfer and thermal storage fluids.

3.2. Mechanisms

Studies have shown that nanofluids exhibit anomalously high thermal conductivity which cannot be predicted accurately by the classical models. In early studies, several heat transfer mechanisms for nanofluids were proposed and analyzed by Wang et al. [62] and Keblinski et al. [63]. Wang et al. [62] suggested that the microscopic motion of nanoparticles, surface properties, and the structural effects might be the reasons for the enhanced thermal conductivity of nanofluids. In nanofluids, the microscopic motion of the nanoparticles due to van der Waals force, stochastic force (causing Brownian motion) and electrostatic force can be significant. Keblinski et al. [63] later elucidated four possible mechanisms which include nanoparticles Brownian motion, interfacial nanolayer at the nanoparticle/fluid interface, nature of heat transport in the nanoparticle, and nanoparticle clustering. However, these mechanisms were mainly proposed for the observed anomalously high thermal conductivity of nanofluids containing spherical shape nanoparticles. Although the contributions of some of these factors particularly Brownian motion and nature of heat transport in nanoparticle to the enhanced the thermal conductivity of nanofluids were demonstrated to be insignificant [62-63], some researchers [64-66] held contrary views.

Besides these aforementioned mechanisms, the effects of nanoparticle surface chemistry and particles interaction could be significant in enhancing the thermal conductivity of nanofluids. Although interfacial nanolayer and nanoparticles clustering are recently considered as key factors for the thermal conductivity of nanofluids, still there remain controversies about their actual heat transfer mechanisms.

Nevertheless, due to different dominant mechanisms, theoretical models for nanofluids containing spherical nanoparticles and cylindrical nanoparticles (nano-rod or tube) are different. With large aspect ratio and complex morphologies of CNT, heat transfer mechanisms for CNT-nanofluids are far more complicated.

Compared to theoretical efforts for nanofluids containing spherical nanoparticle very limited research efforts have been given to identify the heat transfer mechanisms and to develop model for the anomalously high thermal conductivity of CNT-nanofluids. Thus, heat transfer mechanisms for the anomalous thermal conductivity of nanorods or CNT-nanofluids are not yet well-understood and conclusive.

3.3. Thermal Conductivity Models for CNT-Nanofluids

Since the treatise by Maxwell [67] numbers of models have been developed to predict the effective thermal conductivity of solid particle suspensions (k_{eff}). These classical models such as those attributed to Maxwell [67] and Hamilton and Crosser [68] were developed for predicting the effective thermal conductivity of a continuum medium with well-dispersed solid particles.

The Maxwell's [67] model predicts the effective electrical or thermal conductivity of liquid-solid suspensions for very low (dilute) concentration of spherical particles.

Hamilton and Crosser [68] modified Maxwell's [67] model for both the spherical and non-spherical particles by applying a shape factor and it considers the thermal conductivities of both the solid and liquid phases, volume fraction and the shape of the dispersed particles. This model shows that the increase in thermal conductivity for non-spherical particles is higher compared to that of the spherical particles.

Research showed that these classical effective medium theory-based models [67-68] are unable to predict the anomalously high thermal conductivity of nanofluids. This is because these models do not include any nanoscale effects of dispersed nanomaterials. However, the popular Hamilton and Crosser (HC) model [68] is commonly used for the prediction of the effective thermal conductivity of nanofluids having CNT-nanofluids.

Some of the representative existing models used to predict the effective thermal conductivity of nanofluids containing nanorods or nanotubes are elaborated here. Taking a part of their previous model [66] for spherical nanoparticles, Patel et al. [69] later reported a simple model for the thermal conductivity of carbon nanotube nanofluids. Strauss and Pober [70] developed geometric models which considered periodic lattices with equal surface-to-surface distances between the nanotubes and their nearest neighbors in all directions. A combination of series model in the direction of the thermal gradient and a parallel model perpendicular to it is used along with averaging over all orientations to derive an effective bulk thermal conductivity model.

The model developed by the author [21] for the effective thermal conductivity of the suspensions of cylindrical nanoparticles (nanofluids) takes into account the effects of particle size, concentration, and interfacial nanolayer. Unlike two-phase system as used in Hamilton and Crosser model [68], this model was developed considering three phases (nanoparticle, interfacial layer, and base fluid) combination in nanofluids.

Sastry et al. [71] presented a different modeling approach which was based on 3-D CNT chain formation in the base fluid and a thermal resistance network. They used probability density functions for random CNT orientation and CNT-CNT contact determination. Several parameters of this model are to be obtained by random number generators.

Recently Walvekar et al. [72] proposed a simple thermal conductivity model for CNT-based nanofluids. This model has been derived from Kumar et al.´s [66] model, which is valid for only spherical particles. However, the main drawback of this model is that it assumed the liquid medium as particles which are surrounded by other nanoparticles.

A detailed summary of some the selective classical and recently developed models used to predict the effective thermal conductivity of nanofluids containing cylindrical nanoparticles (nanorods) and nanotubes (CNT) is presented in Table 3. A more comprehensive list of available models for the effective thermal conductivity of these nanofluids and analysis of their heat transfer mechanisms can be found elsewhere [78].

Models reported in Table 3 have their limitations and are unable to accurately predict the thermal conductivity of CNT-nanofluids. Except those classical models most of the recent models are validated using very limited data which are mainly from their own experiments. These models also contain unknown parameters which are either used to fit the experimental data or cannot be obtained directly. Thus these models are not widely accepted. There is therefore an urgent need to identify the real heat transfer mechanisms and to develop model for the prediction of the enhanced thermal conductivity of CNT-nanofluids.

Table 3. Existing models for the effective thermal conductivity of nanofluids containing rod or tube shape nanoparticles

Researchers	Models/ Expressions and remarks
Maxwell [67]	$$k_{eff}/k_f = \frac{k_p + 2k_f + 2\phi_p(k_p - k_f)}{k_p + 2k_f - \phi_p(k_p - k_f)}$$ It depends on the thermal conductivities of particle (k_p) and base fluids (k_f) and the volume fraction of solid. It is valid for spherical particles only.
Hamilton and Crosser [68]	$$k_{eff}/k_f = \left[\frac{k_p + (n-1)k_f - (n-1)\phi_p(k_f - k_p)}{k_p + (n-1)k_f + \phi_p(k_f - k_p)}\right]$$ where shape factor $n = 3/\psi$ where ψ is the particle sphericity. It is valid for both the spherical and cylindrical particles. For cylindrical particles sphericity $\psi = 0.5$.
Yu and Choi [73]	$$k_{eff}/k_f = 1 + \frac{n\phi_e A}{1 - \phi_e A}$$ It is a renovated HC [68] model where $A = \frac{1}{3}\sum_{j=a,b,c} \frac{k_{pj} - k_f}{k_{pj} + (n-1)k_f}$, $n = 3\psi^{-\alpha}$, ϕ_e is the equivalent volume fraction of complex particles and α is an empirical parameter.
Kumar et al. [66]	$$k_{eff}/k_f = 1 + c\frac{K_B T}{2\pi\eta r_p^3}\frac{\phi_p r_f}{k_f(1 - \phi_p)}$$ It is developed from the combination of kinetic theory and Fourier's law where η is the viscosity, c is a constant, r_f is the radius of liquid particle, T is the temperature and K_B is the Boltzmann's constant.
Xue [74]	$$9(1-\phi_c)\frac{k_{eff} - k_f}{2k_{eff} + k_f} + \phi_c[\frac{k_{eff} - k_{c,x}}{k_{eff} + B_{2,x}(k_{c,x} - k_{eff})} +$$ $$4\frac{k_{eff} - k_{c,y}}{2k_{eff} + (1 - B_{2,x})(k_{c,y} - k_{eff})}] = 0$$ Maxwell theory [67] and average polarization theory with the interfacial shell effect was the base of this model.
Gao et al. [75]	$$9(1-\phi)\frac{k_{eff} - k_f}{2k_{eff} + k_f} + \phi[\frac{k_{eff} - k_{c,x}}{k_{eff} + L_x(k_{c,x} - k_{eff})} + 4\frac{k_{eff} - k_{c,y}}{2k_{eff} + (1 - L_x)(k_{c,y} - k_{eff})}] = 0$$ This is basically the same as Xue [74] model developed for CNT-nanofluids.
Sabbaghzadeh and Ebrahimi [76]	$$k_{eff} = k_f[1 - \phi(1 + M')] + \phi(k_p + k_{lr}M') + \phi(1 + M')\frac{d_f}{Pr D}(0.35 +$$ $$0.56 Re_f^{0.52})Pr_f^{0.3} k_f$$ where $M' = \left[(\frac{t}{r_p} + 1)^2 - 1\right]$, d_f is the diameter of base fluid molecule, t is nanolayer thickness, and D is the diameter of the complex nanoparticle.

Researchers	Models/ Expressions and remarks
Murshed et al. [21]	$$k_{eff} = \frac{(k_p - k_{lr})\phi_p k_{lr}[\gamma_1^2 - \gamma^2 + 1] + (k_p + k_{lr})\gamma_1^2[\phi_p\gamma^2(k_{lr} - k_f) + k_f]}{\gamma_1^2(k_p + k_{lr}) - (k_p - k_{lr})\phi_p[\gamma_1^2 + \gamma^2 - 1]}$$ It is a three-phase (particle, interfacial layer, and base fluid) static model. Here $\gamma = 1+h/r_p$, $\gamma_1 = 1+h/d_p$ and k_{lr} is the thermal conductivity of nanolayer.
Clancy and Gates [77]	$$k_{eff}/k_f = \frac{3 + \phi(\beta_x + \beta_z)}{2 - \phi\beta_x}$$ This model was derived for SWCNT and considered the effect of interfacial thermal resistance which was determined using molecular dynamic simulations.
Walvekar et al. [72]	$$k_{eff} = k_f \left[1 + \frac{k_p \frac{(2\phi(r_p + l_p))}{r_p l_p}}{k_f \frac{(3(1-\phi))}{r_f}} \right] + \frac{C\phi(T - T_0)}{r_p^2 l_p^2 \mu_f} \ln\left(\frac{l_p}{d_p}\right)$$ This model was derived from Kumar et al.´s [66] model and it assumed the liquid as particles which are surrounded by other nanoparticles. Here constants $C = 2 \times 10^{-27}$ and $T_0 = 273$ K.

4. VISCOSITY AND SPECIFIC HEAT CAPACITY

4.1. Viscosity Results

Viscosity of nanofluids is as crucial as thermal conductivity in any engineering and thermal systems that employ fluids flow. The understanding of convection heat transfer is directly related to the viscosity of the nanofluids. Furthermore, viscosity can also influence the electrical properties of any fluids. However, compared to studies on the viscosity of nanofluids containing oxide and metallic nanoparticles [79-80], limited research works have been performed on this key property of CNT-nanofluids.

In a convective heat transfer study, Ding et al. [28] measured the viscosity of aqueous MWCNT-nanofluids as a function of shear rate and at different concentrations as well as temperatures. At any shear rate, they found that the viscosity of their nanofluids increases with increasing CNT loading and decreases with increasing temperature. They also observed a clear shear shinning phenomenon in their surfactant free nanofluids. However, CNT-nanofluids having Gum Arabic surfactant showed mixed behavior as they found a shear thinning behavior at low shear rates but slight shear thickening at shear rates larger than 200 s^{-1}. They concluded that at low shear rates the presence of this surfactant affects little on the viscosity of the nanofluids but may play a role at high shear rates.

The effects of CNT loading, surfactant concentration and dispersing energy (ultrasonication) on the thermal conductivity and steady shear viscosity of oil (PAO6)-based MWCNT-nanofluids were studied by Yang et al. [53]. Polyisobutene succinimide (PIBSI) was used as surfactant in their study. It was demonstrated that the PIBSI dispersant controls the viscosity of nanofluids particularly at low shear conditions. For example, an astounding six orders of magnitude decrease in viscosity of nanofluids was found due to addition of surfactant and decreasing the shear stress from 0.1 Pa to 0.07 Pa. However, dispersant concentration lower or higher than 3 wt. % yielded higher viscosity of nanofluids. While nanofluids with lowest concentration of CNT can be treated as a Newtonian fluid, nanofluids containing 0.09 and 0.13 vol. % of CNT showed slight shear thinning nature at low shear stress. Nevertheless, viscosity of nanofluids found to increase considerably with increasing CNT concentration.

In a study on thermal and flow characteristics, Ko et al. [81] measured the viscosity of aqueous CNT nanofluids. They used both the functionalization of CNT and surfactant (SDS) addition methods to prepare sable nanofluids. The viscosity results showed that their nanofluids are shear thinning fluids where the viscosity decreased with increasing shear rate. At any shear rate the viscosity of nanofluids was found to increase with increasing the volumetric loading of CNT.

Chen et al. [26] measured viscosity of nanofluids containing MWCNT of 15 nm diameters in distilled water. Effects of volumetric concentration and temperature on the viscosity of this nanofluid were studied. They reported that at low volume fractions (<0.4 vol. %), nanofluids have lower viscosity than that of the base fluid. However, the viscosity only increases with increasing from higher than 0.4 % volumetric loading of MWCNT and there is almost no change in viscosity of their nanofluids with the temperature up to 55°C. Interestingly, a sudden increase in viscosity was observed at temperature higher than 55°C.

The same group [37] later studied the rheology of silicone oil-based treated MWCNT nanofluids at various concentrations and temperatures. They found this nanofluid to remain Newtonian manner in all concentrations and temperatures. Also the enhanced viscosity of this nanofluid further increases significantly with increasing concentration of MWCNT and decreases non-linearly with increasing temperature. It was also reported that addition of hexamethyldisiloxane (HMDS) dispersant in silicone oil can decrease the viscosity of silicone oil but has little effect on the rheology of nanofluids. This group [82] further reported rheological behaviors of similar MWCNT-nanofluids in different base fluids which include water, ethylene glycol, glycerol, and silicone oil. Like their previous study [37], glycerol and silicone oil-based nanofluids were found to be of Newtonian nature. The measured viscosity of all of their nanofluids showed significantly increased and decreased viscosity with concentration (except 0.2 vol. %) of MWCNT and temperature (up to 55 °C), respectively. Interestingly, at MWNT volume fraction of 0.002, nanofluids have lower viscosity than the corresponding base fluids. They attributed such unusual results because of lubricative effect of MWCNT. Nonetheless, for ethylene glycol and glycerol based nanofluids; almost no viscosity augmentation appeared when the temperature was higher than 55 °C.

An investigation of the effect of ultrasonication on viscosity and heat transfer performance of water-based MWCNT nanofluids was conducted by Garg et al. [32]. They added 0.25 wt. % Gum Arabic surfactant in their sample nanofluids. All nanofluids were found to show non-Newtonian behavior especially at 15 °C and they fall into power law fluids. The viscosity of their nanofluids increased with increasing sonication time from 20 minutes to 40 minutes and thereafter decreased with increase in ultrasonication time.

Phuoc et al. [55] measured the viscosity of suspensions of MWCNT in chitosan (surfactant) added water having 0.5 vol.% acetic acid. For the first time, they demonstrated that depending on the concentration MWCNT can be used either to enhance or reduce the base fluid's viscosity. A reduction of viscosity up to 20% was observed at 0.5 wt. % of MWCNT. For all other MWCNT concentrations, nanofluids showed higher viscosity compared to that of the base fluids. For the viscosity-enhancement case, their nanofluids behaved as non-Newtonian shear-thinning fluids.

In a study on temperature-dependent thermal conductivity, Harish et al. [56] reported viscosity data of their SWCNT/water nanofluids and observed that the viscosity of the nanofluids increased and decreased substantially with SWCNT concentration and temperature, respectively. For instance, an increase in viscosity of 30% was reported for SWCNT concentration of 0.3 vol. %.

Hung and Chou [83] studied the effect of chitosan dispersant and MWCNT concentration on few features including thermal conductivity and viscosity of water-based nanofluids. It was concluded that both the chitosan and MWCNT concentrations increase the viscosity of nanofluids. A maximum 233% enhancement of viscosity of water was achieved due to addition of 1.5 wt. % MWCNT and 0.4 wt. % chitosan dispersant in it. An experimental investigation on the thermal conductivity and viscosity of ethylene glycol-based CNT nanofluids was performed by Singh et al. [57]. As usual the viscosity of nanofluids found to increase and decrease significantly with increasing of CNT concentration and temperature, respectively. Their nanofluids showed Newtonian behavior.

Vakili-Nezhaada and Dorany [84] measured the viscosity of suspensions of SWCNT in lube oil cuts at different temperatures ranging from 25°C to 100°C. Like most of the previous studies, they also found that the viscosity of nanofluids increased and decreased with increasing CNT loading and temperature, respectively. A maximum 33% increase in viscosity of nanofluids was found at SWCNT loading of 0.2 wt. %.

Estelle et al. [85] recently studied the effect of shear history on the rheology of aqueous based CNT-nanofluids. They found that at low shear rates this nanofluid behaves as a viscoelastic media and is shear-thinning at higher shear rates. These shear history dependent behaviors were believed to be due to breaking down of the CNT-nanofluids agglomerations and structural network. In another study this same group [86] also reported the viscosity of the same aqueous CNT-nanofluids at various volumetric concentrations of CNT and temperatures ranging from 0 to 40 °C. They revealed that while at low CNT concentration the nanofluids are Newtonian, at high concentration they behave as shear-thinning materials. Although viscosity of their nanofluids found to increase with concentration, no pronounceable effect of temperature on the relative viscosity (viscosity of nanofluids with respect to base fluids) was observed at high shear rate (1000 s^{-1}).

Summary of rheological behavior and key results on the viscosity of various CNT-nanofluids is provided in Table 4. It is clearly seen that although a couple of researchers observed reduction of viscosity with loading of CNT, most of the studies however, reported significantly enhanced viscosity of these nanofluids compared to their base fluids. Surfactant also plays a role in altering the rheology of nanofluids.

Above literature review on this special class of nanofluids revealed that they exhibit both Newtonian and non-Newtonian behaviors and their viscosities are considerably higher than those of the base fluids. In addition, the viscosity increased further with increasing concentration of nanoparticles. A nonlinear decrease in viscosity of these nanofluids with increasing temperature was also observed in the literature. Such substantial decreasing of viscosity with temperature makes these nanofluids even more attractive for their applications at high temperature environments. The existing models are found to severely under-predict the measured viscosity data of these nanofluids.

Table 4. Summary of viscosity results of CNT-nanofluids from the literature

Researchers/ References	CNT/Base fluids	Stabilization	Key results		
			Rheology	CNT loading and conditions	Relative viscosity (η_{nf}/η_f)
Yang et al. [53]	MWCNT/Oil (PAO6)	PIBSI surfactant	Newtonian and shear-thinning	0.34 vol.% (at 10 Pa)	1570
				0.34 vol.% (at 50 Pa)	30.20
Chen et al. [26]	MWCNT/W	surface treated CNT	Not available	<0.4 vol.% and <55°C	<1 (η_{nf} decrease)
				>0.4 vol.% and <55°C	>1 (η_{nf} increase)
				any vol.% above 55°C	>1 (η_{nf} increase)
				0.2vol.% at 65°C	1.7
Chen and Xie [37]	MWCNT/ Silicone oil (SO)	surface treated CNT and HMDS surfactant	Newtonian	1 wt.% (+5.4 wt.% HMDS) at 20 °C	~1 (0.98)
				0.54wt.% (+5.4wt.% HMDS) at 20 °C	<1 (η_{nf} decrease)
Phuoc et al. [55]	MWCNT/W	Chitosan surfactant	non-Newtonian	0.5 wt.%	0.80 (η_{nf} decrease)
Hung and Chou [83]	MWCNT/W	Chitosan surfactant	Not available	0.25 wt.% (+ 0.4 wt.% chitosan)	2.27
				1.5 wt.% (+ 0.4 wt.% chitosan)	3.33
Harish et al. [56]	SWCNT/W	Sodium deoxycholate surfactant	Not available	0.3 vol.%	1.30
Halelfadl et al. [86]	CNT/W	SDBS surfactant	Newtonian and shear-thinning	0.557 vol.%	5.2
				0.055 vol.%	~ 0.92 (decreased η_{nf})

4.2. Viscosity Models

Despite the growing number of research works recently conducted on viscosity of nanofluids [80], very few studies have focused on the development of theoretical models for the prediction of the viscosity of nanofluids and thus no widely accepted model is yet available [80, 87]. Researchers commonly used classical models or empirical models obtained by fitting their own experimental data. The most popular classical viscosity model is the Einstein's model [88] in which he considered a dilute suspension of spherical particles that can be defined as an effective viscosity of suspension (η_{eff}) given by

$$\eta_{eff}/\eta_f = (1 + 2.5\phi)$$ (1)

where η_f is viscosity of base fluid and ϕ is the volume fraction of dispersed particles. Although this model is widely used for any concentrations (low to high) of particles, it is mainly valid for very low concentration of particles (typically less than 1%).

Since the nanofluids are mostly not dilute suspensions, the power law-based models such as attributed to Krieger and Dougherty [89] and Nielsen [90] are more appropriate for the estimation of the effective viscosity compared to the Einstein's[88] and Batchelor's [91] models.

A semi-empirical relationship for the shear viscosity covering the full range of particle volume fraction was obtained by Krieger and Dougherty [89] and it has the form

$$\eta_{eff}/\eta_f = (1 - \phi/\phi_m)^{-[\eta]\phi_m}$$ (2)

where the maximum packing fraction $\phi_m \approx 0.605$ at high shear rates and the intrinsic viscosity $[\eta] = 2.5$ for hard spheres.

Later a generalized equation for the relative elastic moduli of composite materials (also widely known as relative viscosity) was proposed by Nielsen [90] and his power law-based model can be expressed as:

$$\eta_{eff}/\eta_f = (1 + 1.5\phi)e^{\phi/(1-\phi_m)}$$ (3)

where ϕ_m is the maximum packing fraction. For randomly dispersed spheres, the maximum close packing fraction is approximately 0.64.

Nevertheless, all these models are found unable (under-predict) to estimate the measured viscosities of nanofluids [80, 87]. Unless introducing imperial model from fitting the experimental results, no efforts have been devoted to develop a rigorous model for the prediction of viscosity of CNT-nanofluids.

4.3. Specific Heat Capacity

Specific heat capacity is very important in determining other heat transfer properties, heat transfer rates under flow conditions, evaluating heat storage capacity of thermal management systems as well as enthalpy calculations in various processes. However, only a couple of research works were performed to investigate the specific heat capacity of CNT-nanofluids. Using ethylene glycol-based SWCNT nanofluids Amrollahi et al.[31] demonstrated that the volumetric heat capacity of ethylene glycol increased with volume fraction of SWCNT and with temperature as well. Liu et al. [36] measured the specific heat of MWNT/water nanofluid using differential scanning calorimetry (DSC) and found that due to addition of 0.1 vol.% of MWNT the specific heat of city water at 20 °C increased slightly (about 0.4%).

In investigating the heat dissipation performance of MWCNT nanofluids in a motorcycle radiator, Teng and Yu [42] recently measured temperature (80 to 95 °C) and concentration dependence of specific heat of their CNT-nanocoolants. The cationic chitosan dispersant was also added into the base fluid which was a mixture of water and EG (1:1 volumetric ratio). Their results showed that while the specific heat of nanofluids increased with temperature, it however decreased gradually with increasing MWCNT concentration. Results also revealed that adding dispersant to nanofluid increased its specific heat. Such results trends was believed to be because both the dispersant and MWCNT have respectively higher and lower values of specific heat compared to that of the base fluid (W/EG). Although enhanced specific heat was reported for all these CNT-nanofluids, no solid conclusions can be made based on these very limited results.

No other works on specific heat and thermal diffusivity of CNT-nanofluids can be found in the literature. Thus it is imperative to conduct more investigations on these important properties of this particular type of nanofluids.

5. CONVECTIVE HEAT TRANSFER CHARACTERISTICS

Over the last decade, extensive research has been conducted on the convective heat transfer performance nanofluids [13, 15, 92]. However, in spite of having ultra-high thermal conductivity and good dispersion behavior of CNT, very limited investigations have been made on the convective heat transfer of CNT-nanofluids.

Ding et al. [28] was the first to investigate the heat transfer performance in laminar flow of aqueous CNT nanofluids through a horizontal tube. A significant enhancement of the convective heat transfer in comparison with water was reported in their study. For example, at CNT mass fraction of 0.5 % and Reynolds number (*Re*) of 800 an intriguingly high enhancement (maximum 350%) of heat transfer coefficient was observed. The first convective heat transfer experiments with aqueous CNT-nanofluid in a microchannel with hydraulic diameter of 355 µm at Reynolds numbers between 2 to 17 was conducted by Faulkner et al. [93]. They found considerable enhancement in heat transfer coefficient of this nanofluid at CNT concentration of 4.4%. From a different study Garg et al. [32] showed that heat transfer performance of dispersant (GA) added MWCNT/water nanofluid increased until an optimum ultrasonication time was reached and decreased on further ultrasonication. However, the maximum enhancement of convective heat transfer was found to be 32%. Amrollahi et al. [94] studied convective heat transfer of aqueous MWCNT-nanofluids under laminar and turbulent flow conditions and reported up to 40% increase in heat transfer coefficient at 0.25 wt.% of MWCNT concentration.

Another investigation on the convection heat transfer of aqueous CNT nanofluid flowing through a horizontal copper tube under constant heat flux and laminar flow conditions was performed by Rashidi and Nezamabad [95]. A significant increase in heat transfer coefficient of this nanofluid was reported and the increased heat transfer coefficient further increased with increasing concentration of CNT and Reynolds number. The enhancement was found particularly significant at entrance region and decreased with axial distance from the inlet.

Mare´ et al. [96] investigated the thermal performances of two types of nanofluids (water-based oxides of alumina and carbon nanotubes) in two plate heat exchangers. Comparing heat transfer enhancement and pumping power loss nanofluids containing carbon nanotubes showed better thermal- hydraulic performance than base fluid i.e., pure water.

Lotfi et al. [97] experimentally studied the convective heat transfer performance of MWCNT/ water nanofluid in a horizontal shell and tube heat exchanger. Multi-walled carbon nanotubes used in their study were synthesized by the catalytic chemical vapor deposition (CCVD) method over Co–Mo/MgO nanocatalyst. Results showed that the presence of the nanotubes enhances the heat transfer rate in a shell and tube heat exchanger.

In an experimental investigation with functionalized CNT (f-MWCNT)-nanofluids Aravind et al. [38] found almost the same trend of the heat transfer coefficient profiles for EG/f-MWCNT and DIW/f-MWCNT nanofluids used in their study. For a volume concentration of 0.03% and flow rate of 56 mL/s, the maximum enhancements of heat transfer coefficient of MWCNT/ DI water and MWCNT/EG-based nanofluids were respectively 65% and 180%. These enhancements of heat transfer coefficients were several times larger than their thermal conductivity enhancements.

Ruan and Jacobi [98] studied convective heat transfer characteristics of water and ethylene glycol-based MWCNT nanofluids in an intertube falling-film flow system. They reported that while the heat transfer coefficient of water-based nanofluids first decreases and then increases with increasing concentration of CNT, the heat transfer coefficient of ethylene glycol-based nanofluids only decreases with increasing CNT concentration. They also introduced a model for the prediction of the enhanced heat transfer of their nanofluids in the flow systems.

Very recently Kumaresana et al. [99] performed experiments on convective heat transfer characteristics of water/EG mixture (70/30 by volume)-based MWCNT nanofluid in a tubular heat exchanger. The convective heat transfer coefficient of this nanofluid was found to increase to a maximum of 160% at a 0.45 vol. % of MWCNT. They ascribed several factors such as nanotube rearrangement and high aspect ratio as well as delay in boundary layer development for the observed heat transfer enhancement of their nanofluid.

Although numbers of studies have been conducted to develop model for the anomalously high thermal conductivity of CNT-nanofluids [13-14], almost no rigorous work has been performed to understand and identify the real mechanisms for the observed significant enhancement of convective heat transfer of this specific type of nanofluids [15,100-101].

CONCLUSION

Thermophysical properties and heat transfer characteristics CNT-laden nanofluids are presented and critically reviewed in this chapter. Results showed that CNT-nanofluids exhibit significantly higher thermal features such as thermal conductivity, viscosity, and convective heat transfer coefficient compared to their base fluids as well as other nanofluids. These enhanced properties of these nanofluids are significantly influenced by the concentration of CNT and temperature. Other factors also play role in altering these properties in some extent.

These nanofluids with their high thermal conductivity can potentially be used as the next generation heat transfer fluids. However, research works are mainly focusing on their anomalous thermal conductivity whereas other heat transfer and thermophysical properties are also very important in order for their practical applications particularly as coolants in thermal management systems as well as solar energy based applications.

This review clearly demonstrates that despite inconsistent data in the literature, substantial increases in thermal conductivity and convective heat transfer of this particular type of nanofluids compared to their base fluids are undisputed. However, available data are still limited and scattered to clearly understand the underlying heat transfer mechanisms of these nanofluids. It is therefore imperative to conduct more comprehensive studies on the thermophysical properties and convective heat transfer performance of CNT-nanofluids under various important conditions or factors such as concentration, temperature, pressure, flow conditions, heater and tube geometry.

It is also found that most studies used aqueous MWCNT nanofluids. Thus other types of carbon nanotubes such as single- or double-walled nanotube and different types of base fluids like silicone oil and refrigerants-based systems need to be investigated explicitly.

Literature survey showed that despite the observed reduction of viscosity with loading of CNT in a couple of studies, most of the researchers however reported significantly higher viscosity of these nanofluids compared to their base fluids. The enhanced viscosity of nanofluids increases and decreases with increasing nanoparticles concentration and temperature, respectively. Like thermal conductivity, CNT concentration, fluids temperature as well as surfactant also play important role to alter the viscosity of nanofluids. Review also revealed that these nanofluids showed both Newtonian and non-Newtonian shear-thinning behaviors and their rheological behaviors also depend on concentration of CNT, shear rate and temperature as well. The observed nonlinear decrease in viscosity of these nanofluids with increasing temperature makes these nanofluids even more attractive for their applications at elevated temperatures.

Despite numerous studies devoted to the development of models for the prediction of the thermal conductivity of nanofluids containing mainly spherical-shape nanoparticles, very limited attempts have been made to identify the actual heat transfer mechanisms and to develop model for the enhanced thermal conductivity of CNT or cylindrical nanoparticles dispersed nanofluids. Most of the existing models have limitations and are unable to accurately predict the thermal conductivity of CNT-nanofluids. On the other hand, almost no rigorous analytical study has been performed to understand and identify the mechanisms for the enhanced convective heat transfer characteristics of these nanofluids. Thus more research efforts are needed to identify the actual mechanisms and to develop models for the prediction of the enhanced conductive and convective heat transfer properties of this special class of nanofluids.

Besides having superior thermophysical properties, heat transfer characteristics, and benefits compared to the base fluids, these novel fluids show great promises to be used in many important applications such as thermal management systems, advanced coolants as well as solar energy technologies.

ACKNOWLEDGMENT

This research was partially supported by Fundação para a Ciência e a Tecnologia (FCT), Portugal through the projects, PEst-OE/QUI/UI0536/2011 and PTDC/EQU-FTT/104614/2008 attributed to Centro de Ciências Moleculares e Materiais, Faculdade de Ciências da Universidade de Lisboa.

REFERENCES

[1]   Choi, S. U. S. In *Developments and applications of non-Newtonian flows*; Siginer, D. A.; Wang, H. P.; Eds.; ASME Publishing: New York,USA,1995; FED-Vol. 231/MD-Vol. 66, pp 99-105.

[2]   Yang, Y. M.; Maa, J. R. *Int. J. Heat Mass Transfer* 1984, 27, 145–147.

[3]   Masuda, H; Ebata, A.; Teramae, K.; Hishinuma, N. *Netsu Bussei* 1993, 4, 227-233.

[4]   Grimm, A. *Powdered Aluminum-containing Heat Transfer Fluids*, German patent DE 4131516 A, 1993.

[5]   Iijima, S. *Nature* 1991, 354, 56-58.

[6]   Oberlin, A.; Endo, M.; Koyama, T. *J. Cryst. Growth* 1976, 32, 335-349.

[7]   S. Iijima, T. Ichihashi, *Nature* 1993, 363, 603-605.

[8]   Bethune, D. S.; Kiang, C. H.; Devries, M. S.; Gorman, G.; Savoy, R.; Vazquez, J.; Beyers, R.; *Nature* 1993, 363, 605-607.

[9]   Terrones, M. *Annu. Rev. Mater. Res.* 2003, 33, 419-501.

[10] De Volder, M. F. L.; Tawfick, S. H.; Baughman, R. H.; Hart, A. J. *Science* 2013, 339, 535-539.

[11] Coleman, J. N.; Khan, U.; Blau, W. J.; Gun'ko, Y. K. *Carbon* 2006, 44, 1624-1652.

[12] Lee, K. J.; Yoon, S. H.; Jang, J.; *Small* 2007, 3, 1209-1213.

[13] Das, S. K.; Choi, S. U. S.; Patel, H. E. *Heat Transf. Eng.* 2006, 27, 3-19.

[14] Murshed, S. M. S.; Leong, K. C.; Yang, C. *Appl. Therm. Eng.* 2008, 28, 2109-2125.

[15] Murshed, S. M. S.; Nieto de Castro, C. A.; Lourenço, M. J. V.; Lopes, M. L. M.; Santos, F. J. V. *Ren. Sust. En. Rev.* 2011, 15, 2342-2354.

[16] Murshed, S. M. S.; Nieto de Castro, C. A. In *Green Solvents I: Properties and Applications in Chemistry*, Ali M.; Inamuddin; Eds.; Springer, London, UK 2012, Ch. 14, pp.397-415.

[17] Xie, H.; Chen, L. *J. Chem. Eng. Data* 2011, 56, 1030-1041.

[18] Murshed, S. M. S.; Leong, K. C.; Yang, C. *Int. J. Therm. Sci.* 2005, 44, 367-373.

[19] Tan, S. H.; Murshed, S. M. S.; Nguyen, N. T.; Wong, T. N.; Yobas, L. *J. Phys. D: Appl. Phys.* 2008, 41, 165501.

[20] Murshed, S. M. S.; Leong, K. C.; Yang, C.; Nguyen, N. T. *Int. J. Nanosci.* 2008, 7, 325-331.

[21] Murshed, S. M. S.; Leong, K. C.; Yang, C. *Int. J. Therm. Sci.*2008, 47, 560-568.

[22] Murshed, S. M. S.; Leong, K. C.; Yang, C. *J. Phys. D: Appl. Phys.* 2006, 39, 5316-5322.

[23] Murshed, S. M. S.; Nieto de Castro, C. A.; Lourenço, M. J. V.; Lopes, M. L. M.; Santos, F. J. V. *J. Phys.: Conf. Ser.* 2012, 395, 012117.

[24] Choi, S. U. S.; Zhang, Z. G.; Yu, W.; Lockwood, F. E.; Grulke, E. A. *Appl. Phys. Lett.* 2001, 79, 2252-2254.

[25] Xie, H.; Lee, H.; Youn, W.; Choi, M. *J. Appl. Phys.* 2003, 94, 4967-4971.

[26] Chen, L.; Xie, H., Li, Y.; Yu, W. *Thermochim. Acta* 2008, 477, 21-24.

[27] Assael, M. J.; Chen, C. F.; Metaxa, I.; Wakeham, W. A. *Int. J. Thermophys.* 2004, 25, 971-985.

[28] Ding, Y.; Alias, H.; Wen, D.; Williams, R. A. *Int. J. Heat Mass Transf.* 2006, 49, 240-250.

[29] Hwang, Y. J.; Ahn, Y. C.; Shin, H. S.; Lee, C. G., Kim, G. T.; Park, H. S.; Lee, J. K. *Curr. Appl. Phys.* 2006, 6, 1068-1071.

[30] Liu, M. S., Lin, M. C. C.; Huang, I. T.; Wang, C. C. *Int. Comm. Heat Mass Transf.* 2005, 32, 1202-1210.

[31] Amrollahi, A.; Hamidi, A. A.; Rashidi, A. M. *Nanotechnology* 2008, 19, 315701.

[32] Garg, P.; Alvarado, J. L., Marsh, C.; Carlson, T. A., Kessler, D. A.; Annamalai, K. *Int. J. Heat Mass Transf.* 2009, 52, 5090-5101.

[33] Park, K. J.; Jung, D. *Int. J. Heat Mass Transf.* 2007, 50, 4499-4502.

[34] Liu, Z. H.; Yang, X. F.; Xiong, J. G. *Int. J. Therm. Sci.* 2010, 49,1156-1164.

[35] Nasiri, A.; Niasar, M. S.; Rashidi, A.; Amrollahi, A.; Khodafarin, R. *Exp. Therm. Fluid. Sci.* 2011, 35, 717-723.

[36] Liu, M. S.; Lin, M. C. C.; Wang, C. C. *Nanoscal. Res. Lett.* 2011, 6, 297.

[37] Chen, L., Xie, H. *Colloid. Surf. A: Physicochem. Eng. Asp.* 2009, 352, 136-140.

[38] Aravind, S. S. J.; Baskar, P., Baby, T. T.; Sabareesh, R. K.; Das, S.; Ramaprabhu, S. *J. Phys. Chem. C* 2011, 115, 16737-16744.

[39] Park, K. J.; Jung, D., Shim, S. E. *Int. J. Multiphas. Flow* 2009, 35, 525-532.

[40] Kathiravan, R.; Kumar, R.; Gupta, A.; Chandra, R.; Jain, P. K. *Int. J. Heat Mass Transf.* 2011, 54,1289-1296.

[41] Assael, M. J., Metaxa, I. N.; Kakosimos, K.; Constantinou, D. *Int. J. Thermophys.* 2006, 27, 999-1016.

[42] Teng, T. P.; Yu, C. C. *Exp. Therm. Fluid Sci.* 2013, 49, 22-30.

[43] Jiang, W.; Ding, G.; Peng, H. *Int. J. Therm. Sci.* 2009, 48, 1108-1115.

[44] Kim, Y. J.; Ma, H.; Yu, Q. *Nanotechnology* 2010, 21, 295703.

[45] Nasiri, A.; Niasar, M. S.; Rashidi, A.; Khodafarin, M. R. *Int. J. Heat Mass Transf.* 2012, 55, 1529-1535.

[46] Harish, S.; Ishikawa, K.; Einarsson, E.; Aikawa, S.; Chiashi, S.; Shiomi, J.; Maruyama, S. *Int. J. Heat Mass Transf.* 2012, 55, 3885-3890.

[47] Hwang, Y. J.; Park, H. S., Lee, J. K., Jung, W. H. *Curr. Appl. Phys.* 2006, 6S1, e67-71.

[48] Nanda, J.; Maranville, C.; Bollin, S. C.; Sawall, D.; Ohtani, H.; Remillard, J. T.; Ginder, J. M. *J. Phys. Chem. C* 2008, 112, 654-658.

[49] Assael, M. J.; Metaxa, I. N.; Arvanitidis, J.; Christofilos, D.; Lioutas, C. *Int. J. Thermophys.* 2005, 26, 647-664.

[50] Han, Z. H.; Yang, B.; Kim, S. H.; Zachariah, M. R. *Nanotechnology* 2007, 18, 105701.

[51] Glory, J.; Bonetti, M.; Helezen, M.; Hermite, M. M. L.; Reynaud, C. *J. Appl. Phys.* 2008,103, 094309.

[52] Jha, N.; Ramaprabhua, S. *J. Appl. Phys.* 2009, 106, 084317.

[53] Yang, Y.; Grulke, E. A.; Zhang, Z. G.; Wu, G. *J. Appl. Phys.* 2006, 99, 114307.

[54] Jana, S.; Khojin, A. S.; Zhong, W. H. *Thermochim. Acta* 2007, 462, 45-55.

[55] Phuoc, T. X., Massoudi, M.; Chen, R. H. *Int. J. Therm. Sci.* 2011, 50, 12-18.

[56] Harish, S.; Ishikawa, K.; Einarsson, E.; Aikawa, S.; Inoue, T.; Zhao, P.; Watanabe, M.; Chiashi, S.; Shiomi, J.; Maruyama, S. *Mater. Express* 2012, 2, 213-223.

[57] Singh, N.; Chand, G.; Kanagaraj, S. *Heat Transf. Eng.* 2012, 33, 821-827.

[58] Xie, H.; Chen, L. *Phys. Lett. A* 2009, 373, 1861-1864.

[59] Shaikha, S.; Lafdi, K.; Ponnappan, R. *J. Appl. Phys.* 2007,101, 064302.

[60] Hwang, Y.; Lee, J. K.; Lee, C. H.; Jung, Y. M.; Cheong, S. I.; Lee, C. G.; Ku, B. C.; Jang, S. P. *Thermochim. Acta* 2007, 455, 70-74.

[61] Gu, B.; Hou, B.; Lu, Z.; Wang, Z.; Chen, S. *Int. J. Heat Mass Transf.* 2013, 64,108-114.

[62] Wang, X.; Xu, X.; Choi, S. U. S. *J. Thermophys. Heat Transfer* 1999, 13, 474-480.

[63] Keblinski, P.; Phillpot, S.; Choi, S. U. S.; Eastman, J. A. *Int. J. Heat Mass Transfer* 2002, 45, 855-863.

[64] Xuan, Y.; Li, Q.; Hu, W. *AIChE J.* 2003, 49, 1038-1043.

[65] Koo, J.; Kleinstreuer, C. *J. Nanoparticle Res.* 2004, 6, 577-588.

[66] Kumar, D. H.; Patel, H. E.; Kumar, V. R. R.; Sundararajan, T.; Pradeep, T.; Das, S. K. *Phys. Rev. Lett.* 2004, 93, 144301.

[67] Maxwell, J. C. *A Treatise on Electricity and Magnetism*, Clarendon Press: Oxford, U.K., 1891.

[68] Hamilton, R. L.; Crosser, O. K. *Ind. Eng. Chem. Fund.* 1962, 1, 187-191.

[69] Patel, H. E.; Anoop, K. B.; Sundararajan, T.; Das, S. K. *Bull. Mat. Sci.* 2008, 31, 387-390.

[70] Strauss, M. T.; Pober, R. L. *J. Appl. Phys.* 2006, 100, 084328.

[71] Sastry, N. N. V.; Bhunia, A.; Sundararajan, T.; Das, S. K. *Nanotechnology* 2008, 19, 055704.

[72] Walvekar, R.; Faris, I. A.; Khalid, M. *Heat Transfer-Asian Res.* 2012, 41, 145-163.

[73] Yu, W.; Choi, S. U. S. *J. Nanopart. Res.* 2004, 6, 355-361.

[74] Xue, Q-Z. *Phys. Lett. A* 2003, 307, 313-317.

[75] Gao, L.; Zhou, X.; Ding, Y. *Chem. Phys. Lett.* 2007, 434, 297-300.

[76] Sabbaghzadeh, J.; Ebrahimi, S. *Int. J. Nanosci.* 2007, 6, 45-49.

[77] Clancy, T. C.; Gates, T. S. *Polymer* 2006, 47, 5990-5996.

[78] Murshed, S. M. S.; Nieto de Castro C. A. *Ren. Sust. En. Rev.* 2014, 37, 155-167.

[79] Chen, H.; Ding, Y.; Tan, C. *New J. Phys.* 2007, 9, 367-1-24.

[80] Mahbubul, I. M., Saidur, R.; Amalina, M. A. *Int. J. Heat Mass Transf.* 2012, 55, 874-885.

[81] Ko, G. W., Heo, K.; Lee K.; Choi, M. In *Proceeding of PARTEC 2007– Congress on Particle Technology*; Nürnberg, Germany, March 2007.

[82] Chen, L.; Xie, H., Yu, W.; Li, Y. *J. Disper. Sci. Technol.* 2011, 32, 550-554.

[83] Hung, Y. H.; Chou, W. C. *Int. J. Chem. Eng. Appl.* 2012, 3, 5.

[84] Vakili-Nezhaada, G.; Dorany, A. *Energy Proc.* 2012, 14, 512-517.

[85] Estellé, P.; Halelfadl, S.; Doner, N., Maré, T. *Curr. Nanosci.* 2013, 9, 225-230.

[86] Halelfadl, S.; Estellé, P., Aladag, B.; Doner, N., Maré, T. *Int. J. Therm. Sci.* 2013, 71, 111-117.

[87] Sundar, L. S.; Sharma, K. V., Naik, M. T., Singh, M. K. *Ren. Sust. En. Rev.* 2013, 25, 670-686.

[88] Einstein, A. *Investigations on the Theory of the Brownian Movement*, Dover Publications, Inc.: New York, 1956.

[89] Krieger, I. M.; Dougherty, T. *J. Trans. Soc. Rheol.* 1959, 3, 137-152.

[90] Nielsen, L. E. *J. Appl. Phys.* 1970, 41, 4626-4627.

[91] Batchelor, G. K.; *J. Fluid Mech.* 1977, 83, 97-111.

[92] Kakaç, S.; Pramuanjaroenkij, A. *Int. J. Heat Mass Transf.* 2009, 52, 3187-3196.

[93] Faulkner, D.; Rector, D. R.; Davison, J. J.; Shekarriz, R. In *Proceedings of the IMECE*, Vol.375, ASME Heat Transfer Division, California, USA, 2004, Paper: IMECE2004-62147.

[94] Amrollahi, A.; Rashidi, A. M.; Lotfi, R.; Meibodi, M. E.; Kashefi, K. *Int. Comm. Heat Mass Transf.* 2010, 37, 717-723.

[95] Rashidi, F.; Nezamabad, N. M. In *Proceedings of the WCE*, Vol. 3 (Eds: Ao, S. I.; Gelman, L.; Hukins, D. W. L.; Hunter, A.; Korsunsky, A. M.), IA Eng., London, UK, 2011, 2441.

[96] Mare´, T.; Halelfadl, S.; Sow, O.; Estelle, P., Duret, S.; Bazantay, F. *Exp. Therm. Fluid Sci.* 2011, 35, 1535-1543.

[97] Lotfi, R.; Rashidi, A. M., Amrollahi, A. *Int. Comm. Heat Mass Transf.* 2012, 39,108-111.

[98] Ruan, B., Jacobi, A. M. *Int. J. Heat Mass Transf.* 2012, 55, 3186-3195.

[99] Kumaresana, V.; Velraj, R.; Das, S. K. *Int. J. Refrig.* 2012, 35, 2287-2296.

[100] Yang, C.; Li, W.; Sano, Y.; Mochizuki, M.; Nakayama, A. *J. Heat Transfer* 2013, 135, 054504.

[101] Hussein, A. M.; Sharma, K. V.; Bakar, R. A.; Kadirgama, K. *Ren. Sust. En. Rev.* 2014, 29, 734-743.

In: Nanofluids: Synthesis, Properties and Applications
Editors: S.M. Sohel Murshed, C.A. Nieto de Castro

ISBN: 978-1-63321-677-8
© 2014 Nova Science Publishers, Inc.

Chapter 4

THERMAL PROPERTIES OF MAGNETIC NANOFLUIDS

P. D. Shima[1,2], Baldev Raj[3] and John Philip[1,]*

[1]SMARTS, Metallurgy and Materials Group,
Indira Gandhi Centre for Atomic Research, Kalpakkam, India
[2]Laboratoire des Colloïdes et Matériaux divisés, ESPCI, Paris, France
[3]PSG Institutions, Coimbatore, Tamilnadu, India

ABSTRACT

Magnetic nanofluids are suspensions of nanometer sized magnetic particles stabilized aginst agglomeration and sedimentation. The uniqueness of magnetic nanofluid is that its properties and the location can be easily controlled by an external magnetic field, which is being exploited for many scientific, industrial, and commercial applications. During the search for superior coolants with better heat transfer efficiencies, it was found that magnetic nanofluids can produce a dramatic thermal conductivity enhancement (> 300%) due to the efficient transport of heat through the percolating nanoparticle paths. It has also been demonstrated that the field-induced thermal conductivity enhancement can be precisely and reversibly tuned from a low to very high value by varying the magnetic field strength and its orientation. Since the application of magnetic field enhances not only the thermal conductivity but also the rheological properties of the magnetic nanofluid, they find applications in smart cooling cum damping devices. This chapter summarizes the recent research on thermal conductivity of magnetic nanofluids. The effects of volume fraction, magnetic field strength, nanoparticle size, temperature, base fluid material, aggregation and additives on thermal conductivity of magnetic nanofluids are discussed in detail.

1. INTRODUCTION

Magnetic nanofluids, which are suspensions of magnetic nanoparticles, constitute a special class of nanofluids that exhibit both magnetic and fluid properties [1]. Since the

---

[*] E-mail: philip@igcar.gov.in.

properties and the position of these fluids can easily be controlled by an external magnetic field, they have recently attracted many scientific, industrial, and commercial applications [2-4]. These materials have been found to have several fascinating applications such as magneto-optical wavelength filter [5, 6], optical modulators [7], nonlinear optical materials [8], tunable optical fiber filter [9], optical grating [10], cation sensors [11], defect detection sensors [12] and optical switches [13]. In addition, they have been a wonderful model system for fundamental studies [14]. Furthermore, they have found to have applications in magneto-fluidic seals, lubricants, density separation, inkjet printers, refrigeration, diagnostics in medicine, clutches, tunable dampers, etc.

With the depleting hydrocarbon reserves, the demand for reducing power consumption and development of superior coolants with improved performance is increasing [15]. Based on the initial anomalous thermal conductivity enhancement in nanofluids, they have been widely studied and projected as potential candidates for heat transfer applications [16-18]. However, further studies reveal modest thermal conductivity (k) enhancement in conventional nanofluids, which necessitated the need to develop nanofluids with significantly large k, especially for miniature devices such as micro- and nano-electromechanical systems (MEMS and NEMS). Reports showed substantial enhancement in k for magnetic nanofluids in the presence of a magnetic field [19]. The enhancement and control of k of magnetic nanofluids in the presence of a magnetic field offers promising applications in cooling and heat transfer. Consequently, *thermal properties* of magnetic nanofluids have been a topic of intense research in recent years.

Magnetic nanofluids are prepared by dispersing metal or metal oxide magnetic nanoparticles in various carrier fluids. The most commonly used nanoparticles for the preparation of magnetic nanofluids are magnetite (Fe_3O_4), iron (Fe) and hematite (Fe_2O_3) with particle size in the range of 5 − 20 nm. The suspended nanoparticles are often stabilized with a layer of surfactant to prevent aggregation and the subsequent settling of particles, thereby enhancing the stability of the suspension. The choice of surfactant depends on the base fluid used for dispersing the magnetic nanoparticles. The most commonly used base fluids for the preparation of magnetic nanofluids are water, ethylene glycol (EG), kerosene and other heat transfer oils. To prepare stable oil-based magnetic nanofluids, the nanoparticles are often coated with fatty acids like oleic acid.

The carboxylic acid group of fatty acid binds to the surface of magnetic nanoparticles and the aliphatic chain extends into the nonpolar solvent, preventing aggregation of particles by steric hindrance. For stable water-based magnetic nanofluids, the nanoparticles are often coated with a quaternary ammonium salt like tetra methyl ammonium hydroxide (TMAOH) that stabilizes the nanoparticles by electrostatically. When the particles are coated with TMAOH, the surface of particles is charged with hydroxide ions and the tetra methyl ammonium group acts as counter ions in the solution that creates a double layer in an aqueous environment. The net repulsion between two similarly charged particles raises the energy required for the particles to agglomerate and stabilize the nanoparticle suspensions.

Figure 1(a) and Figure 1(b) shows the schematics of steric and electrostatic stabilization of magnetic nanoparticles by fatty acids and TMAOH coating, respectively.

2. FACTORS AFFECTING THERMAL CONDUCTIVITY OF MAGNETIC NANOFLUIDS

The study of heat transport in nanoparticle dispersions is rather new. The thermal conductivity of magnetic nanofluids were reported in the eighties [20-23]. Later, many systematic experimental studies have been carried out on thermal conductivity of magnetic nanofluids using variety of magnetic nanomaterials [24]. Detailed studies over the last one decade show that the k of magnetic nanofluids depend on factors like volume fraction (ϕ), magnetic field strength and its orientation, nanoparticle size, nature of base fluid, nature of nanoparticle material, temperature, presence of additives etc. Studies on each of these factors are discussed briefly.

2.1. Effect of Volume Fraction

The effective medium theory (EMT) or mean field theory of Maxwell is most frequently used to analyze the nanofluid k results [25]. For a nanofluid containing non-interacting spherical nanoparticles, the EMT predicts

$$\frac{k}{k_f} = \frac{1+2\beta\phi}{1-\beta\phi} \qquad (1)$$

where ϕ is the nanofluid volume fraction; k and k_f are the thermal conductivities of the nanofluid and the base fluid, respectively. β is given by $(k_p-k_f)/(k_p+2k_f)$, where k_p is the thermal conductivity of nanoparticle.

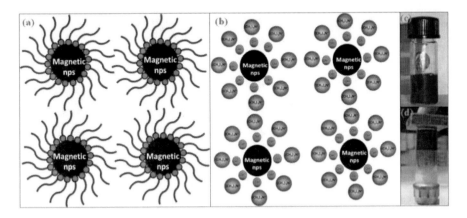

Figure 1. The schematic representation of (a) steric and (b) electrostatic stabilization of magnetic nanoparticles by fatty acids and TMAOH coating, respectively. Photographs of a stable magnetic nanofluids are shown in the panels (c) in absence and (d) in the presence of a magnet.

When a finite temperature discontinuity exists at the nanoparticle-fluid interface, $k_f \rightarrow k_f + \alpha k_p$, where $\alpha = 2R_b k_f /d$, R_b is the interfacial thermal resistance and 'd' is the nanoparticle size. The thermal conductivity enhancements beyond those predicted by Eq. (1) are often reported to be anomalous. Studies on k of nanofluids at different ϕ shows conflicting reports of k enhancement. Thermal conductivity enhancement within [19, 20, 26-31] and beyond EMT predictions are reported [32-42].

2.1.1. k Enhancement within the EMT Predictions

The variation of k with particle concentration in dispersions of Fe_3O_4 particles in diester, hydrocarbon, water and fluorocarbon carriers were well described by Tareef's equation (often known was Maxwell's expression for solid/liquid mixtures) [20]. In cobalt (d = 75 nm)/toluene and cobalt (d = 60 nm)/apiezon ferrofluids, the variation of k with ϕ was within the predictions of Tareef's equation at low particle concentration (ϕ < 0.1) [22]. In another report, the k of Fe_3O_4/kerosene magnetic fluids was described by Tareef's equation up to particle loading of ϕ = 0.1 [21]. In the regions where ϕ was > 0.1, the measured values of k deviated systematically from Tareef's equation and the k of concentrated magnetic fluids (ϕ > 0.10) was described by Zarichnyak equation for thermal conductivity of binary system with a disordered arrangement of the components [43].

The thermal conductivity studies in iron-based magnetorheological (MR) suspension that contains dispersions of carbonyl-grade iron powder (d = 1 − 3 μm) in oil/lithium-grease base fluid, showed a k enhancement in agreement with the Maxwell's and Bruggeman models at low ϕ [Figure 2(a)] [26]. However, at higher ϕ, the measured k values were significantly higher than that of the Maxwell model and were in better agreement with the Bruggeman model. In MR suspensions containing dispersions of carbonyl iron particles (d = 1 − 3 μm) in silicone oil, experimental k data matches with the Bruggeman model at low ϕ and the model underestimates the k at high ϕ (Figure 2(a)) [44].

In microencapsulated phase change material (MMPCM) suspension, k increased with increase in both Fe nanoparticle (d = 25 nm) concentration and MMPCM particle concentration [27]. Here, the enhancement in k with magnetic nanoparticle and MMPCM particle concentration was within EMT predictions. The maximum k enhancement observed in MMPCM suspension was ~ 12 % with 9.21 vol.% of particle concentration. The highest k enhancement with magnetic nanoparticle concentration was 7 % for 4 wt.% of Fe nanoparticle contents in suspension. In another study, Fe (d = 26 nm)/water magnetic nanofluids showed a k enhancement within the prediction of EMT.[28] Maximum k enhancement observed was 14.9% in magnetic fluid having a particle loading ϕ = 0.05.

The enhancement in k with ϕ was within the predictions of EMT for heptane-based Fe_3O_4 nanofluids (d = 10 nm) where a maximum of 5% k enhancement was observed for nanofluid with a particle loading of 7 wt.% (ϕ = 0.01) [29]. The k enhancement in EG-based α-Fe_2O_3 (d = 29 nm) and Fe_3O_4 (d = 15 nm) nanofluids at 30 ^0C was in agreement with Maxwell model prediction with a maximum deviation of 3 % and 6 % for Fe_2O_3 and Fe_3O_4 nanofluids, respectively (Figure 2(b)) [30]. Here, k increases linearly with ϕ reaching the highest values of 11% and 15% for Fe_2O_3 (ϕ = 0.066) and Fe_3O_4 (ϕ = 0.069) nanofluids, respectively. Thermal conductivity enhancement within the predictions of EMT is reported in water:EG-based Fe_2O_3 (d = 20 nm) nanofluids where the maximum k enhancement was ~ 3% for a nanofluid with a particle lading of ϕ = 0.02 (Figure 2(b)) [45].

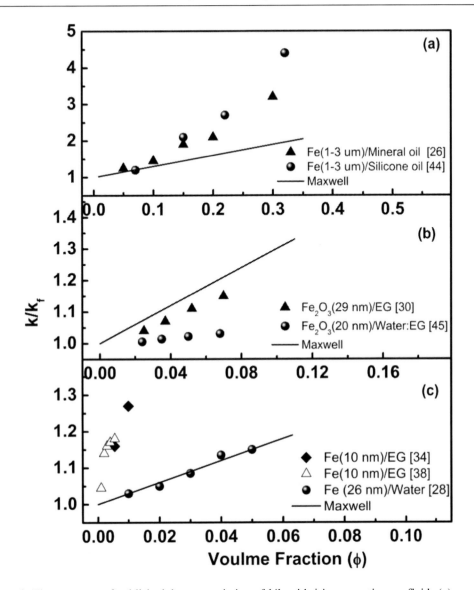

Figure 2. The summary of published data on variation of k/k_f with ϕ in magnetic nanofluids (a): Magnetorheological fluids; (b): Fe$_2$O$_3$ based nanofluid and (c): Fe based nanofluid. The Maxwell fit is shown by the black lines.

Kerosene-based Fe$_3$O$_4$ (d = 9.8 nm) nanofluids showed 17.25% enhancement in k at a particle loading of ϕ = 0.047 that was within the predictions of EMT [46]. Dispersions of Fe$_3$O$_4$ (d = 10 nm) nanoparticles in a mixture of diesel oil and polydimethylsiloxane (PDMS) base fluids showed a k enhancement beyond EMT predictions in low viscous base fluid of 100% diesel oil, which has a viscosity (η) of 4.188 cP [47, 48]. In highly viscous base fluid that consists of 50% diesel oil and 50% PDMS (η = 140.4 cP), the k of nanofluids was within the predictions of the Maxwell equation. It was reported that the k of water-based magnetic fluids do not directly depend on the solid phase concentration when different surfactants were used [49]. In another study, fluorescein isothiocyanate (FITC) labeled water-based magnetite nanofluids (d < 10 nm) showed a insignificant enhancement in k [50].

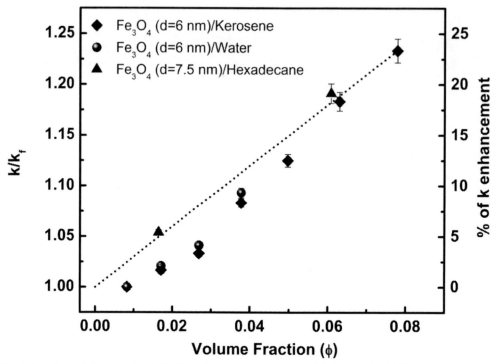

"Reprinted from Advances in Colloid and Interface Science, 183-184, John Philip and P.D. Shima, Thermal properties of nanofluids, 30-45, Copyright (2012), with permission from Elsevier".

Figure 3. The variation of k/k_f and % of k with volume fraction for Fe$_3$O$_4$ (d = 6 nm)/water, Fe$_3$O$_4$ (d = 7.5 nm)/hexadecane and Fe$_3$O$_4$ (d = 6 nm)/kerosene nanofluids. The Maxwell fit is shown by the black dotted line.

A negative k enhancement was observed in samples at low particle concentrations. The observed k variation with ϕ was explained in terms of particle aggregation. The negative k enhancement was explained by assuming that the effective k of aggregates is smaller than that of k of bulk materials and the base fluids. Surfactant-stabilized stable oil (kerosene and hexadecane) and water-based Fe$_3$O$_4$ nanofluids showed a k enhancement within the predictions of Maxwell's effective medium theory [19, 51]. Stable oil and water-based magnetic nanofluids were prepared by coating Fe$_3$O$_4$ nanoparticles with oleic acid and TMAOH, respectively. The oleic acid provides steric stabilization to nanoparticles whereas, TMAOH offers electrostatic stabilization. Figure 3 shows the variation of k/k_f and % of k with volume fraction for Fe$_3$O$_4$ (d = 6 nm)/water, Fe$_3$O$_4$ (d = 7.5 nm)/hexadecane and Fe$_3$O$_4$ (d = 6 nm)/kerosene nanofluids together with the Maxwell's fit. No enhancement in k was observed up to a volume fraction of $\phi = 0.015$, above which k enhancement was linear with the volume fraction of Fe$_3$O$_4$ nanoparticles. The highest k enhancement of 23% was observed in kerosene-based Fe$_3$O$_4$ nanofluids at a particle loading of $\phi = 0.078$. The data fitted with Maxwell model show good agreement, especially at higher volume fractions.

2.1.2. k Enhancement beyond EMT Predictions

Polyethylene glycol-based carbon coated Fe nanofluids are reported to have k enhancement beyond EMT predictions [32]. At a particle loading of 1.5 wt.%, a k enhancement of 28% was observed, which was further increased up to 30% when glycerin dispersion agent was added. An approximately 10% increase in k was observed in heat transfer nanofluids containing Fe_2O_3 nanoparticles and carbon nanotubes (CNTs) even at very low particle loading of 0.02 wt.% [33]. In EG-based Fe nanofluids, at a particle loading of 1.0 vol.%, the k enhancement was in the range of $21 - 33\%$ which was much higher than the EMT predictions [34]. The size distributions of Fe nanoparticles in the carrier fluid measured using DLS was much higher ($\sim$ 500 nm) than the primary crystallite size (10 nm) because of aggregation. Nanofluids containing dispersions of hybrid sphere/CNT particles (consisting of numerous CNTs attached to alumina/iron oxide bimetallic sphere) in polyalphaolefin (PAO), showed a k enhancement of $\sim$ 21% at a low particle loading of 0.2 vol.% [35].

EG-based Fe nanofluids (d = 10 nm) showed a k enhancement beyond EMT predictions where the maximum k enhancement observed was 16.5% for nanofluid with a particle loading of 0.3 vol.% [36]. Large k enhancement was observed in EG-based Fe nanofluids (d = 10 nm) sonicated with high powered pulses [38, 52]. The observed k enhancement in Fe nanofluids was nonlinear with ϕ, which was attributed to the rapid clustering of nanoparticles. The k of EG-based Fe nanofluid (ϕ = 0.055) exhibited a k enhancement of 18%, immediately after sonication and the thermal conductivity was saturated after 30 min of sonication [37]. Figure 2(c) shows the summary of published data on k of Fe nanofluids in water and EG-based fluids together with Maxwell fit.

In EG-based nanofluids, k enhancement was beyond EMT predictions whereas in water-based Fe nanofluids, k enhancement was within EMT predictions. Kerosene-based Fe_3O_4 nanofluids (d = 15 nm) showed a maximum k enhancement of $\sim$ 34% at a particle loading ϕ = 0.01 which was much above the EMT predictions [42]. In the above nanofluids, the average particle size measured using DLS (155 nm) was much higher than the primary nanoparticle size (15 nm) confirming the aggregation of nanoparticle in dispersions. Water-based Fe_3O_4 nanofluids (d = 9.8 nm) showed a k enhancement beyond EMT predictions and the maximum k enhancement observed was 38% for nanofluid with a particle loading of ϕ = 0.04 [41]. The abnormal and nonlinear thermal conductivities of above nanofluids were attributed to the nanoparticle alignment along the heat conduction, which was confirmed by microscopy analysis.

The k of transformer oil-based magnetic nanofluids with a Fe_3O_4 loading of ϕ = 0.20 was 100% higher than that of the base liquid, at room temperature [22]. The thermal conductivities of hexane, heptane and light mineral oil-based Fe_3O_4 (d = 6 and 10 nm) nanofluids showed k enhancement beyond EMT predictions [40]. In the above nanofluids, $\sim$ 28% increase in k was obtained even at low particle loading of 2.5 wt.%. Oil-based Fe_3O_4 nanofluids showed a k enhancement beyond EMT predictions at 80 ^{0}C [39]. For a particle (Fe_3O_4) loading of 4 wt.%, the k enhancement was $\sim$ 26 and 16 % for nanofluids containing 4 and 44 nm, respectively, at 80 ^{0}C.

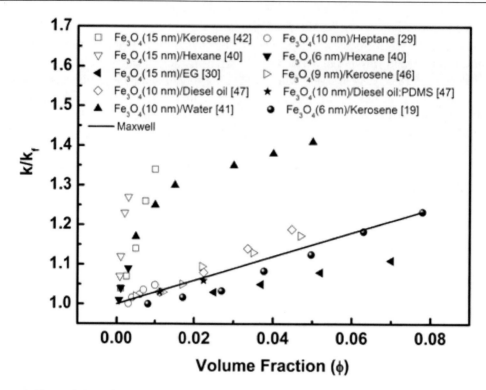

Figure 4. The variation of k/k_f with ϕ for various Fe$_3$O$_4$ nanofluids in different base fluids together with Maxwell fit.

In water-based Fe$_3$O$_4$ (d = 10 nm) nanofluids, the k enhancement was beyond EMT predictions [53]. The highest k enhancement observed was 11.4% in a nanofluid with ϕ = 0.03. Figure 4 shows the summary of published data on k of Fe$_3$O$_4$ nanofluids in various base fluids together with Maxwell fit. Thermal conductivity variations within and beyond EMT predictions can be seen.

2.2. Effect of Magnetic Field Strength and Orientation

Philip et al. [19] observed a dramatic enhancement of k in a nanofluid containing magnetite particles of average diameter of 6.7 nm under the influence of an applied magnetic field. They observed that the k increases with increase in applied field strength up to a critical magnetic field, beyond which the k values decrease a little due to zippering transitions between the field induced chains. With increasing concentration of Fe$_3$O$_4$, the enhancement in k was very significant. Figure 5 shows the k/k_f and the corresponding percentage of enhancement in k as a function of applied magnetic field for kerosene-based Fe$_3$O$_4$ nanofluids with ϕ = 0.063 and 0.049. Here, the k started to increase drastically at very low magnetic fields. The highest value of enhancement in k observed is 300% for a nanofluid of ϕ = 0.063.

The large enhancement in k in presence of magnetic field that is parallel to temperature gradient is explained as follows: Ferrofluids consist of a colloidal suspension of single domain superparamagnetic nanoparticles with a magnetic moment 'm'. The interparticle dipole-dipole interaction U$_d$(ij) between the magnetic particles is given by [54]

$$U_d(ij) = -\left[3\frac{(m_i \cdot r_{ij})(m_j \cdot r_{ij})}{r_{ij}^5} - \frac{(m_i \cdot m_j)}{r_{ij}^3}\right], r_{ij} = r_i - r_j \qquad (2)$$

The dipolar interaction energy depends on the distance r_{ij} between the ith and j^{th} particles and the mutual orientation of their magnetic moments m_i and m_j. When the dipolar interaction energy becomes sufficiently strong, the magnetic particles form chain like structures. The effective attraction between two ferromagnetic particles is described by a coupling constant L = $U_d(ij)/k_BT$, which involves two competing factors: magnetic dipolar interaction energy $U_d(ij)$ and thermal energy k_BT, where, k_B is the Boltzmann constant and T is the temperature. Dipolar structure formation is expected when the dipolar potential exceeds thermal fluctuations; i.e. for a dipolar coupling constant L > 1.

Without any external magnetic field, the magnetic moments of the scatterers are oriented in random direction. In the presence of magnetic field, the nanoparticles align in the direction of magnetic field when the magnetic dipolar interaction energy $U_d(ij)$ dominates over the thermal energy k_BT. The equilibrium chain length and flexibility of the chains depends on the orientational correlations between the magnetic moments of particles inside a chain. The chain flexibility decreases with field strength and in a strong field, the chain aggregate resembles a stiff rod like chain [54].

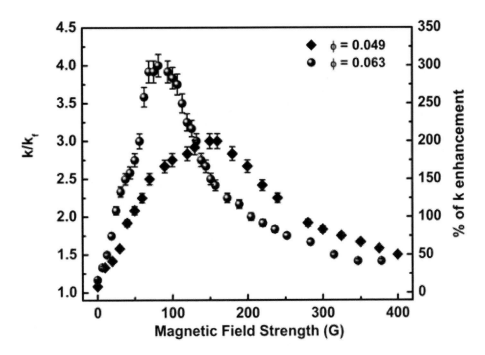

Figure 5. The k/k_f and % of enhancement in k as a function of external magnetic field strength for kerosene-based Fe$_3$O$_4$ nanofluids with ϕ = 0.049 and 0.063. The applied magnetic field is parallel to temperature gradient here.

The particle concentration for hard-sphere suspensions is related to its volume fraction, ϕ = NV_p, where N is the number density of particles and V_p is their volume (= $4\pi r^3/3$, where r is the particle of radius). The extent of chain formation in the presence of an external magnetic field increases with increase in ϕ, since the number of particles per unit volume increases with increase in ϕ. Thus for a given magnetic field strength, the enhancement in k will be higher for the nanofluid with maximum particle loading. Further, the saturation magnetization of Fe_3O_4 nanoparticle dispersions also increases with increase in nanoparticle concentration [55]. The decrease in the k enhancement observed at high magnetic field strengths (Figure 5) is explained as follows: After the formation of linear chains, they come together to form denser chains due to interaction between the adjacent columns [56]. The decrease in k observed above a critical magnetic field strength is expected to be due to 'zippering' of chains. Phase contrast microscopic studies under external magnetic field confirm such lateral overlap (zippering) of chains at sufficiently high magnetic field strengths. Figure 6(a) shows the micrograph of *k*erosene-based Fe_3O_4 nanofluids with $\phi = 0.05$ in the absence of external magnetic field, where no aggregates are visible. Figures 6(b) – (k) shows the micrograph of the nanofluid in presence of increasing magnetic strengths.

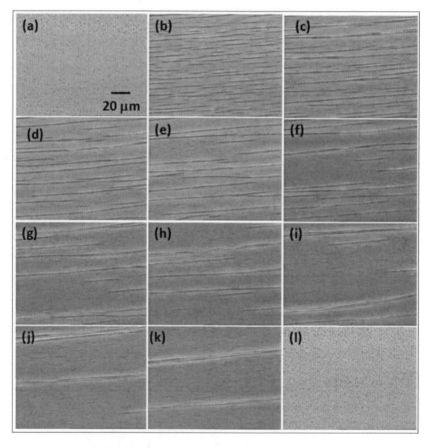

Figure 6. The phase contrast microscopy images of *k*erosene-based Fe_3O_4 nanofluids with $\phi = 0.05$ (a) in the absence of external magnetic field, (b) – (k) in the presence of increasing magnetic field and (l) after removal of magnetic field.

At low magnetic field strength [Figure 6(b)], smaller dipolar chains are formed, whose length increases with increasing magnetic field strength [Figure 6(c)], resulting in an evenly spaced single nanoparticle chains throughout the nanofluid volume. At very high magnetic field strengths [Figures 6(d) – (k)], zippering of the dipolar chain is evident in the microscopic images. Figure 6(l) shows the image after switching off the magnetic field where no permanent aggregates are visible. This confirms the perfect reversibility of the dipolar chains formed in the fluid.

Mean field model predicts series and parallel modes of thermal conduction through nanofluids. The parallel mode has a geometric configuration that allows the most efficient means of heat propagation [57]. Therefore, extremely large k enhancement is possible with parallel modes. Hashin and Shtrikman (HS) bounds for k of a nanofluid, on the basis of ϕ alone is given by [58]

$$k_f \left[1 + \frac{3\phi(k_p - k_f)}{3k_f + (1-\phi)(k_p - k_f)}\right] \leq k \leq \left[1 - \frac{3(1-\phi)(k_p - k_f)}{3k_p - \phi(k_f - k_f)}\right]k_p \qquad (3)$$

In the lower HS limit, nanoparticles are well suspended and conduction is through series modes whereas in the upper HS limit, the conduction path is through dispersed particles. In the absence of magnetic field, the particles are well dispersed, the nanofluids exhibit series mode conduction and the observed variation of k/k_f with ϕ is well within the lower Maxwell limit. In the limit ($\phi k_p/k_f$) >> 1, (k_p and k_f are thermal conductivities of particles and fluid respectively) the predicted values of k/k_f for the upper HS and parallel modes are $(2\phi/3)k_p/k_f$ and $\phi k_p/k_f$, respectively. Figure 7 shows the percentage of enhancement of k without and with magnetic field of different strengths for kerosene-based Fe_3O_4 nanofluids. The upper, lower bounds, series and parallel bound fits are also shown in Figure 7. It can be seen that the experimental data points at the highest magnetic field falls within the parallel mode of conduction.

It was demonstrated that for magnetite nanoparticles, an average particle diameter less than 10 nm is insufficient for significant dipolar structure formation at zero field [59]. Dipolar structure formation is expected when the dipolar potential exceeds thermal fluctuations. The calculated value of L is found to be ~ 0.3 for 10 nm Fe_3O_4 particles coated with a 2 nm thick organic surfactant layer [60]. Since the particles used in this study was of ~ 10 nm size and capped with a surfactant, no dipolar structures occur at zero field, which was confirmed from the TEM data too. Further, the observed k/k_f variation with ϕ at zero field also supports the series mode conduction and was in agreement with the predictions of EMT for well dispersed stable nanofluids. As the magnetic field strength was increased progressively, continuous conduction paths emerge along the nanoparticle chains that result from a series to parallel mode of conduction. At a magnetic field strength of 315 G, the k/k_f data fits fairly well with the parallel mode conduction.

The effect of magnetic field orientation (i.e., the orientation of nano-chains with respect to the heat flow direction) on k enhancement was studied by measuring thermal conductivity under different magnetic field orientations with respect to the thermal gradient [61]. Figure 8 shows the variation of k/k_f with magnetic field strength for kerosene-based Fe_3O_4 nanofluids at ϕ = 0.045 under different field orientations of 0, 20, 70 and 90°. The maximum enhancement in k was observed when the field direction was exactly parallel to the thermal gradient whereas practically no enhancement was observed when field is perpendicular to thermal gradient. A gradual reduction in the k enhancement was observed as the field direction was shifted from parallel to perpendicular direction with respect to thermal gradient. Inset of figure 8 shows the schematics of possible nanoparticle orientation with respect to thermal gradient when the magnetic field direction was varied from parallel to perpendicular direction.

The measurement of k in stable oil-based magnetite nanofluids during rise and decay of magnetic field strength showed that the k enhancement is reversible with a small hysteresis [62]. The k enhancement was reversible even under repeated magnetic cycling. Figure 9 shows the k/k_f and the corresponding k enhancement as a function of applied magnetic field strength for $k$erosene-based Fe_3O_4 with $\phi = 0.045$. The variation of k/k_f at three different magnetic cycles (rise and decay) showed that the enhancement is reversible with a slight hysteresis. The maximum enhancement observed was 216% at an applied magnetic field strength of 101 G. The observed reversible tunable thermal property of nanofluid was shown to have many technological applications. Thermal conductivity studies in iron-based MR suspension that contains dispersions of carbonyl-grade Fe powder (d = 1 − 3 µm) in oil/lithium-grease base fluid, showed an enhancement in k with increase in applied magnetic field parallel to the temperature gradient for different volume fractions and reaches a saturation at high field strengths [26].

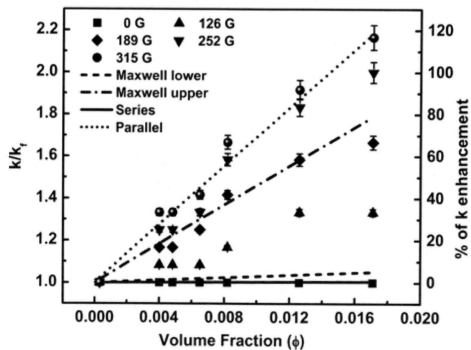

"Reprinted from Advances in Colloid and Interface Science, 183-184, John Philip and P.D. Shima, Thermal properties of nanofluids, 30-45, Copyright (2012), with permission from Elsevier".

Figure 7. The variation of k/k_f with volume fraction in absence and in presence of different magnetic field strengths of 126, 189, 252 and 315 G for $k$erosene-based Fe_3O_4 nanofluids fitted with Maxwell upper and lower bounds, and Maxwell upper and lower bounds.

A maximum of ~ 100% k enhancement was observed in MR suspension with a particle loading of $\phi = 0.191$ when the magnetic field was parallel to the temperature gradient. No k enhancement was observed when the magnetic field direction was perpendicular to temperature gradient. The field induced k enhancements in MR suspension showed a hysteresis.

In suspension with $\phi = 0.30$, upon initial magnetization, the k increased to its saturation value, but when the applied field was decreased, the k value remained at its saturation value even at zero applied field, which is due to the retention of field induced structures against thermal fluctuations due to residual magnetization of the particles. In such cases, k values can be brought to zero by mechanical agitation of the suspension. Thermal conductivity of water, PAO, polyol ester oil-based nickel (Ni) coated CNT nanofluids showed significant enhancement in k (> 60%) under an applied magnetic field [63]. The nanotubes randomly dispersed in the fluid initially and then gradually aligned under the prolonged exposure to a magnetic field. The chain length calculated from the microscopy images was found to be around 30 – 150 µm, which was much longer than the real length of individual nanotubes (5 – 40 µm), which confirms that the nanotubes were aligned and form chains and clusters in the presence of an external magnetic field.

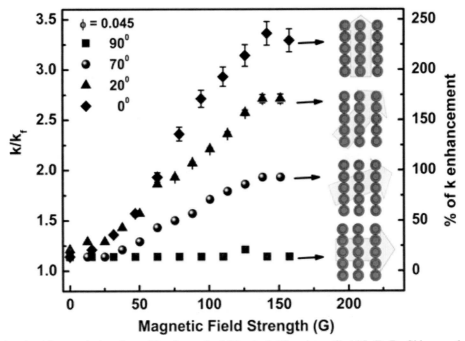

"Reprinted with permission from The Journal of Physical Chemistry C, 115, P. D. Shima and John Philip, Tuning of Thermal Conductivity and Rheology of Nanofluids Using an External Stimulus, 20097–20104. Copyright (2011) American Chemical Society".

Figure 8. The k/k_f and % of enhancement of k as a function of external magnetic field strength for kerosene-based Fe_3O_4 nanofluids with $\phi = 0.045$ in presence of different field orientations of 0, 20, 70 and 90°. Inset shows the schematics of direction of heat and the possible nanoparticle structures for different field directions.

Recent studies showed that k of water and heptane-based Fe$_3$O$_4$ (d = 10 nm) nanofluids can be increased even at low concentrations and low magnetic field strengths where the magnetic field was applied perpendicular to the hot wire [29]. In water-based Fe$_3$O$_4$ nanofluids, at a particle loading of 1.63 wt.%, k enhancements were 1.93% and 5.2% at magnetic field strengths of 0.05 and 0.1 Tesla, respectively. On the other hand, for Fe$_3$O$_4$ nanoparticles dispersed in heptane, k enhancements were 0.6% and 2.8% at a particle loading of 1.28 wt.%, for 0.05 and 0.1 Tesla, respectively. Under such low particle loadings, no k enhancement is expected due to the structure formation. Since the exposure time of external magnetic field was short, the increase in k with applied magnetic field and temperature gradient was attributed to the thermomagnetic convection due to temperature gradient that resulted in a non-uniform magnetic body force. After turning off the magnetic field, the k of both water and hexane nanofluids decreased with time, but it never reached their initial k value (at zero field).

No field-induced k enhancement was observed in *k*erosene-based Mn-Zn ferrite (d = 6.7 nm) nanofluid even at high particle loading whereas, *k*erosene-based Fe$_3$O$_4$ (d = 10 nm) nanofluids showed an enhancement in k with increase in magnetic field strength [64]. This must be due to poor stability and weak magnetization of particles due to impurities and additives. An enhancement in k with increase in magnetic field strength is reported for water-based Fe$_3$O$_4$/MWCNTs and Fe$_3$O$_4$ at SiO$_2$/MWCNTs nanofluids [65]. The maximum field-induced k enhancement observed was 20 and 24% for Fe$_3$O$_4$/MWCNTs and Fe$_3$O$_4$ at SiO$_2$/MWCNTs nanofluids, respectively at particle loading of 0.03 vol.% at a magnetic field strength of 80 G.

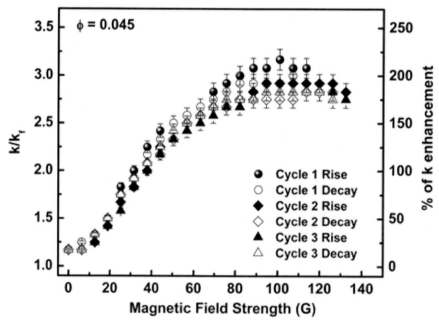

"Reprinted with permission from Applied Physics Letters, 92, John Philip et al., Nanofluid with tunable thermal properties, 043108. Copyright [2008], AIP Publishing LLC".

Figure 9. The k/k_f and % of enhancement in k as a function of increasing and decreasing applied magnetic field strengths at three different magnetic cycles for *k*erosene-based Fe$_3$O$_4$ nanofluids with $\phi = 0.045$.

However, when the magnetic field was increased to a high value of about 130 G, the particles started agglomerating in the base fluid due to poor stability of nanofluids. Moreover, after switching off the field, even though the k decreases a bit, it did not reach the initial zero field k value, which indicates the settling and segregation of nanoparticles in the presence of magnetic field due to poor stability. Engine oil-based magnetic nanofluids containing oleic acid coated Fe_3O_4 (d = 10 nm) nanoparticles showed an enhancement in k with increase in magnetic field strength, when the field was applied parallel to temperature gradient [66]. Maximum k enhancement observed was 200% for a particle loading of $\phi = 0.05$ at a magnetic field strength of 0.0025 Tesla.

A high field-induced k enhancement was observed in untreated Fe_2O_3/water nanofluids compared to sodium dodecyl benzene sulfonate (SDBS) stabilized Fe_2O_3/water nanofluids [67]. The microscopy images showed spontaneous nanoparticle alignments in untreated Fe_2O_3/water nanofluids even without external magnetic field and enhanced particle alignment in presence of a magnetic field. The dispersion stability of above water-based Fe_2O_3 nanofluids was enhanced by the addition of SDBS surfactant where magnetic alignment was not observed. In water-based magnetic nanofluids containing single-walled CNTs (SWCNTs) and Fe_2O_3 nanoparticles, k increases upon exposure to the external magnetic field, reaching a maximum k value at an exposure time of 30 s. Further exposure to magnetic field led to a decrease in k due to clumping of nanoparticle chains [68]. Here, the aggregation of Fe_2O_3 nanoparticles on nanotubes and their alignment in presence of magnetic field led to a significant increase in k [33, 69]. In Fe_2O_3-based nanofluids (0.02 wt.% without carbon nanotube and magnetic field), the k remained reasonably constant with time. With the addition of 0.02 wt.% carbon nanotubes to the above nanofluid at a pH 7, the k increases due to local aggregation of the nanoparticles and k was independent with time. In presence of magnetic field, k of the above nanofluid increased first and reaches a pea$k$ and then decreases with time. The maximum k enhancement observed was about 50% higher than that of the base fluid. At higher magnetic field strengths, the time to reach the maximum k values was shorter and the k of nanofluids under uniform magnetic field was higher than that under non-uniform field. Interestingly, with longer magnetic field exposure time (around $100 - 120$ min), the k value is found to decrease to a value which was even lower than that of the fluid with nanotubes at zero magnetic field [70]. The observed trend in the k was attributed to the precipitation and sedimentation of CNTs together with magnetic particles under a magnetic field for an extended period of time. The k value achieves the original value when the magnetic field was removed and the dispersion was sonicated for a few minutes. Thermal conductivity of magnetic nanofluids containing Ni coated CNTs were found to be enhanced under an applied magnetic field. Under an external magnetic field, k initially increases with time and reaches a pea$k$ value. On prolonged holding of magnetic field, the k was decreased due to formation of bigger clumps of CNTs. At higher magnetic field strengths, the maximum k was reached in a shorter time period due to quic$k$ assembling and alignment of CNTs. A relatively low k value was observed when the above nanofluid was $k$ept under a wea$k$ magnetic field of 0.18 kOe for long time [71].

Water-based Fe_3O_4 (d = 10 nm) nanofluids showed an enhancement of k in the presence of an external magnetic field [72]. In nanofluid having a particle loading of $\phi = 0.05$, a maximum k enhancement of 200% was observed at field strength of 1000 G. At low magnetic field strength of 150 G, the field-induced k enhancement was lower and the maximum k value was reached after a longer exposure time. Increase in the magnetic field intensity (300 G and 600 G) resulted in a higher k value and a shorter time to reach the maximum value. At very high field strength (1000 G), a very high k value was attained in short time interval, which was found to decrease with time. However, after switching off the field, the k of the magnetic nanofluid did not return to its starting value immediately, but followed an exponential decay. The complete return to the starting k value was obtained by ultrasonication of the sample which shows some permanent aggregation of nanoparticle in presence of magnetic field. Moreover, the k of above nanofluids drastically decreased in the presence of a magnetic field with increase in temperature.

Fe (d = 26 nm)/water magnetic nanofluids showed no field-induced k enhancement when the external magnetic field direction was perpendicular to the temperature gradient [28]. On the other hand, thermal conductivity of Fe nanofluid increased with an increase in magnetic field strength when the field direction was parallel to the temperature gradient. In addition, the k increment under the external magnetic field also depend on the concentration of the magnetic particles; the higher the particle concentration, the more remarkable was the field induced k enhancement. At magnetic field strength of 240 G, the k enhancement was 12 and 25% for nanofluids with particle loadings of $\phi = 0.01$ and $\phi = 0.05$, respectively.

Kerosene-based Fe_3O_4 (d = 9.8 nm) nanofluids showed field-induced k enhancement when the magnetic field was applied perpendicular to the temperature gradient [46]. No enhancement in k was observed for nanofluids with $\phi < 0.01$. But with increase in ϕ, there was a significant increase in k in presence of magnetic field. In nanofluids with $\phi > 0.01$, no change in k was observed at lower field strengths whereas for higher magnetic fields, k showed an increase and a saturation at 1 kOe field strength. The maximum k enhancement observed in the above nanofluid was ~ 30% for nanofluid with a particle loading of $\phi = 0.047$. In MR suspensions containing dispersions of carbonyl iron particles (d = 1 − 3 µm) in silicone oil, k enhancement was ~ 30 % along the field direction and was independent of particle concentration [44]. The thermal conductivity of MMPCM suspension increased remarkably with the strength of an external magnetic field that was parallel to the temperature gradient inside the fluid [27]. The maximum field induced k enhancement observed was 22% for a suspension with 9.21 vol.% MMPCM particles. The low k of core/shell materials and the low Fe nanoparticles concentration were the reasons attributed to the weaker response of the suspension to an external magnetic field and a lower field-induced k enhancement.

In synthetic ester-based Fe_3O_4 (d = 13 nm) nanofluids, k was found to increase with increase in magnetic field strength when the field direction was parallel to temperature gradient [73]. Interestingly, when the field direction was perpendicular to temperature gradient, there was a decrease in k with the increase in field strength. The field dependent k studies in water and n-decane-based cobalt ferrite ($CoFe_2O_4$) and γ-Fe_2O_3 nanofluids showed a decrease in k up to a magnetic field strength of $0.2 − 0.3$ mT and a field independent k at higher field strength [74]. Shulman et al. [75] have studied the effect of magnetic fields on the thermal conductivity of suspension of magnetic particles in various oils. They observed that at a constant magnetic field, the suspensions show anisotropy in the thermal conductivity. However, Kronkalns [76] and Blum et al. [77] reported no change in the coefficient of thermal conductivity of magnetic suspensions either in the transverse (with respect to the direction of the thermal flux) or the longitudinal magnetic field direction. No effect of magnetic field on k of magnetic fluids in fields up to H = 200 kA/ m was observed in the presence of a perpendicular magnetic field. Dispersions of Fe$_3$O$_4$ nanoparticles in diester, hydrocarbon, water and fluorocarbon carriers also showed no magnetic field-dependent k in fields up to 0.1 Tesla [20].

2.3. Effect of Nanoparticle Size

Stable kerosene-based Fe$_3$O$_4$ nanofluids showed an increase in k with the increase in nanoparticle size [78]. Figure 10 shows the thermal conductivity ratio (k/k_f) and % of k enhancement as a function of particle size. The thermal conductivity was measured at two different particle loadings of $\phi = 0.01$ and $\phi = 0.055$ as a function of particle size for kerosene-based Fe$_3$O$_4$ nanofluids. With a particle loading of $\phi = 0.01$, no enhancement in k is observed for nanofluid. For nanofluid with $\phi = 0.055$, the enhancement was about 5 and 25% respectively, for 2.8 and 9.5 nm sized particles. Oil-based Fe$_3$O$_4$ nanofluids are reported to have an increase in k with decrease in nanoparticle size [39].

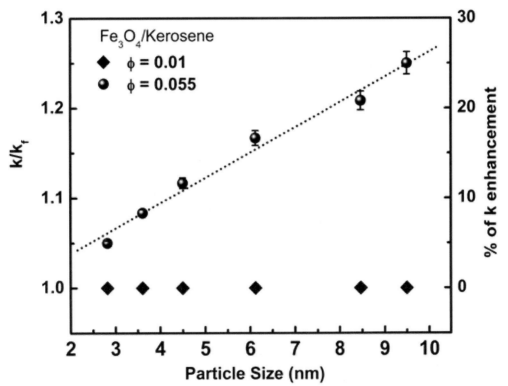

"Reprinted from Advances in Colloid and Interface Science, 183-184, John Philip and P.D. Shima, Thermal properties of nanofluids, 30-45, Copyright (2012), with permission from Elsevier".

Figure 10. The k/k_f and the % of k enhancement of magnetite nanofluids having two different volume fractions $\phi = 0.01$ and $\phi = 0.055$ as a function of nanoparticles size. Best fit is shown by the solid line.

For a particle loading of 4 wt.%, the k enhancement was ~ 26 and 16% for nanofluids containing 4 and 44 nm sized Fe$_3$O$_4$ nanoparticles, respectively. The thermal conductivity studies in hexane and heptane-based Fe$_3$O$_4$ nanofluids showed a high k enhancement for nanofluids with bigger sized particles (d = 10 nm) compared to nanofluids with smaller sized particles (d = 6 nm) [40]. The k enhancement was ~ 22% for hexane-based Fe_3O_4 nanofluids (d = 10 nm) at a particle loading of 1.9 wt.% and the k enhancement was 10% in hexane-based Fe_3O_4 nanofluids containing smaller sized particles (d = 6 nm) at a particle loading of 2.14 wt.%. A similar variation in k with nanoparticle size was observed in the above study for heptane-based Fe_3O_4 nanofluids also.

2.4. Effect of Base Fluid Material

Thermal conductivity studies in stable Fe_3O_4 nanofluids with particle size 6 nm in three different base fluids of kerosene, hexadecane and water showed no dependence of base fluid materials on nanofluid k [51]. The effect of base fluid η on field-induced k enhancement of magnetic nanofluids were studied by following the time dependent k (after 2 − 6 min exposure to the magnetic field) of Ni coated nanotubes in various base fluids (water, PAO, polyol ester oil) with different viscosities [63]. The results show that the viscosities of base fluids do influence the k enhancement; the more viscous the base fluid, the less k enhancement. The highly viscous fluid makes it difficult for the Ni nanotubes to stretch and align. Therefore, the k enhancement was maximum (~ 80%) for water-based nanofluids and minimum for PAO based nanofluids (~ 11%). The effect of base fluid η on the k enhancement of magnetic nanofluids was studied by dispersing Fe_3O_4 (d = 10 nm) nanoparticles in a mixture of diesel oil and PDMS base fluids [48]. When 100% diesel oil was used as a low viscous base fluid (η = 4.188 cP), the k of nanofluids was higher than that of Maxwell prediction. The highest enhancement in k was 18.85% at a particle loading ϕ = 0.0448. The k of nanofluids was found to decrease with increase in viscosity of base fluid. In a highly viscous base fluid (η =140.4 cP), the k of nanofluid was within the predicted values of Maxwell equation. The thermal conductivity of Fe_3O_4 nanofluids showed a high k enhancement for hexane-based nanofluids compared to heptane-based nanofluids [40]. The k enhancement was ~ 22% for hexane-based Fe_3O_4 nanofluids (d = 10 nm) at a particle loading of 1.9 wt.% and the k enhancement was 10% in hexane-based Fe_3O_4 nanofluids (d = 6 nm). However, even at a particle loading > 4 wt.%, the k enhancement was < 5% for heptane-based nanofluids containing both 6 and 10 nm sized particles. The light mineral oil-based nanofluids showed even lesser k enhancement compared to heptane-based nanofluids. The results were contrary to the previous reports on a lower k enhancement for nanofluids with a base fluid of lower k (k of heptane and light mineral oil at 25 ^{0}C are 122 mW/m-K and 125 mW/m-K, respectively). The above study concluded that the difference in the effective k for different nanofluids should not be attributed to the inherent k difference of the base fluids, but more likely to the compatibility of the base fluid with a specific particle type, in which the solvation of the stabilizing layer plays a determining role.

2.5. Effect of Nanoparticle Material

Thermal conductivity studies in iron oxide nanofluids showed a higher k for Fe_2O_3 (d = 29 nm)/EG nanofluids compared to Fe_3O_4 (d = 15 nm)/EG nanofluids [30]. The maximum k

enhancement observed was 15 and 11% for Fe_2O_3 (ϕ = 0.066) and Fe_3O_4 (ϕ = 0.069) nanofluids, respectively. For the same volume fractions, the k of kerosene-based Mn-Zn ferrite (d = 6.7 nm) nanofluid was higher compared to kerosene-based Fe_3O_4 (d = 10 nm) nanofluids [64]. One of the reasons attributed for the above observation was the higher bulk k value of Mn-Zn ferrite (29 W/m-K) compared to that of Fe_3O_4 (5 W/m-K). A thermal valve with a high on-off ratio was demonstrated using magnetic nanofluid that was actuated by a non-uniform magnetic field where a higher switching ratio was observed at lower ϕ in $CoFe_2O_4$ nanofluids compared to Fe_3O_4 nanofluids [79].

2.6. Effect of Temperature

Some reports show an enhancement in k/k_f with increase in temperature [35, 39, 53, 65], whereas other reports show least or no effect of temperature on k/k_f of magnetic nanofluids [30, 31, 40, 42, 46, 80]. There is also a report of reduction in k/k_f with increase in temperature in magnetic nanofluids [22]. An overview of experimental investigations on effect of temperature on magnetic nanofluid k is given below.

2.6.1. Invariant Thermal Conductivity with Increase in Temperature

Aqueous and non-aqueous (kerosene and hexadecane) based stable Fe_3O_4 nanofluids with average particle size of 6 nm showed a temperature independent k over the temperature range of 25 − 50 °C [80]. Figures 11(a) − (c) show the variation of k as a function of temperature for Fe_3O_4-based nanofluids in kerosene, hexadecane, and water based fluids, respectively. Over the temperature range of 25 − 50°C, the absolute k of oil-based nanofluids (kerosene and hexadecane) and base fluids decreases with the increase in temperature. On the contrary, in water-based nanofluid, the thermal conductivity increases with temperature. Figures 12(a) − (c) show the variation of thermal conductivity ratio with base fluid (k/k_f) as a function of temperature for Fe_3O_4-based nanofluids in kerosene, hexadecane, and water based fluids. Interestingly, k/k_f remains constant with increase in temperature, irrespective of the nature of base fluid. These results suggest that the k of the nanofluids simply tracks the k of the base fluid. The DLS measurements in the above nanofluids showed no change in the size with a rise in temperature indicating the absence of aggregation with the rise in temperature.

The invariance of k ratio with rise in temperature over different temperature ranges have been observed by many other groups in kerosene-based Fe_3O_4 nanofluids with varying nanoparticle sizes [42, 46, 81]. Thermal conductivity enhancement in EG-based Fe_2O_3 (d = 29 nm) and Fe_3O_4 (d = 15 nm) were reported to be temperature independent over a temperature range of 10 − 50 ^{0}C [30]. Kerosene-based Fe_3O_4 nanofluids (d = 15 nm) also showed an invariant k ratio with the rise in temperature over the temperature range of 10 − 60 ^{0}C. The thermal conductivity of hexane-based Fe_3O_4 (d = 6 and 10 nm) nanofluids were found to increase linearly with increase in temperature over a temperature range of 30 − 50 ^{0}C. However, a decrease in k with temperature was observed for the base fluid (hexane) as well [40].

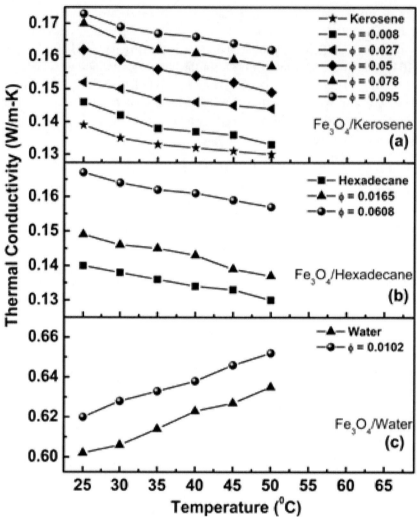

"Reprinted with permission from The Journal of Physical Chemistry C, 114, P. D. Shima et al., Synthesis of Aqueous and Nonaqueous Iron Oxide Nanofluids and Study of Temperature Dependence on Thermal Conductivity and Viscosity, 18825–18833. Copyright (2010) American Chemical Society".

Figure 11. The variation of thermal conductivity with temperature for (a): Kerosene; (b): Hexadecane and (c) Water-based Fe$_3$O$_4$ nanofluids of average particle size 6 nm.

2.6.2. Increase in Thermal Conductivity with Rise in Temperature

Oil-based Fe$_3$O$_4$ nanofluids showed a significantly high *k* enhancement with increasing temperature over the temperature range of 20 – 80 ^{0}C [39]. In comparison with the low *k* values at room temperature, *k* enhancements of ~ 26% and 16 % are observed in nanofluids containing 4 and 44 nm sized Fe$_3$O$_4$ nanoparticles, respectively for a particle loading of 4 wt.%. An enhancement in *k* with increase in temperature was reported for water-based Fe$_3$O$_4$/MWCNTs and Fe$_3$O$_4$ at SiO$_2$/MWCNTs nanofluids.

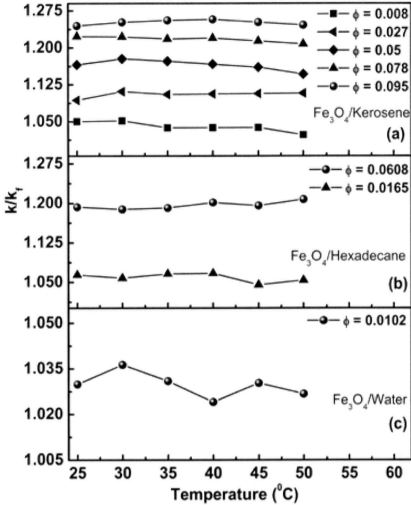

"Reprinted from Advances in Colloid and Interface Science, 183-184, John Philip and P.D. Shima, Thermal properties of nanofluids, 30-45, Copyright (2012), with permission from Elsevier".

Figure 12. The variation of thermal conductivity ratio with temperature for (a): Kerosene; (b): Hexadecane and (c) Water-based Fe$_3$O$_4$ nanofluids of average particle size 6 nm.

For a particle loading of 0.005 vol.%, k enhancement was 3% at 30 ^{0}C and 5% at 50 ^{0}C for water-based Fe$_3$O$_4$/MWCNTs nanofluid [65]. In water-based Fe$_3$O$_4$ at SiO$_2$/MWCNTs nanofluid, for a particle loading of 0.005 vol.%, the k enhancement was 3.7 and 6% at 30 and 50 ^{0}C, respectively. Nanofluids containing dispersions of hybrid sphere/CNT particles in PAO also showed an enhancement in k with rise in temperature over a temperature range of 10 – 90 ^{0}C [35]. For nanofluid with particle loading of 0.2 vol.%, the k enhancement was about 20.5% at 10 ^{0}C and 23.6% at 90 ^{0}C. An enhancement in k with rise in temperature was reported for water-based Fe$_3$O$_4$ nanofluids over the temperature range of 10 – 40 ^{0}C [53].

Studies show that the surfactant dictates the temperature dependence of k of water-based Fe$_3$O$_4$ fluids [49]. For samples with high magnetic solid concentrations (ϕ = 0.025 and 0.028), k was found to increase with rise in temperature over the temperature range of 20 – 60 ^{0}C. In sample with ϕ = 0.024, k do not depend strongly on temperature, and for sample with ϕ =

0.022, the change in k with temperature was similar to that of the surfactant solution. The isolated asymmetrical micelles and the entire micelle layers that appear in concentrated sodium oleate solution undergo a partial destruction when the temperature rises. When the micelle formations are disrupted, the k of the solution seems to rise. Since the rise of k with temperature was observed in the magnetic fluids with the highest concentration of sodium oleate, it was concluded that the k of water-based magnetic fluid mainly depends on the surfactant. Since the volume fraction of surfactant molecules in the water-based magnetic fluid was two orders more than that of magnetite in the same fluid, the contribution of surfactant to the effective k of magnetic fluid was high.

2.6.3. Decrease in Thermal Conductivity with Increase in Temperature

The k of transformer oil-based Fe_3O_4 magnetic nanofluids was found to decrease with the rise in temperature over the temperature range of $20 - 80\ ^0C$ [22]. The k of base fluid (transformer oil) also decreases with rise in temperature. The excess k of the magnetic fluid over that of its base fluid gradually decreases with increase in temperature from 30% at room temperature to 17% at 80 ^0C.

2.7. Effect of Sonication, Aggregation and the Presence of Additives

The stable magnetic nanofluids showed a time independent k after sonication [82]. Figure 13 shows the variation of k/kf with time after sonication for kerosene [Figure 13(a)], hexadecane [Figure 13(b)] and water [Figure 13(c)] based Fe_3O_4 nanofluids. An invariant k/k_f with time after sonication was observed in all the nanofluids. Moreover, the k values of the above nanofluids were invariant even after four months. The particle size remains constant with time for the above nanofluids as observed in the DLS studies (Figure 14), which is consistent with the primary crystallite size obtained from XRD results. Average particle size remains the same in stable magnetic nanofluids even after four months.

The microscopy images showed spontaneous nanoparticle alignments in untreated Fe_2O_3/water nanofluids and the dispersion stability of the nanofluids were enhanced by the addition of SDBS surfactant [67]. In SDBS-stabilized Fe_2O_3 nanofluids (0.02 wt.%), the k remains reasonably constant with time [33]. The dispersion stability of water-based magnetic nanofluids contains SWCNTs and Fe_2O_3 nanoparticles was also enhanced by the addition of SDBS [68]. Thermal conductivity studies in (SWCNTs+Fe_2O_3)/PAO nanofluids showed a relatively high k after sonication, which gradually decreased with time and reached the k value of PAO oil after one hour [83]. From the k measurements at different time intervals in the above nanofluids, it was concluded that increase in k after sonication is because of good dispersion and orientation of the nanoparticles.

Studies have shown that the surfactant sodium oleate determines the temperature dependence of k of water-based Fe_3O_4 fluids [49]. On the other hand, it was also reported that the presence of unbound surfactants has least effect on k of diester and hydrocarbon-based magnetic nanofluids [20]. In EG-based Fe nanofluids, for a particle loading of 1.0 vol.%, the k enhancement was in the range of $21 - 33\%$, which was beyond the EMT predictions [34].

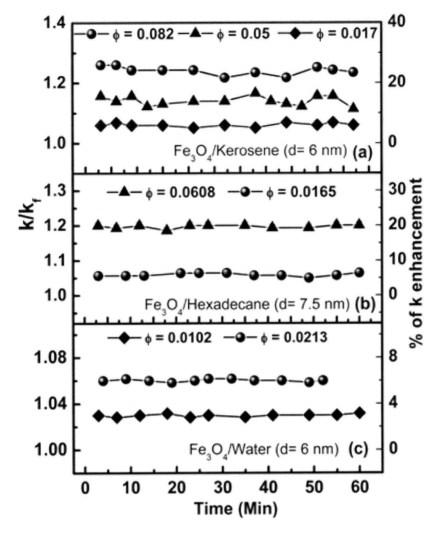

Figure 13. The variation of k/k_f with time after sonication for (a): Kerosene-based Fe_3O_4 nanofluids (d = 8 nm); (b): Hexadecane-based Fe_3O_4 nanofluids (d = 8 nm) and (c): Water-based Fe_3O_4 nanofluids (d = 8 nm).

The size distributions of Fe nanoparticles in the carrier fluid measured using DLS was much higher (~ 500 nm) than the primary crystallite size (10 nm) that confirmed the aggregation in nanofluids. Oleic acid coated kerosene-based Fe_3O_4 nanofluids (d = 15 nm) showed an invariant k enhancement with time and the observed k enhancement was beyond EMT predictions [42]. The average particle size measured using DLS (155 nm) in the above nanofluids was much higher than the primary nanoparticle size of 15 nm that confirms the aggregation of nanoparticle in dispersions.

In polyethylene glycol nanofluids containing carbon coated Fe nanoparticles, the centrifugal sedimentation was ~ 70 min for a particle loading of 0.1 wt.% and sedimentation time remained the same (50 min) when the particle loading is changed from 1 – 1.5 wt.%. To improve the stability of nanofluids, different proportions of glycerin was added to Fe nanofluid that has a particle loading of 1 wt.% where the centrifugal sedimentation time increased when dispersion agent concentration increased from 0 – 1 wt.%.

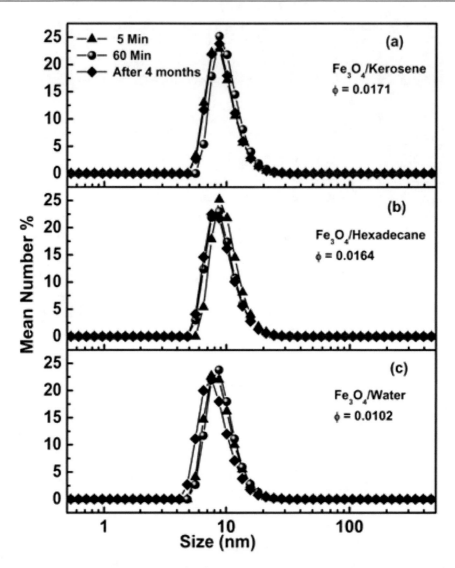

Figure 14. The plot of mean number percentage as a function of particles size at different time intervals after sonication of 5, 60 min and after 4 months for (a): Kerosene-based Fe$_3$O$_4$ (d = 6 nm) nanofluids; (b): Hexadecane-based Fe$_3$O$_4$ nanofluids (d = 7.5 nm) and (c): Water-based Fe$_3$O$_4$ nanofluids (d = 8 nm).

With a particle loading of 1.5 wt.%, the k of carbon coated Fe nanofluid was increased by 28% which was further increased up to 30% by adding glycerin dispersion agent in nanofluid [32]. At a dispersion agent loading of 1%, the centrifugal sedimentation time of carbon coated Fe nanofluids increased from 45 min to 60 min. When dispersion agent loading increase from 1 – 2%, centrifugal sedimentation time show a decline trend and finally reduce to 50 min for carbon coated Fe nanofluids at a dispersion agent loading of 2%. From the above results, it was concluded that there is an optimal value for dispersion agent to improve the stability of nanofluid. In the same report, influence of dispersing methods on stability of nanofluids was studied by dispersing polyethylene glycol based Fe nanofluids with particle loading of 1% by different dispersing ways such as magnetic stirring, ultrasonic dispersion and ball milling.

Nanofluid dispersed by magnetic stirring and ultrasonic dispersion method showed poor stability and a certain amount of sedimentation after 14 days whereas the ball milled samples gave good stability against sedimentation.

Large k enhancement was observed in EG based Fe nanofluids (d = 10 nm) sonicated with high powered pulses [38, 52]. The k of nanofluid showed a gradual increase with the increase in sonication time which saturates at high sonication times. The k of 0.55 vol.% Fe nanofluid exhibited 18% k enhancement with a 30 min sonication and k was saturated after 30 min [37]. Measurement of the cluster size revealed that ultrasonication broke the nanoclusters into smaller clusters. Moreover, Fe nanofluids showed a decrease in k with elapsed time after sonication. From the variations of the cluster size and k as a function of time, it was found that the k enhancement of nanofluids depends on the clustering of nanoparticles. For Fe nanofluid with a particle loading of 0.2 vol.%, a 14.2% enhancement in k is observed immediately after the sonication, which was reduced to 8.5% after an hour. The agglomeration of nanoparticles immediately after sonication led to the formation of larger nanoparticle clusters. The above Fe nanofluids also showed a nonlinear variation in k with ϕ due to rapid clustering of the nanoparticles at high particle concentration.

The variation in k with ϕ beyond EMT predictions, observed in water based Fe_3O_4 nanofluids (d = 9.8 nm), was attributed to the nanoparticle alignment and aggregation [41]. The TEM and DLS analysis of the above nanofluids showed that the aggregation increases with the increasing volume fraction of the nanoparticles. At low particle loading, most nanoparticles were discrete and with the increase in particle concentration, nanoparticle formed clusters.

4. APPLICATIONS OF MAGNETIC NANOFLUIDS FOR HEAT TRANSFER

The large enhancement of k in stable magnetite nanofluids in the presence of a magnetic field and its perfect reversibility was shown to have many technological applications in NEMS and MEMS-based devices [62]. For example, depending upon the cooling requirement, the current or magnetic field can be precisely programmed to obtain the desired level of k enhancement or cooling. The mechanism of heat transport from a cylindrical device with nanofluid coolant around it, without and with magnetic field is depicted in Figure 15. When the field is off, the nanoparticles behave as a normal fluid with random arrangement of particles [Figure 15(a)]. When the field is turned on, the parallel mode conduction leads to a drastic enhancement of k [Figure 15(b)]. Application of magnetic field enhances not only the k of the fluid but also the rheological properties of magnetic fluids. Such field induced enhancements in k and η of ferrofluids can be exploited for a number of technological applications such as damping cum cooling [61, 84].

Figures 16(a) and (b) show the viscous and thermal conductivity changes of hexadecane based Fe_3O_4 nanofluid with a $\phi = 0.067$, where magnetic field strength was varied in a stepwise manner. Here, both k/k_f and η/η_0 measurements were carried out under on–off conditions where the on and off conditions correspond to a magnetic field strength of 120 G and zero, respectively. The shear rate for η measurement was 50 s^{-1}. Though the equilibrium value of viscosity was achieved instantaneously, it was realized after ~ 400 seconds in the case of thermal conductivity.

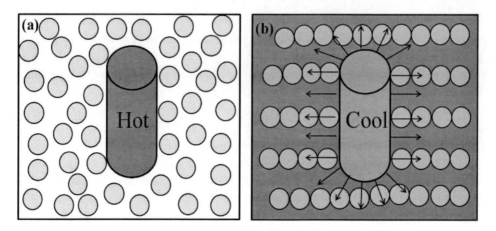

Figure 15. The schematic of the mechanism of heat transport from a cylindrical device immersed in nanofluid (a): without and (b): with magnetic field.

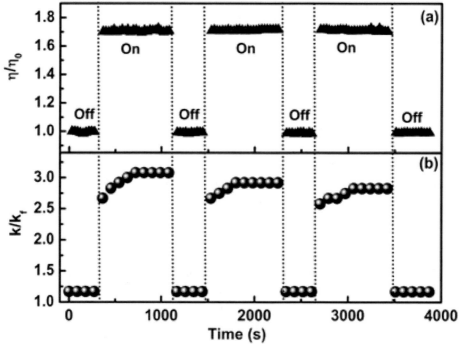

"Reprinted with permission from The Journal of Physical Chemistry C, 115, P. D. Shima and John Philip, Tuning of Thermal Conductivity and Rheology of Nanofluids Using an External Stimulus, 20097–20104. Copyright (2011) American Chemical Society".

Figure 16. The variation of (a) η/η_0 and (b) k/k_f with time during field ON and field OFF condition. The magnetic field strength during ON condition is 120 G. Nanofluid used is hexadecane-based Fe_3O_4 nanofluid with $\phi = 0.067$. The shear rate is 50 s^{-1}.

Upon turning off the magnetic field, both the k/k_f and η/η_0 values drop to zero immediately, showing the perfect reversibility of the observed phenomena and the interesting practical applications is the fluid in smart cooling devices. A uniform perpendicular magnetic field is applied to a flat magnetic fluid layer resulting in the spontaneous surface deformation known as normal field instability [85]. In a recent report, significantly higher on-off conductance ratio was achieved at a device level by exploiting the normal field instability of Fe-based MR fluids across an air gap and also experimentally demonstrated the reversible operation of the above thermal switch using a cyclic on-off test [44].

The experimental setup had a test chamber of thickness 1 mm filled with MR fluid and by applying a strong magnetic field, an array of fluid column was formed across the air gap that provides efficient heat conduction paths. Figure 17 shows the effective thermal conductivity switching ratio obtained from the proposed thermal switch concept. The effective thermal conductivity measurements by filling the chamber with MR fluids to different heights showed a highest switching ratio of 12 at a filling ratio of 65%. The switching mechanism is schematically illustrated in the insets of figure 17. To demonstrate the reversible operation of the thermal switch, a disk-shaped permanent magnet of diameter 1.2 cm was moved horizontally right below the test chamber using a stepper motor to apply periodic magnetic fields of peak amplitude 80 kA/m at frequencies 0.1 – 4 Hz.

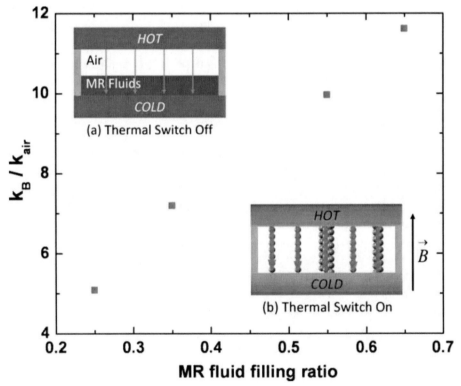

"Reprinted with permission from Journal of Applied Physics, 107, Gilhwan Cha et al., Experimental characterization of thermal conductance switching in magnetorheological fluids, 09B5050. Copyright [2010], AIP Publishing LLC".

Figure 17. The effective thermal conductivity switching ratios obtained from the thermal switch concept. The switching mechanism is schematically illustrated in the insets.

The temporal temperature profile obtained under the periodically varying magnetic field was bound by temperature profiles obtained under static magnetic fields of 0 and 80 kA/m

representing the off and on state, respectively. In another report, the ferromagnetic nanoparticles (d = 20 – 30 nm) was circularly actuated in water-based Fe$_3$O$_4$ nanofluids with the application of rotating magnetic fields and the resulting flow was shown to enhance the cooling capabilities of the system [86]. The heat generated by the miniature film-heater was delivered to a pool filled with nanofluid and surface temperatures are obtained together with constant heat fluxes applied to the system.

Measurements of surface temperature data as a function of heat flux in the absence and presence of magnetic stirrers showed that the heat removal was increased with the introduction of the nanofluid with ferromagnetic particles subject to a rotating magnetic field generated by magnetic stirrers. Figure 18 shows the schematics of heat transfer enhancement mechanism by a magnetically activated nanofluid. The iron oxide nanoparticles motion near the surface and their contact with the heated surface contribute to heat transfer. The magnetic nanoparticles acted as heat carriers, absorbing more heat from the surface of the plate and releasing it to the bul*k* fluid further enhanced the heat transfer and resulted in lower surface temperatures at a fixed heat flux with magnetic stirring. An average heat transfer coefficient enhancement of 37.5% was achieved with magnetic actuation proposing that the magnetic nanofluids can be exploited for thermal management applications. In addition, heat transfer coefficient enhancement was achieved with 5 W of additional power consumption, which ensures that the proposed method could be further exploited to achieve higher energy efficiency. Besides magnetic nanofluids offer several interesting applications in the area of non-contact defect sensors [12, 87-89] that can replace existing Hall probes, cation sensors [90, 91], alcohol sensors [92], ammonia sensors [93], tunable optical filters [5] for lasers, spectro-photometers and other optical devices, light limiting switches [94], liquid hermetic seals [95] etc. Nanofluids are shown to be wonderful model system to probe molecular interactions at nanoscales [96-101], dipolar interaction [102], disorder-order phase transitions [103-105], zippering phenomena [106] etc.

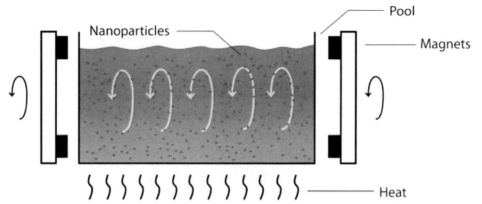

"Reprinted with permission from Journal of Applied Physics, 112, Muhsincan Sesen et al., Heat transfer enhancement with actuation of magnetic nanoparticles suspended in a base fluid, 064320. Copyright [2012], AIP Publishing LLC".

Figure 18. Heat transfer enhancement mechanism by a magnetically activated nanofluid.

CONCLUSION

We presented here the thermal conductivity studies carried out during the last few decades in magnetic nanofluids. Most of the studies show a k enhancement within EMT predictions in stable magnetic nanofluids without magnetic field and a k enhancement beyond EMT predictions in aggregating nanofluids. Studies also reveal that the thermal conductivity of magnetic nanofluids also depend on nanoparticle size, nature of base fluid, nature of nanoparticle material, temperature and presence of additives. Most interestingly, magnetic nanofluids showed significant k enhancement in presence of magnetic field. The enhancement under magnetic field depends on the quality of magnetic nanofluids, the properties of magnetic nanoparticles, carrier fluids, stabilizing moieties, the direction of magnetic field with respect to heat flow direction etc. The increase in k in presence of magnetic field is attributed to the effective conduction of heat through the chainlike structures formed in the nanofluid. The investigations on field induced k enhancement in magnetic nanofluids have shown promising applications in miniature and smart cooling devices.

REFERENCES

[1]   Rosensweig, R. E. Ferrohydrodynamics; Dover: New York, 1997.
[2]   Horng, H. E.; Hong, C. Y.; Yang, S. Y.; Yang, H. C., *Appl. Phys. Lett.,* 2003, 82, 2434-2436.
[3]   Yang, S. Y.; Chieh, J. J.; Horng, H. E.; Hong, C. Y.; Yang, H. C., *Appl. Phys. Lett.,* 2004, 84, 5204-5206.
[4]   Bakuzis, A. F.; Neto, K. S.; Gravina, P. P.; Figueiredo, L. C.; Morais, P. C.; Silva, L. P.; Azevedo, R. B.; Silva, O., *Appl. Phys. Lett.,* 2004, 84, 2355-2357.
[5]   Philip, J.; Jaykumar, T.; Kalyanasundaram, P.; Raj, B., *Meas. Sci. Technol.,* 2003, 14, 1289–1294.
[6]   Liu, T.; Chen, X.; Di, Z.; Zhang, J.; Li, X.; Chen, J., *Appl. Phys. Lett.,* 2007, 91, 121116.
[7]   Chieh, J. J.; Yang, S. Y.; Horng, H. E.; Hong, C. Y.; Yang, H. C., *Appl. Phys. Lett.,* 2007, 90, 133505.
[8]   Huang, J. P.; Yu, K. W., *Appl. Phys. Lett.,* 2005, 86, 041905.
[9]   Liao, W.; Chen, X.; Chen, Y.; Pu, S.; Xia, Y.; Li, Q., *Appl. Phys. Lett.,* 2005, 87, 151122.
[10]   Pu, S.; Chen, X.; Chen, L.; Liao, W.; Chen, Y.; Xia, Y., *Appl. Phys. Lett.,* 2005, 87, 021901.
[11]   Mahendran, V.; Philip, J., *Langmuir,* 2013, 29, 4252-4258.
[12]   Mahendran, V.; Philip, J., *Appl. Phys. Lett.,* 2012, 100, 073104.
[13]   Horng, H. E.; Chen, C. S.; Fang, K. L.; Yang, S. Y.; Chieh, J. J.; Hong, C. Y.; Yang, H. C., *Appl. Phys. Lett.,* 2004, 85, 5592-5594.
[14]   Philip, J.; Prakash, G. G.; Jaykumar, T.; Kalyanasundaram, P.; Raj, B., *Phys. Rev. Lett.,* 2002, 89, 268301.
[15]   Choi, S. U. S., *ASME J. Heat Transfer,* 2009, 131, 033106.
[16]   Keblinski, P.; Eastman, J. A.; Cahill, D. G., *Mater. Today,* 2005, 8, 36-44.

[17] Keblinski, P.; Prasher, R.; Eapen, J., *J. Nanopart. Res.,* 2008, 10, 1089-1097.

[18] Murshed, S. M. S.; Leong, K. C.; Yang, C., *Appl. Therm. Eng.,* 2008, 28, 2109.

[19] Philip, J.; Shima, P. D.; Raj, B., *Appl. Phys. Lett.,* 2007, 91, 203108.

[20] Popplewell, J.; Qenaie, A. A.; Charles, S. W.; Moskowitz, R.; Raj, K., *Colloid Polymer Sci.,* 1982, 260, 333-338.

[21] Fertman, V. E., *J. Eng. Phys. Thermophys.,* 1987, 53, 1097-1105.

[22] Fertman, V. E.; Golovicher, L. E.; Matusevich, N. P., *J. Magn. Magn. Mater.,* 1987, 65, 211-214.

[23] Popplewell, J.; Qenaie, A. A., *J. Magn. Magn. Mater.,* 1987, 65, 215-218.

[24] Nkurikiyimfura, I.; Wang, Y.; Pan, Z., *Renew. Sust. Energy. Rev.,* 2013, 21, 548-561.

[25] Maxwell, J. C. A Treatise on Electricity and Magnetism; Clarendon Press: Oxford, U.K., 1881.

[26] Reinecke, B. N.; Shan, J. W.; Suabedissen, K. K.; Cherkasova, A. S., *J. Appl. Phys.,* 2008, 104, 023507.

[27] Xuan, Y.; Huang, Y.; Li, Q., *Chem. Phys. Lett.,* 2009, 479, 264-269.

[28] Li, Q.; Xuan, Y.; Wang, J., *Exp. Therm. Fluid Sci.,* 2005, 30, 109-116.

[29] Altan, C. L.; Elkatmis, A.; Yuksel, M.; Aslan, N.; Bucak, S., *J. Appl. Phys.,* 2011, 110, 093917.

[30] Gallego, M. J. P.; Lugo, L.; Legido, J. L.; Pineiro, M. M., *J. Appl. Phys.,* 2011, 110, 014309.

[31] Guo, S. Z.; Li, Y.; Jiang, J. S.; Xie, H. Q., *Nanoscale Res. Lett.,* 2010, 5, 1222–1227.

[32] Zhang, H.; Wu, Q.; Lin, J.; Chen, J.; Xu, Z., *J. Appl. Phys.,* 2010, 108, 124304.

[33] Wensel, J.; Wright, B.; Thomas, D.; Douglas, W.; Mannhalter, B.; Cross, W.; Hong, H.; Kellar, J.; Smith, P.; Roy, W., *Appl. Phys. Lett.,* 2008, 92, 023110.

[34] Sinha, K.; Kavlicoglu, B.; Liu, Y.; Gordaninejad, F.; Graeve, O. A., *J. Appl. Phys.,* 2009, 106, 064307.

[35] Han, Z. H.; Yang, B.; Kim, S. H.; Zachariah, M. R., *Nanotechnology,* 2007, 18, 105701.

[36] Yoo, D. H.; Hong, K. S.; Yang, H. S., *Thermochim. Acta,* 2007, 455, 66-69.

[37] Hong, T. K.; Yang, H. S.; Choi, C. J., *J. Appl. Phys.,* 2005, 97, 064311.

[38] Hong, K. S.; Hong, T. K.; Yang, H. S., *Appl. Phys. Lett.,* 2006, 88, 031901.

[39] Wang, B.; Wang, B.; Wei, P.; Wang, X.; Lou, W., *Dalton Trans.,* 2012, 41, 896-899.

[40] Altan, C. L.; Bucak, S., *Nanotechnology,* 2011, 22, 285713.

[41] Zhu, H.; Zhang, C.; Liu, S.; Tang, Y.; Yin, Y., *Appl. Phys. Lett.,* 2006, 89, 023123.

[42] Yu, W.; Xie, H.; Chen, L.; Li, Y., *Colloids Surf. A: Physicochem. Eng. Aspects,* 2010, 355, 109-113.

[43] Zarichnyak, Y. P.; Utkin, A. B., *Powder Metall. Met. Ceram.,* 1985, 24, 307-312.

[44] Cha, G.; Ju, Y. S.; Ahure, L. A.; Wereley, N. M., *J. Appl. Phys.,* 2010, 107, 09B505.

[45] Guo, S. Z.; Li, Y.; Jiang, J. S.; Xie, H. Q., *Nanoscale Res. Lett.,* 2010, 5, 1222-1227.

[46] Parekh, K.; Lee, H. S., *J. Appl. Phys.,* 2010, 107, 09A310.

[47] Tsai, T. H.; Kuo, L. S.; Chen, P. H.; Yang, C. T., *Appl. Phys. Lett.,* 2008, 93, 233121.

[48] Tsai, T. H.; Kuo, L. S.; Chen, P. H.; Yang, C. T., *PIERS Online,* 2009, 5, 231-234.

[49] Safonenko, O. K.; Reks, A. G.; Volkova, N. E., *J. Magn. Magn. Mater.,* 1993, 122, 19-23.

[50] Jiang, W.; Wang, L., *Curr. Nanosci.,* 2011, 7, 480-488.

[51] Philip, J.; Shima, P. D., *Adv. Colloid Interface Sci.,* 2012, 183-184, 30-45.

[52] Hong, T. K.; Yang, H. S., *J. Korean Phys. Soc.,* 2005, 47, S321-S324.

[53] Abareshi, M.; Goharshadi, E. *K.*; Zebarjad, S. M.; Fadafan, H. *K.*; Youssefi, A., *J. Magn. Magn. Mater.,* 2010, 322, 3895-3901.

[54] Mendelev, V. S.; Ivanov, A. O., *Phys. Rev. E,* 2004, 70, 051502.

[55] Bian, P.; Carthy, T. J. M., *Langmuir,* 2010, 26, 6145-6148.

[56] Philip, J.; Shima, P. D.; Raj, B., *Nanotechnology,* 2008, 19, 305706.

[57] Eapen, J.; Williams, W. C.; Buongiorno, J.; Hu, L. W.; Yip, S.; Rusconi, R.; Piazza, R., *Phys. Rev. Lett.,* 2007, 99, 095901.

[58] Hashin, Z.; Shtrikman, S., *J. Appl. Phys.,* 1962, 33, 3125.

[59] Klokkenburg, M.; Vonk, C.; Claesson, E. M.; Meeldijk, J. D.; Erne, B. H.; Philipse, A. P., *J. Am. Chem. Soc.,* 2004, 126, 16706-16707.

[60] Donselaar, L. N.; Frederik, P. M.; Bomans, P.; Buining, P. A.; Humbel, B. M.; Philipse, A. P., *J. Magn. Magn. Mater.,* 1999, 201, 58-61.

[61] Shima, P. D.; Philip, J., *J. Phys. Chem. C,* 2011, 115, 20097–20104.

[62] Philip, J.; Shima, P. D.; Raj, B., *Appl. Phys. Lett.,* 2008, 92, 043108.

[63] Horton, M.; Hong, H.; Li, C.; Shi, B.; Peterson, G. P.; Jin, S., *J. Appl. Phys.,* 2010, 107, 104320.

[64] Parekh, K.; Lee, H. S., *AIP Conf. Proc.,* 2012, 1447, 385-386.

[65] Baby, T. T.; Sundara, R., *J. Appl. Phys.,* 2011, 110, 064325.

[66] Nkurikiyimfura, I.; Wang, Y.; Pan, Z., *Exp. Therm. Fluid Sci.,* 2013, 44, 607-612.

[67] Younes, H.; Christensen, G.; Luan, X.; Hong, H.; Smith, P., *J. Appl. Phys.,* 2012, 111, 064308.

[68] Hong, H.; Luan, X.; Horton, M.; Li, C.; Peterson, G. P., *Thermochim. Acta,* 2011, 525, 87-92.

[69] Hong, H.; Wensel, J.; Smith, P., *J. Nanofluids,* 2013, 2, 38–44.

[70] Hong, H.; Wright, B.; Wensel, J.; Jin, S.; Ye, X. R.; Roy, W., *Synth. Met.,* 2007, 157, 437-440.

[71] Wright, B.; Thomas, D.; Hong, H.; Groven, L.; Puszynski, J.; Duke, E.; Ye, X.; Jin, S., *Appl. Phys. Lett.,* 2007, 91, 173116.

[72] Gavili, A.; Zabihi, F.; Isfahani, T. D.; Sabbaghzadeh, J., *Exp. Therm. Fluid Sci.,* 2012, 41, 94-98.

[73] Krichler, M.; S. Odenbach, *J. Magn. Magn. Mater.,* 2013, 326, 85-90.

[74] Djurek, I.; Znidarsic, A.; Kosak, A.; Djurekc, D., *Croat. Chem. Acta,* 2007, 80, 529-532.

[75] Shul'man, Z. P.; Kordonskii, V. I.; Demchuk, S. A., *Magn. Gidrodin.,* 1977, 4, 30-34.

[76] Kronkalns, G. E., *Magn. Gidrodin.,* 1977, 3, 138-140.

[77] Blum, E. Y.; Kronkalns, G. E.; Fedin, A. G., *Magn. Gidrodin.,* 1977, 1, 28-34.

[78] Shima, P. D.; Philip, J.; Raj, B., *Appl. Phys. Lett.,* 2009, 94, 223101.

[79] Seshadri, I.; Gardner, A.; Mehta, R. J.; Swartwout, R.; Keblinski, P.; Tasciuc, T. B.; Ramanath, G., *Appl. Phys. Lett.,* 2013, 102, 203111.

[80] Shima, P. D.; Philip, J.; Raj, B., *J. Phys. Chem. C,* 2010, 114, 18825-18833.

[81] Xie, H.; Yu, W.; Li, Y.; Chen, L., *Nanoscale Res. Lett.,* 2011, 6, 124.

[82] Shima, P. D.; Philip, J.; Raj, B., *Appl. Phys. Lett.,* 2010, 97, 153113.

[83] Wright, B.; Thomas, D.; Hong, H.; Smith, P., *J. Nanofluids,* 2013, 2, 45–49.

[84] Shima, P. D.; Philip, J.; Raj, B., *Appl. Phys. Lett.,* 2009, 95, 133112.

[85] Rosensweig, R. E. Ferrohydrodynamics; Cambridge Univ. Press: Cambridge, 1985.

[86] Sesen, M.; Teksen, Y.; Sendur, K.; Menguc, M. P.; Ozturk, H.; Acar, H. F. Y.; Kosa, A., *J. Appl. Phys.,* 2012, 112, 064320.

[87] Philip, J.; Rao, C. B.; Jayakumar, T.; Raj, B., *NDT & E Int.,* 2000, 33, 289-295.

[88] Mahendran, V.; Philip, J., *NDT&E International,* 2013, 60, 100-109.

[89] Mahendran, V.; A. Beautlin; Philip, J., *J. Nanofluids,* 2013, 2, 165-174.

[90] Philip, J.; Mahendran, V.; Felicia, L. J., *J. Nanofluids,* 2013, 2, 112-119.

[91] Mahendran, V.; Philip, J., *Appl. Phys. Lett.,* 2013, 102, 163109.

[92] Mahendran, V.; Philip, J., *Sensors and Actuators: B,* 2013, 185, 488-495.

[93] Mahendran, V.; Philip, J., *Appl. Phys. Lett.,* 2013, 102, 063107.

[94] Singh, C. P.; Bindra, K. S.; Shukla, V.; Philip, J.; A.K. Kar; McCarthy, J. E.; Bookey, H. T., *Adv. Sci. Eng. Med.,* 2014, 6 1-5.

[95] Sreedhar, B. K.; Kumar, R. N.; P. Sharma; Ruhela, S.; J. Philip; Sundarraj, S. I.; Chakraborty, N.; Mohana, M.; Sharma, V.; Padmakumar, G.; Nashine, B. K.; Rajan, K. K., *Nucl. Sci. Eng.,* 2013, (In Press).

[96] Mondain-Monval, O.; Espert, A.; Omarjee, P.; Bibette, J.; Leal-Calderon, F.; Philip, J.; Joanny, J. F., *Phys. Rev. Lett.,* 1998, 80, 1778-1781.

[97] Philip, J.; Gnanaprakash, G.; Jayakumar, T.; Kalyanasundaram, P.; Raj, B., *Macromolecules,* 2003, 36, 9230-9236.

[98] Philip, J.; Gnanaprakash, G.; Jaykumar, T.; Kalyanasundaram, P.; Mondain-Monval, O.; Raj, B., *Langmuir,* 2002, 18, 4625–4631.

[99] Philip, J.; Gnanaprakash, G.; Jaykumar, T.; Kalyanasundaram, P.; Raj, B., *Phys. Rev. Lett.,* 2002, 89, 268301.

[100] Philip, J.; Jaykumar, T.; Kalyanasundaram, P.; Raj, B.; Monval, O. M., *Phys. Rev. E,* 2002, 66, 011406.

[101] Philip, J.; Monval, O. M.; Calderon, F. L.; Bibette, J., *J. Phys. D: Appl. Phys.,* 1997, 30, 2798.

[102] Laskar, J. M.; Philip, J.; Raj, B., *Phys. Rev. E,* 2008, 78, 031404.

[103] Laskar, J. M.; Philip, J.; Raj, B., *Phys. Rev. E,* 2010, 82, 021402.

[104] Laskar, J. M.; Raj, B.; Philip, J., *Phys. Rev. E,* 2011, 84, 051403.

[105] Laskar, J. M.; S. Borbjosi; Philip, J.; Raj, B., *Opt. Commn.,* 2012, 285, 1242-1247.

[106] Laskar, J. M.; Philip, J.; Raj, B., *Phys. Rev. E,* 2009, 80, 041401.

In: Nanofluids: Synthesis, Properties and Applications
ISBN: 978-1-63321-677-8
Editors: S.M. Sohel Murshed, C.A. Nieto de Castro © 2014 Nova Science Publishers, Inc.

Chapter 5

VISCOSITY OF NANOFLUIDS CONTAINING METAL OXIDE NANOPARTICLES

S. M. Sohel Murshed[*]*, F. J. V. Santos and C. A. Nieto de Castro*
Centro de Ciências Moleculares e Materiais
Faculdade de Ciências, Universidade de Lisboa,
Lisboa, Portugal

ABSTRACT

In recent years, researchers have shown a great interest in studying the viscosity of nanofluids, particularly nanofluids having various types of metal oxide nanoparticles dispersed in base fluids and they have found anomalous results on this crucial property of nanofluids. This chapter critically reviews the literature and recent research progress in the viscosity of metal oxide based nanofluids. Experimental investigations on the viscosity of nanofluids containing titania and silica nanoparticles in silicone oil conducted by the authors are also reported. Like thermal conductivity, viscosity data of nanofluids vary considerably among the various studies. Nanoparticles concentration and fluid temperature play an important role in changing the viscosity of nanofluids and the effects of these factors on this property of nanofluids are demonstrated. Studies show that nanofluids exhibit both Newtonian and non-Newtonian shear-thinning flow natures and their rheological behavior depends on various factors such as nanoparticle concentration, shear rate and temperature. The review revealed that all these nanofluids have significantly higher viscosity when compared to their base fluids and increasing the concentration of nanoparticles increases the viscosity of nanofluids. Although viscosities of both the base fluids and the nanofluids decrease mostly nonlinearly with increasing temperature, temperature has little influence on the increase or decrease of the relative viscosity of nanofluids. Despite numerous efforts devoted to develop a model for the viscosity of nanofluids, success has not yet been achieved. None of the existing models can accurately predict the viscosity of nanofluids and can explain the underlying mechanisms. Most of the reported models in the literature are mainly obtained from the regression of researchers` own viscosity data.

---

[*] E-mail: smmurshed@fc.ul.pt

1. INTRODUCTION

Nanofluids are suspensions of nanoparticles in conventional fluids [1] and they have potential applications in numerous important fields such as microelectronics, micro-electromechanical systems, microfluidics, transportation, manufacturing, instrumentation, medical, and HVAC systems [2-7]. Viscosity is a very important property of fluids for their thermal performance in practical applications that are particularly related to fluid flow. For instance, the understanding of convective heat transfer and the pumping power are directly related to the viscosity of the fluids. Viscosity of fluids also influences their electrical properties. Compared to the extensive studies on thermal conductivity [3-4, 8-13], viscosity of nanofluids has not received similar attention from researchers until recent years [8, 10, 13-16]. Among the studies on the viscosity of nanofluids, researchers mainly focused on studying the effect of concentration of nanoparticles on the viscosity of aqueous base nanofluids. However, in addition to concentration of nanoparticles it is also important to investigate the effects of other factors such as temperature, base fluids, and shear rate on the viscosity of nanofluids in order to exploit their potential and wide range applications. Compared to carbon or metallic nanoparticles most studies used nanofluids containing metal oxide nanoparticles in base fluids. Thus, it is timely to critically review the findings and research progress on the viscosity of these metal oxide based nanofluids.

Although nanofluids showed higher viscosity when compared to their base fluids, other significantly enhanced thermal properties and characteristics such as thermal conductivity, heat transfer coefficient, and critical heat flux fully compensate that drawback, making this novel class of fluids a strong candidate for the next generation of coolants [2-5]. A survey of published viscosity data reveals a very significant scattering mainly due to the use of different sized particles, equipment, and sample preparation methods [4, 13-16], but quite less controversial when compared to the thermal conductivity results [2-4,9,12]. Nanofluids were found to have both Newtonian and non-Newtonian shear-thinning flow natures [13-17] which depend on several factors such as concentration of particles, temperature and shear-rate. Most of the reported studies are either aqueous or ethylene glycol based [13-16] and few studies have used oil or other fluids as the base fluids [18-21].

As part of the International Nanofluid Property Benchmark Exercise (INPBE) a round-robin exercise [22] involving dozens of different research laboratories around the world on the viscosity of nanofluids containing metal oxide nanoparticles in water and synthetic oil was carried out and the influences of nanoparticle concentration and shape were examined. The viscosity results obtained from the exercise revealed that while some nanofluids showed shear-thinning behavior, others (e.g. spherical Al_2O_3/PAO + surfactant) are Newtonian over the full range of shear rates considered. It was also concluded that for both spherical and cylindrical shaped nanoparticles, concentration is the most significant variable for the linear dependence of the viscosity of these nanofluids and that the classical theory for dilute suspensions (e.g., Einstein´s model [23]) considerably under predicts these viscosity data. The degree of agglomeration of nanoparticles was considered to be the main reason for the observed differences between the prediction and measured data.

This chapter aims to critically review the recent research progress and available findings on viscosity of nanofluids containing metal oxide nanoparticles. Effects of various parameters such as the concentration of nanoparticles and fluid temperature on this important property of these nanofluids are demonstrated. Existing classical models are used in an effort to confirm whether those models are suitable for the prediction of the viscosity of these nanofluids or not. In addition, our recent study on rheological properties of nanofluids containing TiO_2 and SiO_2 nanoparticles in silicone oil is reported and an empirical model for the estimation of viscosity of these silicone oil based nanofluids is also proposed.

2. MODELS FOR VISCOSITY OF NANOFLUIDS

Available literature studies agree that the enhancement of viscosity of nanofluids cannot be predicted by the existing classical model used for mostly dilute suspensions of particles. Researchers have given considerable efforts to develop models for the viscosity of nanofluids but success is not yet achieved as till today no model is available that can accurately predict the viscosity of nanofluids. Thus researchers commonly use classical viscosity models and/or proposed empirical models based on their own data. Among the classical models the well-known Einstein's viscosity model [23] came first, being the most used for the prediction of viscosity of nanofluids. Einstein considered that a dilute suspension of spherical particles can be defined as presenting an effective viscosity of suspension (η_{eff}) given by

$$\eta_{eff} = \eta_f (1 + 2.5\phi) \tag{1}$$

where η_f is the viscosity of the base fluid and ϕ is the volume fraction of the dispersed particles. Even though this model is mainly valid for very low concentration of particles (typically less than 1 vol. %), it is commonly used for any concentrations of particles. Most of the other classical models [24-26] were mainly developed by modifying Einstein´s [23] model and they yield similar increase in viscosity with increasing the volume fraction of nanoparticles.

Brinkman [24] extended Einstein's formula for the viscosity of even less dilute suspensions of particles and his model has the form:

$$\eta_{eff} = \eta_f / (1 - \phi)^{2.3} \tag{2}$$

As the nanofluids studied are mostly not dilute suspensions, it is considered that the power law-based models [26-27] are more appropriate for the prediction of the viscosity of nanofluids when compared to other models such as those attributed to Einstein [23], Batchelor [25], and Brinkman [24].

Considering the possibility of some structures creation in disperse systems Krieger and Dougherty (K-D) [27] developed a semi-empirical power law-based expression for the effective viscosity of suspensions containing any volumetric concentration of particle and expressed it as:

$$\eta_{eff} = \eta_f (1 - \phi / \phi_m)^{-[\eta]\phi_m} \tag{3}$$

where the intrinsic viscosity $[\eta] = 2.5$ for hard spheres and the maximum packing fraction ϕ_m is c.a. 0.605 at high shear rates and varies between 0.495 to 0.54 under quiescent conditions.

Another power law-based model was proposed by Nielsen [26] for composite materials and his model can be written as:

$$\eta_{\text{eff}} = \eta_f (1+1.5\phi)e^{(\phi/(1-\phi_m))} \tag{4}$$

where similarly to Krieger and Dougherty [27] model, ϕ_m is the maximum packing fraction which is approximately 0.64 for randomly dispersed spheres.

For computing the viscosity of aqueous Al_2O_3 (47 nm) nanofluid an Arrhenius type equation was introduced by Nguyen et al. [28]:

$$\eta_{\text{eff}} = 0.904\eta_f e^{0.1482\phi} \tag{5}$$

Most of these models are employed for the predictions and comparisons of the effective viscosity of nanofluids containing metal oxide nanoparticles in various base fluids and they are found unable to predict the measured viscosities of nanofluids which are shown in later sections.

3. LITERATURE STUDIES

3.1. Overview of Selected Studies

Although some review articles [3-4, 13, 15] emphasized the significance of investigating the viscosity of nanofluids, limited research works have been performed on the viscosity of metal oxide based nanofluids. An overview of the findings from some key studies on viscosity of these nanofluids is presented here.

Even before coining the term "nanofluids" Masuda et al. [29] measured the viscosity of suspensions of different types of nanoparticles in water (W), systems that are now recognized as aqueous based nanofluids. They found that while at a volumetric loading of 4.3% TiO_2 (27 nm) nanoparticles increased the viscosity of water by 60%, the viscosity of Al_2O_3 (80 nm)/deionized water (DIW) based nanofluids increased by nearly 82% at 5 vol. % of Al_2O_3 nanoparticle. Whereas, Wang et al. [30] reported that the effective viscosity of Al_2O_3 (28 nm)/distilled water based nanofluid was increased by about 86% at the same 5 vol. % concentration. In their case [30], a mechanical blending technique was used for dispersion of Al_2O_3 nanoparticles in distilled water. About 40% increase in the viscosity of ethylene glycol at a volumetric loading of 3.5% of their Al_2O_3 nanoparticle was reported and results showed that the viscosity of nanofluids depends on the dispersion methods [30]. In contrast, Pak and Cho [8] found that at 10% volumetric concentration of nanoparticles, the viscosities of water-based Al_2O_3 (13 nm) and TiO_2 (27 nm) nanofluids were several times greater than that of the base water. This large discrepancy could be due to the differences in dispersion techniques and sizes of particles used. Pak and Cho [8] also adjusted pH values of their sample and employed an electrostatic repulsion technique. As anticipated, the viscosity of nanofluids depends on the methods used to disperse and stabilize the nanoparticle suspensions. The viscosity results of Pak and Cho [8] were significantly larger than the predictions by the classical Einstein's model [23]. The groups of Das et al. [31] and Putra et al. [32] measured the viscosities of Al_2O_3/W and CuO/W based nanofluids as a function of shear rate for nanoparticle volume concentration ranging from 1% to 4% and they reported Newtonian behavior and a considerable increase in the viscosity with increasing volumetric concentration of the nanoparticles [31-32].

Prasher et al. [33] reported the viscosity of alumina based nanofluids for various shear rates, temperatures and nanoparticle volume concentrations demonstrating that viscosity was independent of shear rate thus confirming their nanofluids are Newtonian in nature. Data also showed that while viscosity increases with increasing nanoparticle volume fraction, the relative increase in viscosity was found independent of temperature.

In a study on natural convection heat transfer of water based nanofluids Wen and Ding [34] reported about 20% increase in the effective viscosity at 2.4% weight concentration of TiO_2 (34 nm) nanoparticles in water. However, a much higher viscosity increase was observed under low shear rate conditions (i.e., 25-100 1/s).

The viscosities of DIW based TiO_2 and Al_2O_3 nanofluids were reported in a study by the lead author [13]. At 5 vol. % of TiO_2 (15 nm) in DIW, the maximum increase in viscosity of DIW was found to be 86%. For the same TiO_2 based nanofluid Masuda et al. [29] showed about 70 % increase in viscosity of 4.3 vol. % of TiO_2 (27 nm) nanoparticles in water. It is noted that Masuda et al. [29] used 27 nm sized TiO_2 nanoparticles and adjusted their suspensions to a high pH value. On the other hand, the viscosity of DIW was increased by 58% at the same volumetric loading (5%) of the cylindrical shape TiO_2 (10×40 nm) nanoparticles [13]. This indicates that the shapes of the particles can influence the viscosity of suspensions and one possible reason for such an increase in viscosity is the particle shape-dependent electrostatic repulsive force which plays a significant role in altering the viscosity of nanofluids through their dispersion and sedimentation. The particle shape can affect clustering and adsorption, which may also change the viscosity of nanofluids. In the same study [13], the maximum increase in viscosity of DIW was found to be nearly 82% at a 5% volumetric concentration of Al_2O_3 (80 nm) nanoparticles. A similar increment (86%) of the viscosity of the Al_2O_3 based nanofluids was also observed by Wang et al. [30].

Comparisons between the experimental results [13] and the predictions by classical models such as Krieger and Dougherty's (K-D) [27] and Nielsen's [26] showed that these models severely under predict the viscosity of nanofluids.

A summary of the selected viscosity results from literature is presented in Table 1 which demonstrates that these metal oxide nanofluids possess higher viscosity when compared to their base fluids. In addition, the enhancement in viscosity further increased with increasing the concentration of nanoparticles. Some studies observed anomalously high viscosity of nanofluids. For example, Nguyen et al. [28] reported viscosities of aqueous nanofluids containing 9 vol. % of CuO and Al_2O_3 nanoparticles to be respectively about 10 times and 5 times that of the base fluid (water). Most of the other studies also reported between 50 to 200 % increase in viscosity of nanofluids at less than 5 % volumetric loading of nanoparticles (Table 1).

Table 1. Summary of nanofluid viscosity data

Nanofluids (NP/BF)	Surfactant (Conditions)	Maximum increase (%) in viscosity over base fluid	References
Al_2O_3 (13 nm)/W	Adjusted pH (at 32 °C)	η increased 248% at 4.3 vol. % of Al_2O_3	Masuda et al. [29]
TiO_2 (27 nm)/W		η increased 69% at 4.3 vol. % of TiO_2	
SiO_2 (12 nm)/W		η increased 158% at 2.3 vol. % of SiO_2	
Al_2O_3 (13 nm)/W	Adjusted pH (at 25 °C and 395s^{-1})	η increased 152% at 2.78 vol. % of Al_2O_3	Pak and Cho [8]
TiO_2 (27 nm)/W		η increased 36% at 3.16 vol. % of TiO_2	
Al_2O_3 (40 nm)/PG	None (30°C)	η increased 38% at 3 vol. % of Al_2O_3	Prasher et al. [33]
CuO (29 nm)/EG+W (EG/W:60/40)	None (at 20 °C)	η increased 259% at 6.12 vol. % of CuO	Numburu et al. [35]
Al_2O_3 (47 nm)/W	None (at 20 °C)	η increased 431% at 9 vol. % of Al_2O_3	Nguyen et al. [28]
CuO (29 nm)/W		η increased 880% at 9 vol. % of CuO	
TiO_2 (15 nm)/DIW	CTAB (at 25 °C)	η increased 86% at 5 vol.% of TiO_2 (sphere)	Murshed et al. [13]
TiO_2 (10×40 nm)/DIW		η increased 58% at 5 vol. % of TiO_2 (rod)	
Al_2O_3 (150 nm)/DIW		η increased 180% at 5 vol. % of Al_2O_3	
TiO_2 (21 nm)/DIW	None (at 21 °C)	η increased 133% at 3 vol. % of TiO_2	Turgut et al. [36]
Al_2O_3 (10 nm)/PAO	certain surfactant (at 25 °C)	η increased 59% at 3 vol. % of Al_2O_3 (sphere)	Zhou et al. [37]
Al_2O_3 (10×80 nm) /PAO		η increased 200% at 3 vol. % of Al_2O_3 (rod)	
CuO (40 nm)/GO	oleic acid (at 20 °C)	η increased 168% at 2.5 vol. % of CuO	Kole and Dey [38]
SiC (100 nm)/DIW	None (at 30 °C)	η increased 102% at 3 vol. % of SiC	Lee et al. [39]
Fe_3O_4 (13 nm)/W	CTAB (at 60 °C)	η increased 197% at 2 vol. % of Fe_3O_4	Sundar et al. [40]
SiO_2 (28.3nm)/EG	None (at 25 °C)	η increased 85% at 8.2 vol. % of SiO_2	Rudyak et al. [41]

Literature review of this class (metal oxide) of nanofluids revealed that these nanofluids exhibit both Newtonian and non-Newtonian behaviors and their viscosities are considerably higher than those of their base fluids. A nonlinear decrease in the viscosity of these nanofluids with increasing temperature was also observed in the literature. Such substantial viscosity decrease with temperature makes nanofluids even more attractive for their applications at elevated temperatures. All reported studies showed that the existing models severely under predict the observed anomalously high viscosity of these nanofluids.

Nevertheless, through a thermal and hydraulic analysis by comparing pressure drops at constant Nusselt number Prasher et al. [33] showed that the relative increase in the viscosity has to be within four times larger than the relative increase in the thermal conductivity for nanofluids to be better than the base liquid for heat transfer applications. This translates that if the thermal conductivity of a certain nanofluid increases 25% at certain concentration of nanoparticles, the increase in viscosity of that nanofluid at that concentration should be within 100 % in order for its heat transfer performance to be really better when compared to just the base fluid. Therefore, despite of their high viscosity, most nanofluids demonstrate better heat transfer than the conventional heat transfer fluids used as base fluids. This is further supported by the findings from numerous studies on the convective heat transfer of nanofluids which showed that nanofluids exhibit significantly higher convective heat transfer coefficients when compared to their base conventional fluids [6].

3.2. Effect of Nanoparticle Concentration on Viscosity

Here nanofluids are classified based on the types of nanoparticles used and the viscosity results from the literature on each type of nanofluids are plotted and analyzed separately in order to demonstrate the effect of the concentration of that nanoparticle on the viscosity. Existing models are also used to predict the viscosity of nanofluids and are compared with experimental data. First, results for the viscosity of nanofluids having Al_2O_3 nanoparticles in various base fluids are presented as a function of Al_2O_3 volume fraction (%) followed by TiO_2 and other metal oxide nanoparticles based nanofluids. It is noted that the enhancements of viscosity of these nanofluids are characterized with the relative viscosity (η_{nf}/η_f) which is the ratio of viscosity of nanofluids and viscosity of their base fluids at the corresponding concentration and at the same temperature.

Figure 1 presents the viscosity of Al_2O_3 nanofluids as a function of volumetric concentration of this nanoparticle and its comparison with the predictions of selected classical models. It is found that most of the studies used water as base fluid, very few other types of base fluids such as PG, PAO and EG/W mixture being also reported in the literature. A previous study by the lead author [13] showed that while the maximum increase in viscosity of DIW was about 180% at 5% volumetric loading of Al_2O_3 (150 nm) nanoparticles, about 82% enhancement in viscosity of DIW was obtained for the same volumetric loading of 80 nm sized Al_2O_3 nanoparticles. Whereas in spite of using smaller sized Al_2O_3 (13 nm) nanoparticles, Masuda et al.'s [29] observed a much larger increase (3.5 times that of water) in the viscosity of their Al_2O_3/W nanofluid at 4.3 volume % of nanoparticles. Interestingly, Pak and Cho [8] reported the viscosity of the same nanofluid to be about 2.5 times that of water for only 2.78 volume % of the same Al_2O_3 (13 nm) nanoparticles at room temperature. For W/EG (80:20 w/w) mixture based nanofluid, Yiamsawas et al. [42] recently found about

2.23 times larger viscosity of their nanofluid compared to the base fluid at 4 vol.% loading of Al$_2$O$_3$ (120 nm) nanoparticle in W/EG.

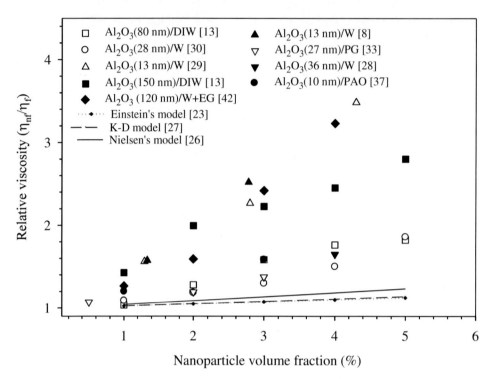

Figure 1. Viscosity of various nanofluids as a function of volumetric concentration of Al$_2$O$_3$ nanoparticles. The sizes of the nanoparticles are identified inside parentheses in the inner caption.

The effect of particle size on the viscosity of this nanofluid can also be evidenced from Figure 1, which shows that the larger the particle size, the higher the viscosity increase. This is because of the effect of particle size and thus its weight on its agglomeration in the base fluid. Figure 1 also shows that the measured viscosities are severely under-predicted by the classical models. In addition to the progress in research on stability, thermal conductivity, and heat transfer characteristics, a brief discussion on viscosity results of this (alumina) nanofluid can be found in a review article by Sridhara and Satapathy [43].

Despite their comparatively low thermal conductivity, TiO$_2$ nanoparticles are widely used in nanofluids mainly due to their easy availability in large quantities and its relatively lower cost. The literature survey indicates that TiO$_2$ dispersed nanofluid is one of the most popular nanofluids used by the researchers. Figure 2 displays literature data on volumetric concentration dependence of viscosity of nanofluids containing TiO$_2$ nanoparticles in different base fluids at room temperature. It can be seen that the enhanced viscosity of these nanofluids increase significantly with increasing the volume fraction of this nanoparticle and the maximum increase in viscosity was 133% at 3 vol.% of TiO$_2$ (21 nm) in water [36]. The overall increase in the viscosity of TiO$_2$ nanofluids with volume fraction is smaller as compared to viscosity of Al$_2$O$_3$ nanofluids (Figure 1). The size and shape of nanoparticle were also found to influence the viscosity of this nanofluid. Nevertheless the data from various research groups vary considerably as well, mainly because of different sample preparation and by the different measurement techniques used by different groups. Except for very low concentration of nanoparticles, the existing models are found to severely underpredict the viscosity of this nanofluid.

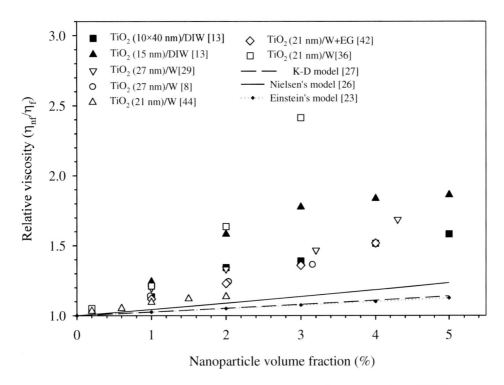

Figure 2. Viscosity of various nanofluids as a function of volumetric concentration of TiO2 nanoparticles. The sizes of the nanoparticles are identified inside parentheses in the inner caption.

Figure 3 presents viscosity results as a function of volumetric concentration of nanoparticles at room temperature, for non-metallic nanofluids using SiO_2, CuO, Fe_3O_4, titanate nanotube (TNT) and SiC based nanofluids but excluding Al_2O_3 and TiO_2 nanoparticles based nanofluids.

Researchers used various types of nanoparticles and different types of base fluids as well as mixtures of fluids (e.g., W/EG). Like other nanoparticles based nanofluids, these nanofluids also showed substantially higher viscosity when compared to their base fluids and the enhanced viscosity further increases with the volumetric loading of nanoparticles. For example, Kole and Dey [38] reported that the viscosity of gear oil increased 168% at 2.5 vol. % CuO (40 nm) nanoparticles. Some nanofluids showed almost linear increase in the viscosity with nanoparticle volume fraction. Once again, except for very low concentration, both the popular Einstein´s [23] model and the power-law based Nelsen´s [26] model are unable to predict the viscosity of these nanofluids.

Based on the above review and the results presented in Table 1 and Figures 1-3, it is unquestionable that the viscosities of all these nanofluids are significantly higher than those of their base fluids and that the enhanced viscosities further increase with increasing concentration of nanoparticles. However, there are discrepancies among the results from various research groups even for the same nanofluids. The reasons for such discrepancies could be due to the differences in the size of the nanoparticle clusters, to differences in the dispersion techniques, and to the use of surfactants. Indeed, the viscosity of nanofluids greatly depends on the methods used to disperse and stabilize the nanoparticle suspension [30].

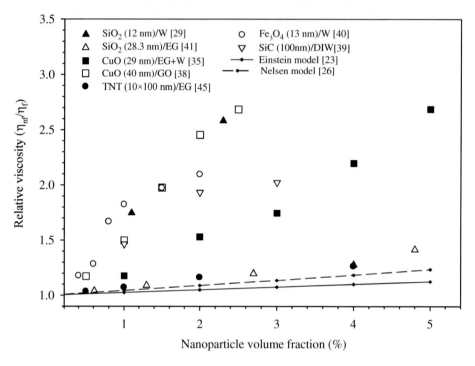

Figure 3. Viscosity of various nanofluids as a function of volumetric concentration of different nanoparticles. The sizes of the nanoparticles are identified inside parentheses in the inner caption.

The classical models were also found to be unable to predict the effective viscosity of these nanofluids. This is because these models considered only particle volume fraction, whereas the nanoparticles in fluids can easily form clusters and can experience surface adsorption due to surface forces. Clustering and adsorption increase the hydrodynamic diameter of nanoparticles leading to the increase of relative viscosity. Besides the particle volume fraction and size, many other factors such as the nature of the particle surface, ionic strength of the base fluid, surfactants, pH value, and particle interaction forces may play a considerable role to change the viscosity of nanofluids. In addition to nanoparticle concentration and the above mentioned factors, temperature also plays a significant role in the viscosity of nanofluids and the effect of temperature on the viscosity of these nanofluids will be presented in the next section.

3.3. Effect of Temperature on Viscosity

Like in the previous section temperature dependent viscosities of different types of metal-oxide nanoparticle based nanofluids are presented and discussed in this section. Viscosity of nanofluids containing only Al_2O_3 nanoparticles in various base fluids are presented first (Figure 4), followed by TiO_2 nanoparticle based nanofluids (Figure 5) and finally all other oxides based nanofluids (Figure 6). Here the relative viscosity term is used to quantify the increase or decrease of viscosity of these nanofluids over their base fluids with respect to temperature. The viscosities of nanofluids and their respective base fluids at the same temperature were used to obtain the relative viscosity.

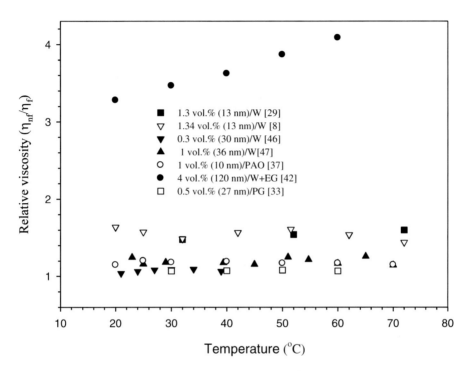

Figure 4. Temperature dependence of viscosity of Al$_2$O$_3$ based nanofluids. The sizes of the nanoparticles are identified inside parentheses in the inner caption.

Figure 4 presents the relative viscosity of Al$_2$O$_3$ based nanofluids as a function of temperature. Except for the results from a recent study [42], no significant effect of temperature on the enhancement of relative viscosity of Al$_2$O$_3$ nanofluids were found (Figure 4). This demonstrates that the effect of temperature on the viscosities of both the nanofluids and base fluids are of similar magnitude and there is no special temperature-based mechanism that alters the viscosity of nanofluids.

Nevertheless, the viscosity data from different studies are scattered as different studies used different sizes of nanoparticles, base fluids, surfactants, sample preparation techniques, and measurement techniques. Interestingly, and except at high temperature (72 °C), two early studies [8, 29] using the same aqueous Al$_2$O$_3$ (27 nm) nanofluids showed very similar influence of temperature. On the other hand, the effect of temperature on the viscosity of TiO$_2$ nanofluids is demonstrated in Figure 5, which showed a similar trend of results as found for Al$_2$O$_3$ nanofluids. Once again unlike results from other groups, Yiamsawas et al. [42] showed considerable effect of temperature on the viscosity of this nanofluid (Figure 5).

Figure 6 depicts temperature dependence of viscosity of nanofluids containing other available metal oxide nanoparticles (except Al$_2$O$_3$ and TiO$_2$) in different base fluids and it shows a very mixed influence of temperature on the enhancement of viscosity of these nanofluids.

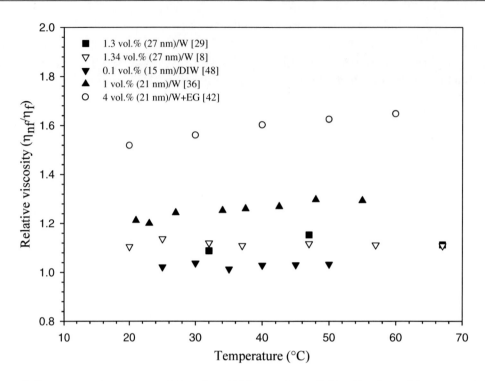

Figure 5. Temperature dependence of viscosity of TiO$_2$ based nanofluids. The sizes of the nanoparticles are identified inside parentheses in the inner caption.

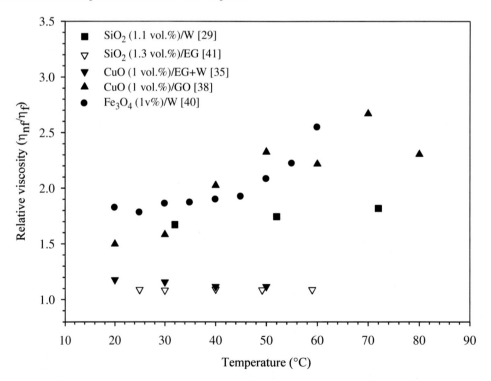

Figure 6. Temperature dependence of viscosity of other metal oxide based nanofluids. The sizes of the nanoparticles are identified inside parentheses in the inner caption.

While several studies found no considerable influence of temperature on the relative viscosity of their nanofluids, two studies [38, 40] reported a scattered but considerable effect of temperature on the viscosity of their nanofluids particularly at higher temperatures (Figure 6). For instance, Sundar et al. [40] recently showed that until about 45°C the temperature had no noticeable effect on the viscosity of their Fe_3O_4 (13 nm)/W nanofluid but the effect of temperature became significant at temperatures above 45°C as there was a sharp increase in viscosity of these nanofluids with the temperature increasing from 45 to 60°C.

In another interesting study, Kole and Dey [38] found a very fluctuating effect of temperature on the enhancement of viscosity of their gear oil based CuO nanofluid particularly above 50°C (Figure 6). However the viscosity of this CuO nanofluid was found to increase almost linearly with temperature up to 50°C. This indicates that the magnitude of influence of temperature on the enhancement of viscosity of nanofluids depends on range of temperature, types and sizes of nanoparticles as well as their base fluids. Nonetheless, data presented in Figures 4-6 reveals that, except for a handful of studies, most of the studies found no significant effect of temperature on the enhancement of viscosity of these metal oxides nanofluids.

In order to be conclusive on the effect of temperature in changing the viscosity of nanofluids more studies are to be conducted for various types of nanofluids for a wide range of temperatures.

4. STUDY BY THE AUTHORS

Recently we have conducted an experimental investigation on the rheological properties of nanofluids containing SiO_2 and TiO_2 nanoparticles in silicone oil (SO) [49] and the results obtained are briefly presented here. To the best of our knowledge, there was no previous work on the viscosity of silicone oil based SiO_2 and TiO_2 nanofluids. These nanoparticles were used because of their numerous potential applications, cheaper price and mass scale availability. On the other hand, besides the wide range of viscosity, silicone oil at room temperature has similar thermal characteristics as the engine oil at its high operating temperature [18] and thus can be used as base fluid for nanofluids. It was important to study rheological properties of silicone oil based nanofluids for exploring their potential applications.

4.1. Experiments and Results

4.1.1. Sample Nanofluids

Two types of sample nanofluids were prepared by dispersing different concentrations of TiO_2 (20 nm) and SiO_2 (10-20 nm) nanoparticles in silicone oil (XIAMETER® PMX-200). The silicone oil (SO) and nanoparticles were purchased from Dow Corning Corporation (USA) and Io-Li-Tec nanomaterials (Germany), respectively. Proper dispersion of nanoparticles and stability of these nanofluids were ensured by undertaking long time magnetic stirrer and ultra-sonication.

Figure 7 depicts photos of both types of sample nanofluids. These nanoparticles were found to be well dispersed in the silicone oil as the samples were stable for more than a week.

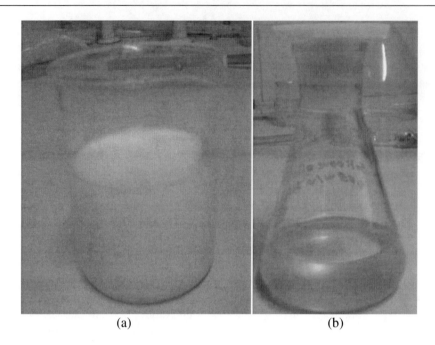

Figure 7. Pictures of nanofluids: (a) TiO$_2$/SO and (b) SiO$_2$/SO.

4.1.2. Measurements

Measurements of viscosity of both the base fluids and nanofluids were performed using a TA Instrument controlled shear stress/strain/rate rheometer (AR 1500ex). The plates of the rheometer have a diameter of 40 mm. The resolution of temperature control of this rheometer is ±0.01°C. Because the heating source (Peltier plate) of this rheometer is located just below the lower plate, a small temperature gradient between the lower plate and the upper plate can generate that the actual sample temperature can be slightly higher or lower than the set point temperature. Thus, a very thin type S thermocouple was inserted into the sample to directly measure the actual sample temperature and to verify the rheometer set point temperature. Measurements of each sample were started after the desired set point temperature was reached.

4.2. Results and Discussion

4.2.1. Rheology of Base Fluid and Nanofluids

In order to verify the rheological behavior of these SO based nanofluids, the relation between shear stress and shear rate for different volumetric loadings of nanoparticles at different temperatures (20°C to 60°C) was studied and results at three representative temperatures are presented for both the base fluid and nanofluids. With linear relationship between shear stress and shear rate Figure 8 demonstrates that the silicone oil (Figure.8a) and its nanofluids containing small concentrations (0.01vol. % and 0.05vol. %) of SiO$_2$ and TiO$_2$ nanoparticles (Figures 8b and 8c) are Newtonian.

Viscosity of Nanofluids Containing Metal Oxide Nanoparticles

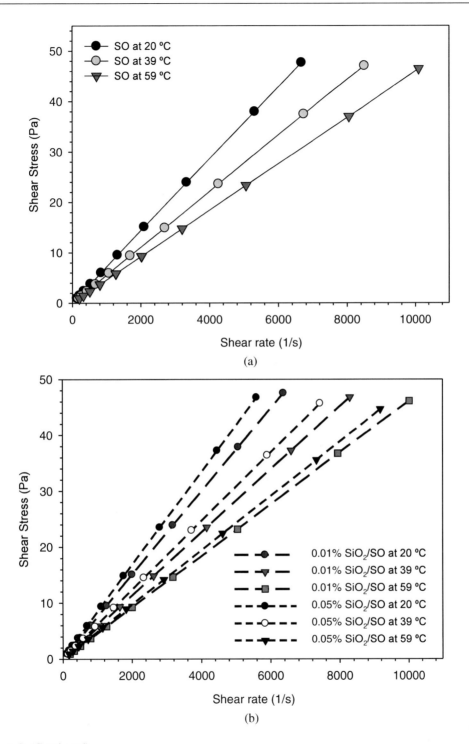

Figure 8. (Continued)

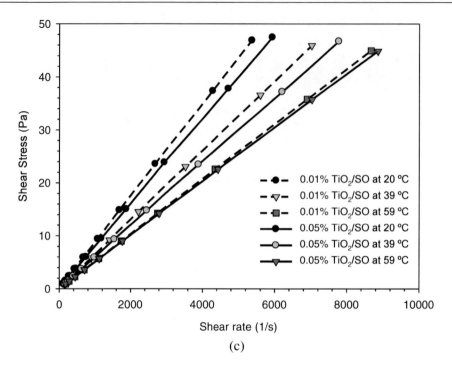
(c)

Figure 8. Shear stress versus shear rate of (a) base fluid (SO), (b) SiO$_2$/SO and (c) TiO$_2$/SO nanofluids at different volumetric concentrations of nanoparticles and temperatures.

Regardless of temperature, the apparent shear viscosities of these SO based nanofluids are independent of shear rate (within the range of the experiment) as can be seen from Figure 9 and this further confirms their Newtonian nature.

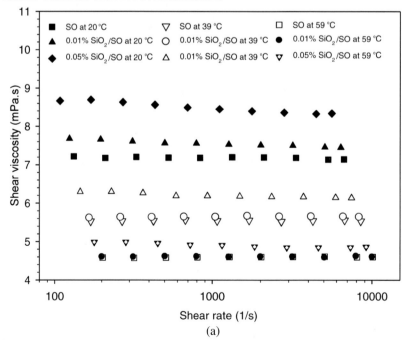
(a)

Figure 9. (Continued)

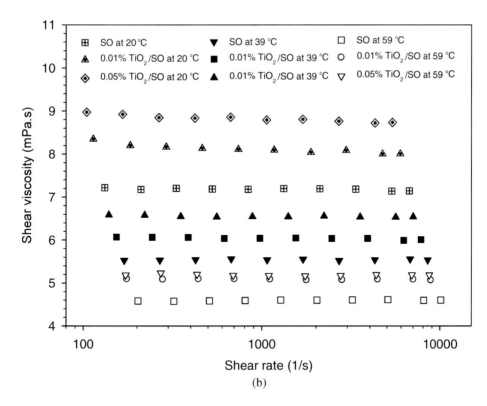

Figure 9. Apparent shear viscosity as a function of shear rate for (a) SO and SiO$_2$/SO and (b) SO and TiO$_2$/SO nanofluids at different temperatures and concentrations.

It is noticeable from Figure 9 that at 20 °C the apparent viscosity of both nanofluids showed slight decreasing trend with shear rate, whereas at higher temperatures they showed more constant trend with shear rate. These results of shear rate independent viscosities of these nanofluids are in agreement with the literature data on silicone oil based TCNT nanofluid [19]. Similar Newtonian nature and shear rate independent viscosity of ethylene glycol (EG) based TiO$_2$ nanofluids was also reported by Chen et al. [14]. It is also noted that at low shear rate (less than 100 s^{-1}) the same research group [50] previously reported a strong non-Newtonian behavior of water based nanofluid containing the same TiO$_2$ nanoparticles. At low shear rates Newtonian behavior of silicone oil (Syltherm 800) based diamond-graphene nanofluid was observed by Yang et al. [20].

Figure 9 also illustrates that while the apparent shear viscosity decreases with increasing temperature, the corresponding shear rate increases as the temperature increases. For SO/TCNT nanofluid similar results was reported by Chen and Xie [19].

4.2.2. Effect of Temperature and Concentration on Viscosity of These Nanofluids

As anticipated, temperature and concentration of nanoparticles were found to have significant influence on the viscosity of these nanofluids. Results of temperature and concentration dependence of viscosity of both SiO$_2$ and TiO$_2$ nanofluids are presented in Figure 10, which clearly shows a substantial decrease in viscosity with increasing temperature and increase with increasing nanoparticles concentration. For instance, due to rising temperature from 20 °C to 59 °C the viscosity of TiO$_2$/SO nanofluid at 0.01 vol. % decreases about 37% and the viscosity of SiO₂/SO nanofluid at 0.05 vol. % also decreases about 42%. Also the maximum increase in viscosity of SO was about 22% for a 0.05 % volumetric loading of TiO₂ nanoparticles. It is noted that the viscosity of silicone oil also showed similar decreasing trend with temperature as the nanofluids.

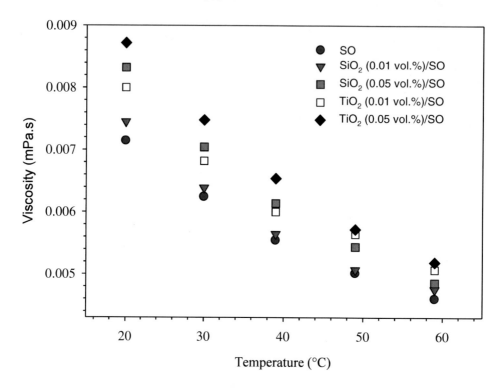

Figure 10. Effect of temperature and concentration on viscosity of SO based SiO₂ and TiO₂ nanofluids.

Figure 10 also shows that at any temperature and volume fraction, TiO₂/SO nanofluid exhibits considerably higher viscosity than SiO₂/SO. This could be due to the fact that besides a larger particle size, TiO₂ nanoparticles form larger agglomerations in SO compared to SiO₂ nanoparticles. Since no data on the viscosity of nanofluids containing these nanoparticles in silicone oil is available in the literature, present results cannot be compared. However, these results are consistent with the literature results for the same nanoparticles based nanofluids. For nanofluids containing SiO₂ and TiO₂ in water, mineral oil or other base fluids researchers found similar increase and nonlinear decrease in viscosity with increasing nanoparticle concentration and temperature, respectively [17, 29, 36, 44, 51]. Recently, Jamshidi et al. [52] investigated the effect of addition of SiO₂ nanoparticles on viscosity of different base fluids (ethylene glycol, transformer oil and water) and reported similar temperature-dependent viscosities of their nanofluids. If the viscosity of nanofluids decreases more rapidly compared to base fluid, it makes nanofluids more fascinating as they can be used in high temperature applications.

4.2.3. Comparisons of Measured and Theoretical Predictions

Apart from the popular Einstein model [23], the classical power-law based model developed by Nielsen [26] and a recent Arrhenius type equation reported by Nguyen et al.

[28] for nanofluids were used to predict and compare the viscosity of these nanofluids. The viscosity of base fluid (SO) at the corresponding temperature was used to calculate the viscosity of these nanofluids using these models.

Figure 11 shows that all these models failed to predict the viscosity of the nanofluids, although the concentration of these nanofluids are suitable for both the Einstein's [23] and Nielsen's [26] models. It is seen that while the under-predictions by both the Einstein [23] and Nielsen [26] models are almost identical, Nguyen et al. [28] equation severely underestimates the results. It is also noted that due to similar under-predictions of viscosity data by these models Figure 11 presents the predicted data for only 0.05 vol. % of nanoparticles. Nevertheless such under-predictions of viscosity results by existing models are commonly reported in the literature [3, 11, 15, and 16].

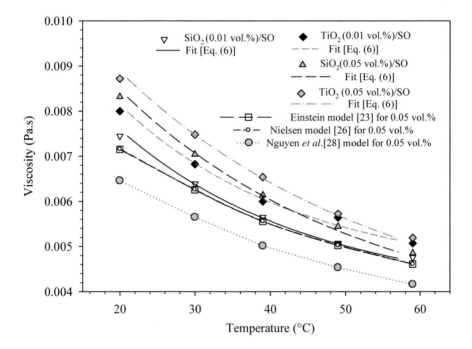

Figure 11. Comparisons experimental data with predictions by the selected existing models for these SO-based nanofluids.

Since the existing models were found unable to predict the temperature dependent viscosity of these nanofluids, a non-linear regression technique was employed to fit the present results and an exponential decay-based empirical correlation as a function of temperature and nanoparticles concentration was developed for the viscosity of these nanofluids (η_{nf}). By incorporating a reference viscosity (i.e., viscosity of base fluid at room temperature) the following generalized empirical correlation was deduced for the viscosity of these nanofluids:

$$\eta_{nf} = \eta_0[A + Be^{-CT}] \qquad (6)$$

where η_0 is the reference viscosity (Pa.s) of base fluids at room temperature (20°C in this case) and T is the temperature (°C). The dimensionless empirical coefficients A, B, and C are related to several factors such as concentration, type of base fluid, nanoparticles materials and agglomeration.

This equation (6) fits the data with a correlation coefficient of $R^2 >0.99$. The coefficients obtained from the data fitting of each nanofluid depend on the concentration of nanoparticles. The values of these coefficients and conditions for both nanofluids are provided in Table 2, which clearly shows that while values of A and C decrease with increasing nanoparticles concentration, value of B increases with the concentration. At this moment it is however not known how nanoparticle concentration or agglomeration is related with these coefficients. Thus more studies need to be conducted to identify the relationship and to understand the influence of concentration, agglomeration, and temperature on the viscosity of nanofluids.

Table 2. Values of empirical coefficients in Eq. (6) for both nanofluids

Coefficients	SiO$_2$/SO		TiO$_2$/SO		Conditions
	0.01 vol.%	0.05 vol.%	0.01 vol.%	0.05 vol.%	
A	0.546	0.448	0.602	0.434	$A <1$
B	1.148	1.372	1.232	1.414	$B >1$
C	0.040	0.031	0.042	0.028	$C <<1$

CONCLUSION

This chapter thoroughly reviews recent research on viscosity of nanofluids containing metal oxide nanoparticles. Our recent investigations on the viscosity of two oxide nanofluids are also presented. The influences of several important parameters such as concentration and temperature on the viscosity of these nanofluids are demonstrated and analyzed.

This review also reveals that these metal oxide nanofluids show both Newtonian and non-Newtonian shear-thinning behaviors and that their rheological behaviors also depend on concentration, shear rate, and temperature as well. All studies found considerably higher viscosity of nanofluids compared to their base fluids and the enhanced viscosity further increases with increasing concentration of nanoparticles. Although viscosity of both the base fluids and nanofluids decrease mostly nonlinearly with increasing temperature, temperature has little influence on the enhancement or decrease of viscosity of nanofluids over their base fluids. While some studies reported no increase or decrease in the relative viscosity of nanofluids, others found an increasing trend with the temperature. Nevertheless, the observed nonlinear decrease in viscosity of these nanofluids with increasing temperature makes these nanofluids more attractive for their applications at elevated temperatures. The existing viscosity models have been tried to predict the observed viscosity of nanofluids and were found to severely under predict the results.

Our recent study on temperature and concentration dependence of viscosity of silicone oil based TiO$_2$ and SiO$_2$ nanofluids is presented. Results showed that both the base fluid (SO) and nanofluids are of Newtonian nature. The viscosity of these nanofluids was considerably higher than that of the base silicone oil and it increased further with increasing volumetric concentration of nanoparticles. At any temperature and volume fraction, TiO$_2$/SO nanofluid showed considerably higher viscosity than that of the SiO$_2$/SO nanofluid. A nonlinear decrease in viscosity of these nanofluids with increasing temperature was observed. Based on the present results an empirical correlation for the temperature dependence of viscosity of these nanofluids was proposed.

Although some research efforts have been devoted on the model development for the viscosity of nanofluids, no noticeable success is achieved as till today no model is available that can predict the viscosity of nanofluids and can explain the underlying mechanisms. Most of the reported models are mainly obtained from the regression of researchers own viscosity data. Thus more research work need to be performed to identify the real mechanisms and to develop models for predicting the viscosity of nanofluids. It is also imperative to conduct more systematic experiments on various types of nanofluids under various conditions to elucidate more on this key property of nanofluids.

ACKNOWLEDGMENT

This research was partially supported by Fundação para a Ciência e a Tecnologia (FCT), Portugal through projects PEst-OE/QUI/UI0536/2011 and PTDC/EQU-FTT/104614/2008 attributed to Centro de Ciências Moleculares e Materiais (CCMM), Faculdade de Ciências, Universidade de Lisboa (FCUL).

NOMENCLATURE

A,B,C	Dimensionless coefficient (-)
T	Temperature (C)
η	Viscosity (Pa.s)
ϕ	Volume fraction (-)

Abbreviations

CNT	Carbon nanotube
DIW	Deionized water
EG	Ethylene glycol
GO	Gear oil
PAO	Poly-alphaolefins oil
PG	Propylene glycol
SO	Silicone oil
TNT	Titanate nanotube
W	Water

REFERENCES

[1] Choi, S. U. S. In *Developments and applications of non-Newtonian flows*; Siginer, D. A.; Wang, H. P.; Eds.; ASME Publishing: New York,USA,1995; FED-Vol. 231/MD-Vol. 66, pp 99-105.

[2] Choi, S. U. S.; Zhang, Z. G.; Keblinski, P. In *Encyclopedia of Nanoscience and Nanotechnology*, Nalwa, H. S.; Ed.; American Scientific Publishers, Los Angeles, USA 2004, Vol.6, pp.757-773.

[3] Das, S. K.; Choi, S. U. S.; Patel, H. E. *Heat Transf. Eng.* 2006, 27, 3-19.

[4] Murshed, S. M. S.; Leong, K. C.; Yang, C. *Appl. Therm. Eng.* 2008, 28, 2109-2125.

[5] Wong, K. V.; De Leon O. *Adv. Mech. Eng.* 2010, 2010, 519659.

[6] Murshed, S. M. S.; Nieto de Castro, C. A.; Lourenço, M. J. V.; Lopes, M. L. M.; Santos, F. J. V. *Ren. Sust. En. Rev.* 2011, 15, 2342-2354.

[7] Murshed, S. M. S.; Nieto de Castro, C. A. In *Green Solvents I: Properties and Applications in Chemistry*, Ali M.; Inamuddin; Eds.; Springer, London, UK 2012, Ch. 14, pp.397-415.

[8] Pak, B. C.; Cho Y. I. *Exp. Heat Transfer* 1998, 11, 151-170.

[9] Yu, W.; France, D. M.; Routbort, J. L.; Choi, S. U. S. *Heat Transf. Eng.* 2008, 29, 432–460.

[10] Murshed, S. M. S.; Leong, K. C.; Yang, C. *Int. J. Therm. Sci.* 2005, 44, 367-373.

[11] Murshed, S. M. S.; Leong, K. C.; Yang, C. *Int. J. Therm. Sci.* 2008, 47, 560-568.

[12] Lee, J.-H.; Lee, S.-H.; Choi, C. J.; Jang, S. P.; Choi, S. U. S. *Int. J. Micro-Nano Scal. Transp.* 2010, 1, 269-322.

[13] Murshed, S. M. S.; Leong, K. C.; Yang, C. In *Handbook of Nanophysics: Nanoparticles and Quantum Dots*; Sattler, K. D.; Ed.; CRC Press: Boca Raton, USA, 2010; Vol.3, pp. 32.1-32.14.

[14] Chen, H.; Ding, Y.; Tan, C. *New J. Phys.* 2007, 9, 367.

[15] Mahbubul, I. M., Saidur, R.; Amalina, M. A. *Int. J. Heat Mass Transf.* 2012, 55, 874-885.

[16] Sundar, L. S.; Sharma, K. V., Naik, M. T., Singh, M. K. *Ren. Sust. En. Rev.* 2013, 25, 670-686.

[17] Naina, H. K.; Gupta, R.; Setia, H.; Wanchoo, R. K. *J. Nanofluid.* 2012, 1, 161-165.

[18] Kolade, B.; Goodson, K. E.; Eaton, J. K. *J. Heat Transfer* 2009, 131, 052402.

[19] Chen, L.; Xie, H. *Colloid. Surf. A: Physicochem. Eng. Aspects* 2009, 352, 136-140.

[20] Yang, Y.; Oztekin, A.; Neti, S.; Mohapatra, S. *J. Nanopart. Res.* 2012, 14, 852.

[21] Felicia, L. J.; John, R.; Philip, J. *J. Nanofluid.* 2013, 2, 75-84.

[22] Venerus, D. C.; Buongiorno, J. et al. *Appl. Rheol.* 2010, 20, 44582 (7 pp).

[23] Einstein, *Investigations on the Theory of the Brownian Movement*, Dover Publications, Inc.: New York, USA, 1956.

[24] Brinkman, H.C. *J. Chem. Phys.* 1952, 20, 571–581.

[25] Batchelor, G. K.; *J. Fluid Mech.* 1977, 83, 97-111.

[26] Nielsen, L. E. *J. Appl. Phys.* 1970, 41, 4626-4627.

[27] Krieger, I. M.; Dougherty, T. *J. Trans. Soc. Rheol.* 1959, 3, 137-152.

[28] Nguyen, C.; Desgranges, F.; Roy, G.; Galanis, N.; Mare, T.; Boucher, S.; Anguemintsa, H. *Int. J. Heat Fluid Flow* 2007, 28, 1492-1506.

[29] Masuda, H; Ebata, A.; Teramae, K.; Hishinuma, N. *Netsu Bussei* 1993, 4, 227-233.

[30] Wang, X.; Xu, X.; Choi, S. U. S. *J. Thermophys. Heat Transfer* 1999, 13, 474-480.

[31] Das, S. K.; Putra, N, Roetzel, W. *Int. J. Heat Mass Transf.* 2003, 46, 851-862.

[32] Putra, N.; Roetzel, W.; Das, S. K. *Heat Mass Transfer* 2003, 39, 775-784.

[33] Prasher, R.; Song, D.; Wang, J.; Phelan, P. E. *Appl. Phys. Lett.* 2006, 89, 133108.

[34] Wen, D.; Ding, Y. *IEEE Trans. Nanotechnol.* 2006, 5, 220-227.

[35] Namburu, P. K.; Kulkarni, D. P.; Misra, D.; Das, D. K. *Exp. Therm. Fluid Sci.* 2007, 32, 397-402.

[36] Turgut, A.; Tavman, I.; Chirtoc, M.; Schuchmann, H. P.; Sauter, C.; Tavman, S. *Int. J. Thermophys.* 2009, 30, 1213-1226.

[37] Zhou, S.-Q.; Ni, R.; Funfschilling, D. *J. Appl. Phys.* 2010, 107, 054317.

[38] Kole, M.; Dey T. K. *Int. J. Therm. Sci.* 2011, 50, 1741-1747.

[39] Lee, S. W.; Park, S. D.; Kang S.; Bang, I. C.; Kim, J. H. *Int. J. Heat Mass Transf.* 2011, 54, 433-438.

[40] Sundar, L. S.; Singh, M. K.; Sousa, A. C. M. *Int. Comm. Heat Mass Transfer* 2013, 44, 7-14.

[41] Rudyak, V. Y. *Adv. Nanoparticles* 2013, 2, 266-279.

[42] Yiamsawas, T.; Mahian, O.; Dalkilic, A. S.; Kaewnai, S.; Wongwises, S. *Appl. Energy* 2013,111, 40-45.

[43] Sridhara, V.; Satapathy, L. N. *Nanoscal. Res. Lett.* 2011, 6, 456 (16 pp).

[44] Duangthongsuk W.; Wongwises, S. *Exp. Therm. Fluid Sci.* 2009, 33, 706-714.

[45] Chen, H.; Ding, Y.; Lapkin, A.; Fan, X. *J. Nanopart. Res.* 2009, 11, 1513-1520.

[46] Lee, J.-H.; Hwang K. S.; Jang, S. P.; Lee, B. H.; Kim, J. H.; Choi, S. U. S.; Choi, C. J. *Int. J. Heat Mass Transf.* 2008, 51, 2651-2656.

[47] Nguyen, C.; Desgranges, F.; Galanis, N.; Roy, G.; Mare, T.; Boucher, S.; Minstsa, H. A. *Int. J. Therm. Sci.* 2008, 47, 103-111.

[48] Murshed, S. M. S.; Tan S. H.; Nguyen, N. T. *J. Phys. D: Appl. Phys.* 41, 2008, 085502.

[49] Murshed, S. M. S.; Santos, F. J. V.; Nieto de Castro, C. A. *J. Nanofluid.* 2013, 2, 161-166.

[50] He, Y. R.; Jin, Y.; Chen, H. S.; Ding, Y. L.; Cang, D. Q.; Lu, H. L. *Int. J. Heat Mass Transfer* 2007, 50, 2272-2281.

[51] Jin, H.; Andritsch, T.; Morshuis, P. H. F.; Smit, J. J. In *Proceedings of the Annual Conference on Electrical Insulation and Dielectric Phenomena (CEIDP)*; Montreal, Canada, October 2012.

[52] Jamshidi, N.; Farhadi, M.; Ganji, D. D.; Sedighi, K. *IJE Trans. B: Appl.* 2012, 25, 201-209.

In: Nanofluids: Synthesis, Properties and Applications ISBN: 978-1-63321-677-8
Editors: S.M. Sohel Murshed, C.A. Nieto de Castro © 2014 Nova Science Publishers, Inc.

Chapter 6

CAPILLARY WETTING OF NANOFLUIDS

Milad Radiom[1], Chun Yang[2,] and Weng Kong Chan[2]*
[1]Department of Chemical Engineering, Virginia Tech.,
Blacksburg, VA, US
[2]School of Mechanical and Aerospace Engineering,
Nanyang Technological University, Singapore

ABSTRACT

Spreading dynamics, equilibrium contact angle and wettability of nanofluids have attracted attention of many researchers in the recent years. This interest stems from the fact that nanofluids, due to their more desirable thermophysical properties, are gradually replacing simple liquids in many applications: cooling of electronic devices, ventilation and air conditioning, and biomedical practices. Important applications such as oil extraction from porous rocks, spin coating, and underfill flow process in flip chip technology directly involve interfacial (capillary) flow of nanofluids driven by capillary forces. These wide-ranging applications make study of nanofluid spreading dynamics and wettability vital from engineering and scientific standpoints. In capillary flow of nanofluids effects from inter-nanoparticle interactions (e.g. electrostatic repulsion, van der Waals attraction, and hydrodynamic coupling), surface adsorption and deposition, aggolomeration and breaking of clusters play complex roles. Experimental investigation into spreading dynamics and dynamic contact angle of nanofluids reveals the contribution of each effect. A theoretical model that incorporates all the physics involved is scarce; however, available theories for simple liquid capillary flow help better understand this complex phenomenon. In this chapter we will present a short preview of capillary flow and our experiments in the field of nanofluid spreading dynamics.

---

[*] Tel.: (+65) 6790-4883, Fax: (+65) 6792-4062, E-mail: mcyang@ntu.edu.sg.

1. INTRODUCTION

The interesting phenomenon of a liquid wetting surface of a solid has received attention for more than a century. Many practical processes (e.g., dip-coating of wire for electrical insulation, flow of oil through porous rocks in petroleum reservoirs, surface cleaning and detergent systems, and some clinical, chemical and biomedical analysis) in fact rely on this phenomenon. Many surfaces are designed to enhance wetting, while others are modified to avoid it. For liquids, a wide range of wetting affinities are attained by adding particles (e.g., surfactants, macromolecules and nanoparticles), and thus modifying the thermo-physical properties of the liquid such as surface tension and viscosity. The wetting of the solid by the liquid is connected to physical chemistry of solid-liquid interactions, surface forces and fluid dynamics; and thus to a very well degree, is controllable. Several aspects of wetting process have been revealed [1]:

1. The out of equilibrium interfacial energy of the connecting interfaces is the driving force of wetting; and wetting spontaneously progresses towards minimization of the total interfacial energy.
2. Interfacial energies are very sensitive to variations in composition or concentration of interface material which might change due to contamination or adsorption.
3. Physical roughness and chemical inhomogeneity of the solid surface are the main sources of hysteresis (for more information on hysteresis the reader is referred to Ref. [2]).

Many practical applications such as underfill flow process in flip chip technology, spin coating, electrowetting and structuring hydrophilic, hydrophobic, and superhydrophobic surfaces involve wetting of the solid surface by nanofluids. Nanofluids differ from simple liquids in that the nanoparticles result in new types of solid-liquid interaction energies, interfacial properties and fluid flow characteristics. These differences along with their important presence in many applications have made capillary wetting of nanofluids an area of research for researchers and scientists in colloids and surface science and nano science communities. There are several theoretical studies [3-6] and experimental investigations [7-11]. Dynamics of nanofluids spreading on solid surfaces is also studied (see Ref. [12] for silica nanoparticles and Ref. [13] for aluminum nanoparticles). Nevertheless, with the constantly renewing applications of nanofluids and the variety of nanoparticles in shape, concentration, and chemistry, etc. this subject requirs greater attention.

When a liquid droplet is put in contact with a solid surface, a spontaneous process starts during which the droplet spreads over the surface and wets it. The time evolution of this process is described by variation in dynamic contact angle (apparent contact angle), θ_a (see Figure 1). Upon achieving a steady state, an equilibrium contact angle θ_e forms. The interface of the liquid and solid forms a line at the boundary (contact line) ℓ_r where three phases are in mutual contact: solid S, liquid L and corresponding equilibrium vapor V (see Figure 1). Hence, there are three interfaces, SL, SV, and LV, where each interface has a certain free energy per unit area (interfacial energy): γ_{SL}, γ_{SV} and γ_{LV}.

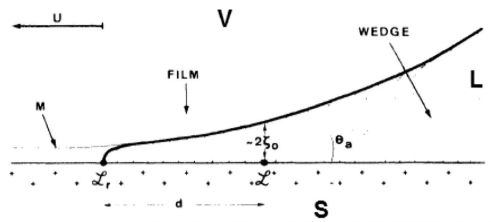

Reprinted with permission from [1]: De Gennes P.G., Rev. Mod. Phys. 57, 827–863 (1985); Copyright 1985 by the American Physical Society.

Figure 1. Schematic of a liquid droplet spreading over a solid surface. The dynamic contact angle θ_a approaches the equilibrium contact angle θ_e at the end of the wetting process. The distance between the real contact line ℓ_r and the nominal contact line ℓ is about 100 Å.

These interfacial energies however only describe the energy content of the interface in the far field from ℓ_r and close to the nominal contact line ℓ. In the vicinity of ℓ_r the energy content is more complex and depends on the local pressure difference due to curvature, and various surface and film interaction energies [14]. The distance between ℓ_r and ℓ is about 100 Å. It is possible to relate θ_e to the far field interfacial energies from balance of interfacial forces at the nominal contact line ℓ and at equilibrium:

$$\gamma_{LV}\cos(\theta_e) = \gamma_{SV} - \gamma_{SL} \tag{1}$$

This equation was developed by the British scientist Thomas Young (1773-1829) [15] and is known as Young's equation.

In practice, common techniques of θ_e measurement (e.g. direct photograph of sessile droplet, or rise of a liquid column in a fine capillary) can only measure it for distances larger than 100 Å from line ℓ_r. The major experimental obstacle in accurate determination of θ_e is hysteresis. Hysteresis is mainly due to solid surface roughness and chemical inhomogeneity, and can affect measurement of θ_e to about or even more than 10°. For more details on this subject the reader is referred to the excellent review by de Gennes [1].

Common techniques to measure γ_{LV} (e.g. direct photograph of pendant droplet, [16-18]) usually give precise values. It is however prone to contamination as γ_{LV} is a strong function of interface composition and concentration [19]. The difference $\gamma_{SV} - \gamma_{SL}$ is also an important parameter that is calculated from equation (1) knowning θ_e and γ_{LV}.

For separations between 30 Å to 1 μm from ℓ_r, the role of surface and film interactions becomes significant in shape and local curvature of the LV interface. For a liquid film of thickness ξ on solid surface, the film energy per unit area of solid-liquid interaction $P(\xi)$ is related to disjoining pressure $\Pi(\xi)$ via [8]:

$$P(\xi) = \int_{\xi}^{\infty} \Pi(\xi') \, d\xi'$$

(2)

There are known to be three contributions to the disjoining pressure: molecular (or van der Waals) Π_m, electrostatic Π_e, and structural Π_s [8]:

$$\Pi(\xi) = \Pi_m(\xi) + \Pi_e(\xi) + \Pi_s(\xi)$$

(3)

The molecular and electrostatic components are included in the well-known DLVO theory. The structural component is explained in various work by Wasan's group (see e.g., [3, 20]) and is related to layering and patterning of nanoparticles in the thin nanofluid film (nano to micron thick) abutting the contact line ℓ_r. For a discussion on the form of disjoining pressure at various length scales, the reader is referred to Matar et al. [6] and Trokhymchuk et al. [3].

It is known that the out-of-equilibrium interfacial energy, $\gamma_{LV}(\cos\theta_e - \cos\theta_a)$, provides the free energy of dynamic spreading. While spreading, this free energy is dissipated by two mechanisms [1]: (i) contact line friction ($T\Sigma_l$) which occurs in proximity of three-phase contact line due to interactions between solid molecules and liquid molecules, and (ii) wedge film viscosity ($T\Sigma_W$) which occurs in the wedge film region due to lubricating and rolling flow patterns. For each mechanism of energy dissipation, a theory is developed: (i) molecular kinetic theory (MKT) [21, 22] models the contact line friction and (ii) hydrodynamic theory (HDT) [23, 24] models the wedge film viscosity.

For partial wetting systems ($\theta_e > 10°$), it is assumed that the mechanisms of energy dissipation coexist and models that combine MKT and HDT are developed by Petrov [25] and De Ruijter [26]. In Petrov's model, it is assumed that the equilibrium contact angle θ_e is not constant and its variation is described by MKT. In De Ruijter's model, it is assumed that θ_e is constant and the mechanisms of energy dissipation are added together to form the total dissipation function, $T\Sigma_{tot} = T\Sigma_l + T\Sigma_W$. These models show generally good agreement with experimental data of simple liquid spreading dynamics [27]. For nanofluids, these models are good starting points paving the path to more sophisticated models that include various effects from nanoparticles in nanofluid spreading dynamics. Such models are yet to be developed.

2. MEASUREMENT OF SURFACE TENSION AND CONTACT ANGLE OF TiO₂-WATER NANOFLUIDS

Fluid wetting capability relies on two properties: surface tension (liquid-vapor interfacial energy γ_{LV}) and contact angle. The objective of this section is to investigate the effect of nanoparticle concentration on surface tension and contact angle of nanofluids. Experiments were carried out using water based TiO_2 nanofluids (15 nm anatase, 99%, Nanostructured and Amorphous Materials Inc., TX, USA) of various nanoparticle volume concentrations 0.05% to 2%. With a FTA200 system (First Ten Angstroms, Inc., VA, USA), surface tension of these nanofluids was measured using photography of pendant droplet. Equilibrium contact angle was determined using photography of sessile droplet on a borosilicate glass slide.

The choice of TiO_2 nanoparticles for the purpose of these experiments was that these nanoparticles are cheap and commercially available. They are spherical, and thus do not exhibit the geometrical difficulties associated with cylindrical nanoparticles. They are also found to be a very good nanoparticle choice for heat transfer applications.

TiO_2 nanoparticles were mixed with deionized (DI) water. Oleic acid surfactant (Riedel-de Haën) was added to stabilize the solution [28]. The solution was stirred for 8 hours and then sonicated for 100 min (Sonicator 3000, MISONIX). The sonication program was 5-sec pulsation and 20-sec off. To avoid solution overheating, and thus evaporation of water content, water-ice bath was used. To understand the stability and durability of these nanofluids, the solution was sonicated for various time periods, namely 50 min to 150 min, and the viscosity of the solution was measured as a function of sonication time [29]. With this method, a saturation sonication time can be obtained. Saturation sonication time refers to a sonication period after which there is no more significant change in a particular thermophysical property. In this work, viscosity was chosen as the reference thermophysical property. Figure 2 shows viscosity of 2% TiO_2 nanofluid as a function of sonication time. After a sonication period of 100 min, no obvious change in the viscosity of the solution was seen.

It is a normal practice to use Transmission Electron Microscope (TEM) to monitor clustering and dispersity of nanoparticles. An aliquot of 0.05% TiO_2 nanofluid was dried on a carbon coated copper grid in air. The TEM nanograph of the sample was then taken immediately. The TEM nanograph showed some levels of clustering (see Figure 3). The sizes of single particles were found to be close to 15 nm as specified by the supplier. The morphology of the monodispersed particles was spherical. This asserted qualitatively that the sonication period and addition of oleic acid surfactant had been successful in producing a well dispersed and close to homogeneous solution. The effective nanoparticle size was 260 nm measured with Particle Size Analyzer (Brookhaven Instruments Corporation, NY, USA).

Photography of pendant droplet is a common surface tension measurement technique (see Figure 4-a). The FTA200 system was employed for this purpose. The volume of the pendant droplet should reach a possible maximum for gravitational force and surface tension force to balance each other. To minimize the effects of evaporation, measurements were performed immediately after the pendant droplet reached its maximum volume.

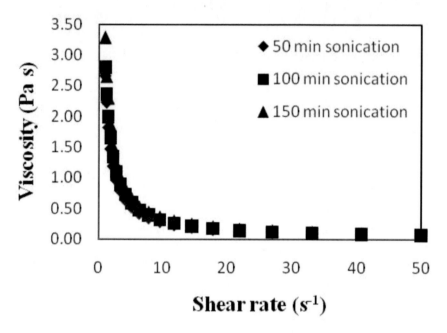

Figure 2. Viscosity of 2% TiO$_2$ nanofluid as a function of sonication time.

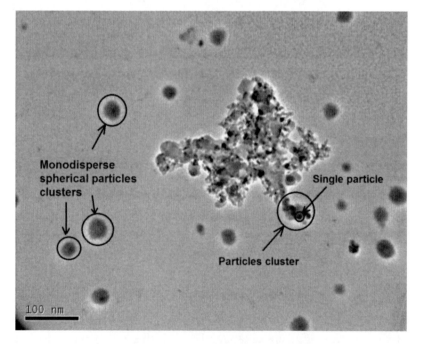

Figure 3. TEM nanographs of 15 nm TiO$_2$ nanoparticles taken immediately after an aliquot of dilute solution was dropped and dried on a carbon coated copper grid [30].

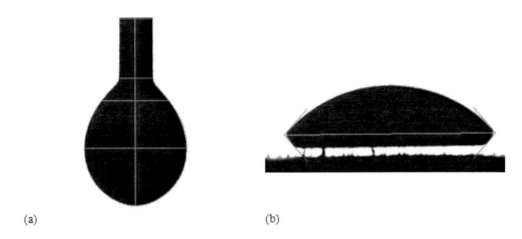

Figure 4. Images of (a) a pendant droplet and (b) a sessile droplet.

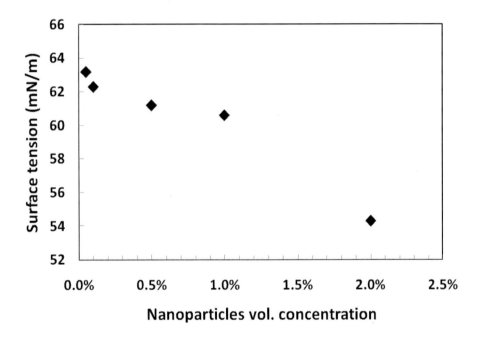

Figure 5. Surface tension of TiO$_2$-DI water nanofluids versus nanoparticle concentration.

Figure 5 shows variation of the surface tension of these nanofluids with nanoparticle concentration. At higher concentrations, the surface tension is lower. This agrees with Gibbs adsorption isotherm [31]. The reduction in surface tension of nanofluids is attributed to the reduction in cohesive energy at fluid-air interface where nanoparticles are added [32-34]. The Brownian motion at interface disperses the nanoparticles to locations at the interface with lower level of LV interfacial energy [35].

Photography of sessile droplet of nanofluids was used to measure the contact angle (see Figure 4-b). The sessile droplet was placed on borosilicate glass slides. The nanofluid was initially pumped out of the syringe at a very low rate (e.g., 1 μL s^{-1}) and detached from the syringe needle tip as soon as it touched the surface of the solid. The sessile droplet volume was less than 6 μl. At this volume, the sessile droplet forms a spherical cap with the radius of its equivalent sphere being smaller than the capillary length $L_C = \sqrt{\gamma_{LV}/\rho g}$ where ρ is the density of the nanofluid and g is the gravitational acceleration constant. The capillary length denotes the ratio of surface tension force to gravity. For small values of this ratio, the macroscopic shape of the droplet is close to a spherical cap [1]. The assumption of a spherical cap is important in fitting for the contact angle. Evaporation can affect these measurements. Evaporation of the sessile droplet can drastically lower the value of the contact angle as compared to the actual equilibrium wetting contact angle. For a sessile droplet of spherical cap, the base radius (r) and volume (V) are related to the contact angle (θ) via:

$$ r^3 = \left(\frac{3V}{\pi} \frac{\sin^3 \theta}{2 - 3\cos\theta + \cos^3 \theta} \right) $$

(4)

A change in the droplet volume less than 5% due to evaporation will result in less than 5% error in the measurement of contact angle. Figure 6 shows variation of the equilibrium contact angle of these nanofluids with nanoparticle concentration. It is shown that the equilibrium contact angle increases with the nanoparticle concentration. Similar trend was partially observed in contact angle of bismuth telluride-water nanofluid on glass and silicon substrates by Vafaei et al. [7] and aluminum-ethanol nanofluid on Teflon-AF coated substrate by Sefiane et al. [13]. In order to understand the observed phenomena, it is necessary to consider the underlying mechanisms responsible for the shape and contact angle of droplets on solid surface. For nanofluids, in addition to the far field relation between the interfacial energies as mentioned in Young's equation (1), the equilibrium contact angle is related to stability of the thin film containing nanoparticles abutting on the wedge film [3]. When a nanofluid wets a solid surface, the surface forces together with structural and depletion effects move the contact line and form the equilibrium contact angle (Eq.(3)). The disjoining pressure results in an excess energy in the film which contributes to the spreading in the same manner as the interfacial tension energies. It is likely that larger clusters and monodispersed particles are formed in the nanofluids with higher concentration. The effective nanoparticle size for the higher concentrations can thus be larger, and thereby the number of nanoparticles in the thin film can be lower. This reduces the wetting affinity of the nanofluid and increases the contact angle at higher concentrations.

3. MEASUREMENT OF DYNAMIC CONTACT ANGLE OF TIO₂-WATER NANOFLUIDS

In this section we present an investigation into spreading dynamics of TiO_2-DI water nanofluids. It is noted that the denser solutions exhibit non-Newtonian viscosity. Since the dynamics of wetting is associated with fluid dynamics, viscosity of these solutions should be accounted for.

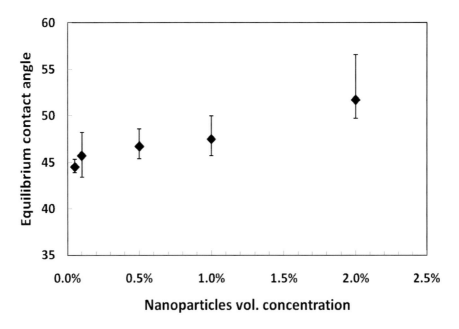

Figure 6. Equilibrium contact angle of TiO$_2$-DI water nanofluids versus nanoparticles concentration.

To model experimental data, a theoretical model based on combination of MKT and HDT similar to De Ruijter's model is used. The non-Newtonian viscosity of the solutions is incorporated in the model.

3.1. Viscosity

Viscosity of the solutions was measured using a controllable low shear rate concentric cylinders rheometer (Contraves, Low Shear 40). The viscosity was measured at shear rates ranging from 0 to 50 s^{-1}. This range corresponds to the shear rates that are common to capillary flow. From Figure 7 it is obvious that 0.5%, 1% and 2% solutions exhibit shear thinning viscosity at shear rates below 20 s^{-1}. At higher shear rates, Newtonian behavior was observed for all solutions. For dilute solutions, 0.05% and 0.1% vol., a weak shear thinning behavior was observed at very low shear rates [36]. A power-law equation is used to model the shear-rate and nanoparticle-concentration dependent viscosity:

$$\frac{\eta_n}{\eta_b} = F(\phi) K \dot{\gamma}^{n-1} \tag{5}$$

where η_n is the viscosity of nanofluid, η_b is the viscosity of DI water equal to 0.927 mPa.s, $F(\phi)$ is a function of nanoparticle volume concentration (ϕ), $K\dot{\gamma}^{n-1}$ is an indicator of shear thinning viscosity with K as the proportionality factor and n as the power-law index. $F(\phi)$ is calculated using Krieger's formula [37]:

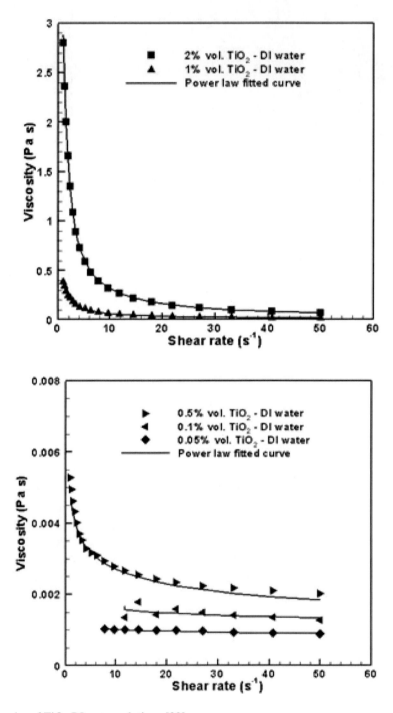

Figure 7. Viscosity of TiO$_2$-DI water solutions [30].

$$F(\phi) = \left(1 - \frac{\phi}{\phi_{max}}\right)^{-2.5\phi_{max}}$$

(6)

where ϕ_{max} is the fluidity limit that is empirically equal to 0.68 for hard spherical particles. In Eq. (5), n and K are empirical constants which are obtained by fitting this equation to the experimental data shown in Figure 7. Table 1 shows values of K and n for various nanoparticle volume concentrations.

Table 1. Power-law viscosity, surface tension and equilibrium contact angle of TiO$_2$-DI water solutions [30]

TiO$_2$ vol. concentration (ϕ)	Power-law index (n)	Proportionality factor (K)	Surface tension (γ_{LV} [N/m])	Equilibrium contact angle (θ_e)
2%	0.04	2932	0.0543	51.7
1%	0.18	432	0.0606	47.5
0.5%	0.76	5	0.0612	46.7
0.1%	0.89	2	0.0623	45.7
0.05%	0.92	1	0.0632	44.5

It is obvious that higher nanoparticle concentration results in a larger non-Newtonian behavior. Figure 7 also shows that the power-law Eq.(5) is in good agreement with experimental data.

3.2. Molecular Kinetics Theory

Schematic of a spreading droplet that is inspired from De Gennes [1] and Blake [22] is depicted in Figure 8. Based on MKT [22], the rate of displacement of the three-phase contact line over adsorption sites on solid surface is equal to the net frequency of molecular movements ($K_W = K^+ - K^-$, where K^+ is the frequency of forward motion and K^- is the frequency of backward motion) multiplied by the average distance between the adsorption sites, λ:

$$U = \lambda K_W$$

(7)

The equilibrium frequency of the three-phase contact line motion (K_W^0) is obtained from Eyring's theory of absolute reaction rates [22]:

$$K_W^0 = \left(\frac{k_B}{\hbar}\right)\exp\left(\frac{-\Delta G_W^*}{Nk_BT}\right) \quad \left(\text{At equilibrium } K_W^0 = K^+ = K^-\right)$$

(8)

where $k_B, \hbar, N$ and T are the Boltzmann constant, Planck constant, Avogadro's number and absolute temperature, respectively. In this equation, ΔG_W^* is the equilibrium free energy of capillary flow. The out-of-equilibrium interfacial energy $\gamma_{LV}\left(\cos\theta_e - \cos\theta_a\right)$ alters the free energy of capillary flow. The net frequency of contact line motion can be obtained as follows [22]:

$$K_W = K^+ - K^- = \left(\frac{2k_B}{\hbar}\right)\exp\left(\frac{-\Delta G_W^*}{Nk_BT}\right)\sinh\left(\frac{\gamma_{LV}\left(\cos\theta_e - \cos\theta_a\right)}{2nk_BT}\right)$$

(9)

where n is the number of adsorption sites per unit area. For small $\sinh$ arguments (i.e., for θ_a close to θ_e) it is easy to assign a friction coefficient ζ for the rate of three-phase contact line motion: $\zeta = nk_BT/K_W^0\lambda$. We thus get $\gamma_{LV}\left(\cos\theta_e - \cos\theta_a\right) = \zeta U$ [22]. De Ruijter et al. [38] showed that the corresponding dissipation function ($T\Sigma_l$) is:

$$T\Sigma_l = \frac{\zeta U^2}{2}$$

(10)

3.3. Hydrodynamic Theory

To calculate the wedge film viscous dissipation ($T\Sigma_W$), the Navier-Stokes equation of motion is solved in the wedge film region. For film thickness (H) much smaller than the radial distance ρ (see Figure 8), and capillary number $Ca = \eta_n U/\gamma_{LV} \ll 1$, the lubrication approximation can be used:

$$\frac{\partial p}{\partial x} = \frac{\partial}{\partial z}\left(\eta_n \frac{\partial u}{\partial z}\right)$$

(11)

where p is pressure, x is the direction of flow and z is normal to direction of flow. For no stress boundary condition at the free fluid-air interface and no slip boundary condition at the solid surface, solution to Eq. (11) gives:

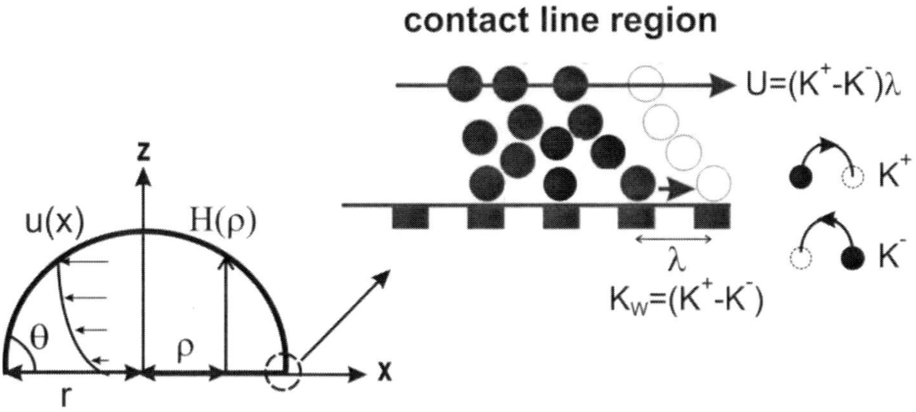

Capillary Wetting of Nanofluids

contact line region

Figure 8. (Left) Schematic of a spreading droplet of radius r and contact angle θ. H is the height of a point on outer layer of the droplet at distance ρ from center. $u(x)$ is the velocity distribution inside the droplet. (Right) K^+, K^- and K_W are the forward, backward and net frequencies of the three-phase contact line motion over adsorption sites on solid surface. λ is the average distance between the adsorption sites. U is the three-phase contact line velocity [30].

$$u = \frac{1}{\left(\eta_b F(\phi) K\right)^{\frac{1}{n}}} \left(\frac{\partial p}{\partial x}\right)^{\frac{1}{n}} \frac{n}{n+1} \left(H^{\frac{1}{n}+1} - (H-z)^{\frac{1}{n}+1}\right)$$

(12)

where η_n is replaced by its expression in equation (5). The average cross sectional fluid velocity in the wedge film ($\bar{u} = \int_0^H u dz \Big/ H$) is equal to the three-phase contact line velocity (i.e., $\bar{u} = U$). This results in:

$$u = \frac{2n+1}{n+1} U \left(1 - \left(1 - \frac{z}{H}\right)^{\frac{1}{n}+1}\right)$$

(13)

The viscous dissipation function in the wedge film can be obtained as follows [1]:

$$T\Sigma_W = \int_0^{r-x_m} \left[\int_0^H \tau\left(\frac{\partial u}{\partial z}\right) dz\right] d\rho = \eta_b F(\phi) K \left(\frac{2n+1}{n}\right)^n U^{n+1} \left\{\int_0^{r-x_m} \left[\left(\frac{1}{H}\right)^n\right] d\rho\right\}$$

(14)

where r is the droplet base radius, τ is the shear stress ($= \eta_n \partial u/\partial z$) and x_m is the cut-off length [23, 24]. Without consideration of x_m, the dissipation of energy at the wedge film grows infinitely close to the three-phase contact line [1]. For a thin wedge film Eq. (14) simplifies to:

$$T\Sigma_W = \eta_b F(\phi) K \left(\frac{2n+1}{n}\right)^n U^{n+1} \frac{1}{\theta^n}\left[\frac{r^{1-n} - x_m^{1-n}}{1-n}\right]$$

(15)

3.4. Dynamic Contact Angle

Combining equations (10) and (15) we get the total dissipation function [38]:

$$T\Sigma_l + T\Sigma_W = \frac{\zeta U^2}{2} + \eta_b F(\phi) K \left(\frac{2n+1}{n}\right)^n U^{n+1} \frac{1}{\theta^n}\left[\frac{r^{1-n} - x_m^{1-n}}{1-n}\right]$$

(16)

The derivative of the total dissipation function with respect to contact line velocity (i.e., $\partial\left[T\Sigma_l + T\Sigma_W\right]/\partial U$) results in the total drag force [1]:

$$f_{drag} = \zeta U + \eta_b F(\phi) K \left(\frac{2n+1}{n}\right)^n \left(\frac{1+n}{1-n}\right)\frac{1}{\theta^n}\left[r^{1-n} - x_m^{1-n}\right]U^n$$

(17)

Finally equating Eq. (17) with the out of equilibrium interfacial energy we get:

$$\gamma_{LV}\left(\cos\theta_e - \cos\theta_a\right) = \zeta U + \eta_b F(\phi) K \left(\frac{2n+1}{n}\right)^n \left(\frac{1+n}{1-n}\right)\frac{1}{\theta_a^n}\left[r^{1-n} - x_m^{1-n}\right]U^n$$

(18)

It is noted that for $n = 1$ (Newtonian fluid), integral of Eq. (14) results in logarithm $\ln\left(r/x_m\right)$. In this case, the final form of Eq. (18) is similar to De Ruijter's model [38] ($\gamma_{LV}\left(\cos\theta_e - \cos\theta_a\right) = \zeta U + 6\eta\Phi(\theta)U\ln\left(r/a\right)$) where $\Phi = \sin^3\theta/2 - 3\cos\theta + \cos^3\theta$ and a is the cut-off length in De Ruijter's model).

In Eq. (15) the base radius (r) is in millimeter length scale while the cut-off length (x_m) is in nanometer length scale. Thus, $r >> x_m$, and consequently $r^{1-n} >> x_m^{1-n}$ for n ranging from 0.04 to 0.92 (see Table 1). Also, for a sessile droplet of spherical geometry, the base radius is geometrically related to the contact angle via Eq. (4). Neglecting x_m^{1-n} and substituting for r gives:

$$\gamma_{LV}\left(\cos\theta_e - \cos\theta_a\right) =$$

$$\zeta U + \eta_b F(\phi) K \left(\frac{2n+1}{n}\right)^n \left(\frac{1+n}{1-n}\right) \frac{\left(\frac{3V}{\pi}\frac{\sin^3\theta_a}{2-3\cos\theta_a+\cos^3\theta_a}\right)^{\frac{1-n}{3}} U^n}{\theta_a^n} \tag{19}$$

Equation (19) shows the dynamic contact angle (θ_a) as a function of contact line velocity (U), solid-liquid molecular interactions (ζ) and non-Newtonian viscosity (n, K). Finally, substituting U with $dr/dt = (dr/d\theta) \times (d\theta/dt)$ the following equation can be obtained for the time evolution of the dynamic contact angle:

$$\sigma\left(\cos\alpha - \cos\alpha_e\right) = \zeta\left[\left(\frac{3V}{\pi}\frac{(1+\cos\alpha)^6}{(2+3\cos\alpha-\cos^3\alpha)^4}\right)^{\frac{1}{3}}\right]\frac{d\alpha}{dt}$$

$$+\eta_b F(\phi) K \left(\frac{2n+1}{n}\right)^n \left(\frac{1+n}{1-n}\right)\left[\left(\frac{3V}{\pi}\frac{\sin^3\alpha}{2+3\cos\alpha-\cos^3\alpha}\right)^{\frac{1}{3}}\right]^{1-n}$$

$$\times\left[\left(\frac{3V}{\pi}\frac{(1+\cos\alpha)^6}{(2+3\cos\alpha-\cos^3\alpha)^4}\right)^{\frac{1}{3}}\right]^n \left(\frac{d\alpha/dt}{\pi-\alpha}\right)^n \tag{20}$$

in which the dynamic contact angle $\theta_a = \pi - \alpha$.

3.5. Results and Discussion

The effective diameter of nanoparticles was equal to 260 nm at the lowest solution concentration 0.05% vol. At higher particle concentrations, the increased interparticle interactions result in larger clusters. This increases the possibility of clusters to deposit on the surface of solid and form a new hydrophilic surface. Due to their larger size these clusters are less possible to deposit on the three-phase contact line and thus a heterogeneous surface forms: within the wedge film and away from the three-phase contact line, deposition of TiO_2 clusters results in a hydrophilic surface with higher surface energy (~ 2.2 J/m^2 [39]) than the three-phase contact line where the bare borosilicate glass is present (~ 0.11 J/m^2 [40]). The higher surface energy inside the droplet shrinks the wetted area by increasing the equilibrium contact angle. The solid-liquid interfacial tension increases which on the other hand enhances the equilibrium contact angle [7].

The shear thinning viscosity of the solutions is due to strong interparticle interactions of the nanoparticle clusters [36, 41, 42]. Other nanofluids such as ethylene glycol based ZnO nanofluid [41] and CuO nanofluid [43] also exhibited shear thinning viscosity at low shear rates.

Equation (20) suggests that the contact line friction dissipation (first term on the RHS of Eq.(20)) and the wedge film viscous dissipation (second term on the RHS of Eq.(20)) can occur at different time scales [44]. The time dependence of these dissipations is shown in Figure 9 where the three-phase contact line velocity ($U = dr/dt$) is plotted versus $\gamma_{LV} \cos\theta_a$. Figure 9 shows a linear trend that is in accordance with the contact line friction dissipation, and a nonlinear trend (see inset) that is in accordance with the wedge film viscous dissipation. This suggests that at start of capillary flow the contact line friction is the dominant dissipative mechanism. As capillary flow slows down, the wedge film viscous dissipation becomes more dominant. This corresponds to the solution higher viscosity at lower shear rates. Transition to wedge film viscous dissipation dominant regime occurs earlier in the dilute solutions; for example Figure 10 shows that for 0.05% concentration the viscous forces start to dominate at times around 4-8 seconds while for 2% concentration at times around 25-32 seconds.

Figure 10 shows the dynamic contact angle of TiO$_2$-DI water nanofluids at various nanoparticle volume concentrations ranging from 0.05% to 2%. Due to limitation in camera frame per second speed (30 fps), the onset of pendant droplet touching the surface of solid cannot be determined accurately. Hence, the time axis in Figure 10 was shifted to where all of the captured images were readable to the FTA200 software. From Figure 10, it is obvious that for higher nanoparticle concentrations, the contact angles are higher. Figure 10 also shows that the spreading of these nanofluids starts from a primary region where the contact angle changes rapidly followed by a region where the contact angle changes more gradually (note that in a very short period of time (less than 300 milliseconds) the contact angle evolves from 180° at point of initial contact to angles that are readable to our software and are plotted on Figure 10 at the shifted zero time).

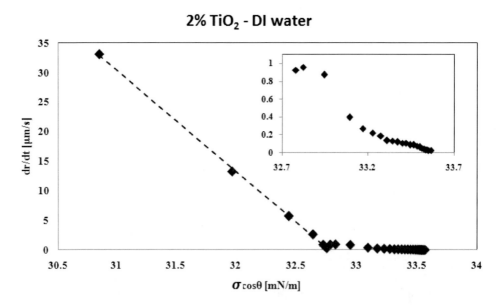

Figure 9. Experimental three-phase contact line velocity ($U = dr/dt$) plotted versus $\gamma_{LV} \cos\theta_a$. Dashed line shows a linear trend that is in accordance with contact line friction dissipation. Inset of the figure shows a non-linear trend that is in accordance with wedge film viscous dissipation [30].

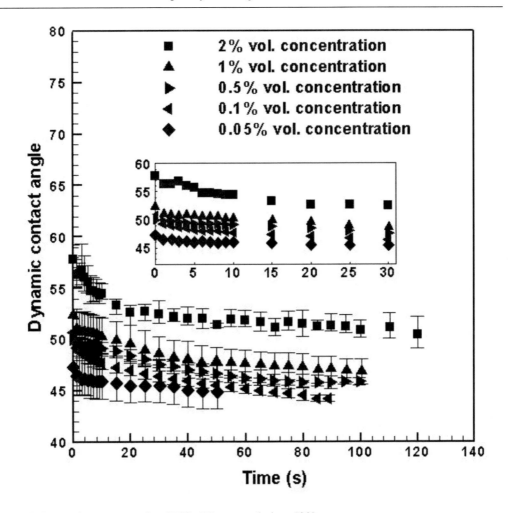

Figure 10. Dynamic contact angle of TiO$_2$-DI water solutions [30].

Table 2. The coefficient of contact line firction ζ, theoretical equlibirum contact angle $\theta_{e,theory}$ and the error of comparison between theory and experiment [30]

Nanoparticle concentration	ζ [Pa s]	$\theta_{e,theory}$	Error
2%	32	52.1	1.1
1%	99	48.2	1
0.5%	464	46.4	0.65
0.1%	483	45.3	0.54
0.05%	486	44.8	0.34

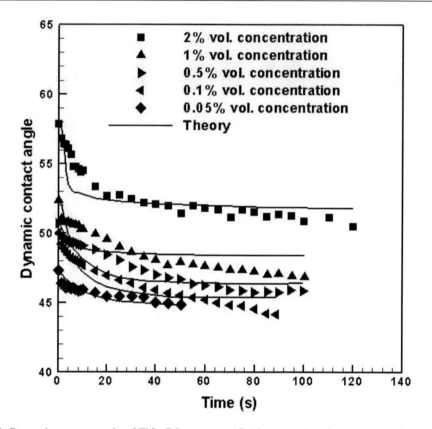

Figure 11. Dynamic contact angle of TiO$_2$-DI water nanofluid, comparison between experiment and theory [30].

In the primary region the contact line friction dissipation predominates the wedge film viscous dissipation causing fast reduction in the contact angle, then the wedge film viscous dissipation controls the droplet spreading [27].

Using Eq.(20) ζ is obtained for the best fit of theory to experimental data that gives the least squared error. Figure 11 shows a reasonable comparison between experimental data and theory.

Table 2 shows values of ζ for various nanoparticle vol. concentrations. From solution concentration 0.05% to 0.5% ζ only changes by 5%, however it drops rapidly for denser solutions. It is possible that the relative higher hydrophobicity at the three-phase contact line for denser solutions lowers the affinity of surface molecules to water molecules, thereby lowering the friction. At dense concentrations, the presence of large amount of nanoparticles in the wedge film varies the flow field structure. Without nanoparticles, it has been stated that there are two flow patterns in the wedge film: rolling and lubricating patterns [1]. Nanoparticles in the wedge film can change these flow patterns and result in more complex flow structures. As a result, dissipation is more pronounced in the wedge film.

Equation (20) gives better results at lower nanoparticle concentrations since complex interparticular interactions are less frequent in dilute solutions. Other sources of disagreement between experiment and theory can be local variations in the concentration of the nanoparticles in the nanofluid [7], pinning of the contact line and variations in solid-liquid interfacial tension [7, 13].

It is not possible to model all these effects in theory, and only simple models which can accommodate some of these effects can be developed. Also shown in Table 2 are the theoretical equilibrium contact angles, $\theta_{e,\text{theory}}$, which are in reasonable agreement with the experimental equilibrium contact angles, θ_e (see Table 1).

CONCLUSION

Due to a wide range of industrial applications, studying capillary flow of liquids laden with nanoparticles is important. TiO_2 nanoparticles are especially interesting to enhance heat removal applications. Agglomeration of TiO_2 nanoparticles results in clusters that have larger effective diameter than the actual nanoparticle size. These clusters can deposit on the surface of solid substrates and thus form a heterogeneous surface condition inside the droplet away from the three-phase contact line that increases the equilibrium contact angle. Dynamic contact angle of TiO_2 nanoparticles dispersed in DI water reveals two stages of spreading: rapid reduction in contact angle coincides with contact line friction dissipation governed by MKT while gradual reduction in contact angle coincides with wedge film viscous dissipation governed by HDT. Non-Newtonian viscosity of the solution is incorporated in HDT model to give reasonable comparison with experimental data. Nanoparticles in the wedge film change lubricating and rolling flow patterns and result in complex flow field structures. Including all physical aspects of such complex flow in theory is not feasible at the current stage. Simple theoretical equations can only give reasonable comparisons with experimental data.

ACKNOWLEDGMENT

The authors gratefully acknowledge the financial support of the research grant (MOE2009-T2-2-102) from the Ministry of Education of Singapore to CY and the Singapore A*STAR scholarship to MR.

REFERENCES

[1]   De Gennes, P. G. *Rev. Mod. Phys.,* 1985, 57, 827-863.
[2]   Pittoni, P. G.; Chang, C. C.; Yu, T. S.; Lin, S. Y. *Colloid. Surf. A,* 2013, 432, 89-98.
[3]   Trokhymchuk, A.; Henderson, D.; Nikolov, A.; Wasan, D. T. *Langmuir,* 2001, 17, 4940-4947.
[4]   Wasan, D. T.; Nikolov A. D. *Nature,* 2003, 423, 156-159.
[5]   Chengara, A.; Nikolov, A. D.; Wasan, D. T.; Trokhymchuk, A.; Henderson, D. *J. Colloid Interface Sci.,* 2004, 280, 192-201.
[6]   Matar, O. K.; Craster, R. V.; Sefiane, K. *Phys. Rev. E,* 2007, 76, 056315-9.

[7] Vafaei, S.; Borca-Tasciuc, T.; Podowski, M. Z.; Purkayastha, A.; Ramanath, G.; Ajayan, P. M. *Nanotech.,* 2006, 17, 2523-2527.

[8] Sefiane, K.; Bennacer, R. *Adv. Colloid Interface Science,* 2009, 147-148, 263-271.

[9] Orejon, D.; Sefiane, K.; Shanahan, M. E. R. *J. Colloid Interface Science,* 2013, 407, 29-38.

[10] Trybala, A.; Okoye, A.; Semenov, S.; Agogo, H.; Rubio, R. G.; Ortega, F.; Starov, V. M. *J. Colloid Interface Sci.,* 2013, 403, 49-57.

[11] Vafaei, S.; Wen, D. S.; Borca-Tasciuc, T. *Langmuir,* 2011, 27, 2211-2218.

[12] Kondiparty, K.; Nikolov, A. D.; Wasan, D.; Liu, K. L. *Langmuir,* 2012, 28, 14618-14623.

[13] Sefiane, K.; Skilling, J.; MacGillivray, J. *Adv. Colloid Interface Sci.,* 2008, 138, 101-120.

[14] Miller, C. A.; Neogi, P. Interfacial phenomena; CRC Press: Boca Raton, USA, 2008.

[15] Young, T. *Phil. Trans. R. Soc.,* 1805, 95, 65-87.

[16] Morita, A.; Carastan, D.; Demarquette, N. *Colloid. Polym. Sci.,* 2002, 280, 857-864.

[17] Yeow, Y. L.; Pepperell, C. J.; Sabturani, F. M.; Leong, Y. K. *Colloid. Surf. A,* 2008, 315, 136-146.

[18] Rai, P. K.; Denn, M. M.; Maldarelli, C. *Langmuir,* 2003, 19, 7370-7373.

[19] de Meijere, K.; Brezesinski, G.; Mohwald, H. *Macromol.,* 1997, 30, 2337-2342.

[20] Wasan, D.; Nikolov, A.; Kondiparty, K. *Curr. Opin. Colloid Interface Sci.,* 2011, 16, 344-349.

[21] Blake, T. D.; Haynes, J. M. *J. Colloid Interface Sci.,* 1969, 30, 421–423.

[22] Blake, T. D. *J. Colloid Interface Sci.,* 2006, 299, 1-13.

[23] Voinov, O. V. *Fluid Dyn.,* 1976, 11, 714-721.

[24] Cox, R. G. *J. Fluid Mech.,* 1986, 168, 169-194.

[25] Petrov, P.; Petrov, I. *Langmuir,* 1992, 8, 1762-1767.

[26] de Ruijter, M. J.; Blake, T. D.; De Coninck, J. *Langmuir,* 1999, 15, 7836-7847.

[27] Seveno, D.; Ledauphin, V.; Martic, G.; Voue, M.; De Coninck, J. *Langmuir,* 2009, 25, 13034-13044.

[28] Murshed, S. M. S.; Leong, K. C.; Yang, C. *Int. J. Thermal Sci.,* 2005, 44, 367-373.

[29] Hong, K. S.; Hong, T. K.; Yang, H. S. *Appl. Phys. Lett.,* 2006, 88, 031901-3.

[30] Radiom, M.; Yang, C.; Chan, W. K. *Nanoscale Research Lett.,* 2013, 8, 282-290.

[31] Hunter, R. J. Foundations of Colloid Science, Oxford University Press: USA, 2001.

[32] Schonhorn, H. *J. Chem. Phys.,* 1965, 43, 2041-2043.

[33] Becher, P. *J. Colloid Interface Sci.,* 1972, 38, 291-293.

[34] Vavruca, I. *J. Colloid Interface Sci.,* 1978, 63, 600-601.

[35] Murshed, S. M. S.; Tan, S. H.; Nguyen, N. T. *J. Phys. D,* 2008, 41, 085502-5.

[36] He, Y.; Jin, Y.; Chen, H.; Ding, Y.; Cang, D.; Lu, H. *Int. J. Heat Mass Tran.,* 2007, 50, 2272-2281.

[37] Phillips, R. J.; Armstrong, R. C.; Brown, R. A.; Graham, A. L.; Abbott, J. R. *Phys. Fluid. A,* 1992, 4, 30-40.

[38] de Ruijter, M. J.; De Coninck, J.; Oshanin, G. *Langmuir,* 1999, 15, 2209-2216.

[39] Naicker, P. K.; Cummings, P. T.; Zhang, H. Z.; Banfield, J. F. *J. Phys. Chem. B,* 2005, 109, 15243-15249.

[40] Rhee, S. K. *J. Mater. Sci.,* 1977, 12, 823-824.

[41] Yu, W.; Xie, H.; Chen, L.; Li, Y. *Thermochim. Acta,* 2009, 491, 92-96.

[42] Chen, H.; Ding, Y.; Tan, C. *New J. Phys.,* 2007, 9, 367.

[43] Kwak, K.; Kim, C. *Korea-Australia Rheology J.,* 2005, 17, 35–40.

[44] de Ruijter, M. J.; Charlot, M.; Voué, M.; De Coninck, J. *Langmuir,* 2000, 16, 2363-2368.

In: Nanofluids: Synthesis, Properties and Applications   ISBN: 978-1-63321-677-8
Editors: S.M. Sohel Murshed, C.A. Nieto de Castro   © 2014 Nova Science Publishers, Inc.

Chapter 7

CONVECTIVE HEAT TRANSFER OF NANOFLUIDS IN TUBES

J. P. Meyer[1], and C. C. Tang[2]*
[1]Department of Mechanical and Aeronautical Engineering,
University of Pretoria, Pretoria, South Africa
[2]Department of Mechanical Engineering,
University of North Dakota, Grand Forks, US

ABSTRACT

The purpose of this chapter is to experimentally investigate the heat transfer enhancement of two types of nanofluids during convective heat transfer in circular tubes. The nanofluids that were considered are carbon nanotubes and aluminium oxide. The enhancements during convective heat transfer of the nanofluids were investigated on two different experimental set-ups in two laboratories by two separate groups of researchers. The one group was at the University of Pretoria in South Africa and the other group at the University of North Dakota in the USA. Both experiments were conducted in smooth circular tubes at constant heat fluxes although the magnitude of the heat fluxes differed and different tube diameters were also used. The Pretoria group conducted experiments using mixtures of water with multi-walled carbon nanotubes (MWCNT's) at concentrations of 0.33%, 0.75%, and 1%. The average outside diameter of the MWCNT's was 10-20 nm with lengths of 10-30 µm. The test section was 1.5 m in length of which the last 1.0 m was heated and the first 500 mm was used for the flow to hydro dynamically fully developed. The internal diameter of the tube was 5.16 mm. It was operated at a constant heat flux of 13 kW/m^2. The North Dakota group conducted their experiments with mixtures of water and aluminium oxide at a concentration of 5.93%. The average diameter of the aluminium oxide particles was 50 nm. They used three different test sections with lengths of 460, 863 and 888 mm. Two tube inner diameters were 2.97 mm and one tube diameter was 4.45 mm. They operated their experiments at a constant heat flux of 18 kW/m^2 for D=4.45 mm, and 47 kW/m^2 for D=2.97 mm and at Reynolds numbers of 40 to 2000. It has been found that the effect of thermal entry for aluminium oxide nanofluid flow is more significant for lower L/D tube, which resulted in

---

* E-mail: josua.meyer@up.ac.za, Tel: 27 12 420 3104.

higher value of average Nusselt number. In laminar flow, the friction factors for both MWCNT (0.33% vol.) and aluminium oxide nanofluids are in good agreement; however the onset of transition observed from the friction factor appears to be influenced by the type of nanofluids and nanoparticle concentration. The comparison in heat transfer results is not as clear cut, as the effects of thermal entry, nanoparticle shape and concentration are at play. Nanofluid flow in test section with longer thermal entry region has higher Nusselt number due to the developing thermal boundary layer. And the type of nanofluids and nanoparticle concentration may be contributing to the development of thermal boundary layer.

1. INTRODUCTION

The physical hardware that makes technologies such as communication, electronic systems and high speed computing possible is usually relatively small in size and is becoming smaller and smaller. The result is that the cooling fluxes which is required by material limitations becomes more challenging as the surface areas for heat transfer decrease while the heat that needs to be removed by improved performance increases.

On a larger physical scale, mechanical systems used for power generation, air-conditioning, refrigeration, transportation, mining, manufacturing, chemical processing, etc., requires more efficient cooling and heating systems. The challenge for these systems is not only to increase the cooling and/or heating capacities but also to reduce cost by using less material which reduces the size and this can be done by decreasing the effective area for heat transfer.

The common problem in both small and large scale components and equipment is that the thermal management is becoming more challenging as the size of the components decreases. The use of fins and microchannels to enhance the heat transfer offer in many cases acceptable solutions but it seems to be inadequate for next generation technologies. Furthermore, conventional cooling and heating fluids such as air and water are becoming inadequate to achieve the heat transfer capacities of the future [1]. The reason is that they have relatively low thermal conductivities with air having a thermal conductivity of 0.025 W/mK and water 0.58 W/mK at 25°C.

Many studies, have investigated methods of improving the heat transfer conductivities of fluids and specifically water by using mixtures of micrometer-sized particles in base fluids such as water and oil. It has been found that although higher conductivities and therefore higher heat transfer capabilities are possible in practice problems are experienced with clogging, sedimentation and increased potential damage of systems due to abrasion [2].

Over the past 25 years, a lot of research has been conducted on colloidal dispersions of solid nanoparticles in liquid base fluids (nanofluids) as a new class of heat transfer fluid [3]. It has been shown that by adding nanoparticles, such as aluminium oxide, titanium dioxide or copper oxide to a fluid the thermal conductivity of the fluid can be enhanced [4-5]. The potential enhancements are high as the thermal conductivities of aluminium oxide (30 W/mK) and MWCNT (3000 W/mK) are approximately 50 and 5000 times higher than that of water.

In recent papers by the authors of this chapter from the laboratories of the University of Pretoria and the University of North Dakota the relevant literature [1, 6, 7] on previous work on the use of nanofluids in convective heat transfer in tubes were summarized and is therefore not discussed here. Another recent paper is by Elnajjar et al. [8] that is complementary to the work of Meyer et al. [6] in which they experimentally investigated the effect of MWCNT suspensions in developing channel flow. They have used a much smaller tube diameter than Meyer et al. and lower mass fractions.

The work of Meyer et al. [6] at the University of Pretoria focused on the measurement of average heat transfer and pressure drop characteristics in the transitional flow regime using different concentrations of MWCNT and they also took measurements in both the laminar and turbulent flow regimes, however, they also have local heat transfer coefficient data for developing flow. The work of Tang et al. [1, 7] at the University of North Dakota focused on using aluminium oxide for laminar thermal developing flow. When the work was conducted independently at the two labs it was not the intention to compare data and therefore the experimental set-ups, methodologies and operating conditions are different. However, comparisons are possible although in many cases not exact comparisons but trends. The purpose of this chapter is to make such comparisons where possible; in some cases the comparisons are done qualitatively and not quantitatively.

2. EXPERIMENTAL SETUP

2.1. Nanofluids Preparation

2.1.1. MWCNT–Water Nanofluid

Three different volume concentrations, measured according to the volume of the base fluid, of MWCNT–water nanofluids were prepared. The MWCNTs have an outside diameter of 10–20 nm, an inside diameter of 3–5 nm and a length of 10–30 μm. The three different volume concentrations were 0.33%, 0.75% and 1.0%, which were dispersed into 10 litres of distilled water. In order to stabilise the three different mixtures, gum arabic (GA) powder was dissolved and added into the distilled water first. Garg et al. [9] used 0.25 wt% GA with 1 wt% MWCNT, or a 1:4 ratio. A similar approach was used here. Garg et al. [9] sonicated various nanofluid samples for different time lengths. They found that the optimum sonication time for a 1 wt% MWCNT and 0.25 wt% GA mixture was 40 min using a 130 W, 20 kHz ultrasonicator. In the current study, the nanofluids were sonicated using an ultrasonicator that had an operating frequency of 24 kHz and a maximum power output of 200 W. The sonication times for the current nanofluids were adjusted to match the optimum sonication time of Garg et al. [9]. Hence the sonication times for the volume concentrations of 0.33%, 0.75% and 1.0% were 30, 80 and 120 min respectively.

For a stable nanoparticle-dispersion, the pH is a key parameter, which is related to the electrostatic charge on the particles' surface and is known as the zeta potential. At the iso-electric point, which is the point where the nanoparticle carries no net electrical charge, the nanoparticles will form agglomerations since there are no sufficient repulsive forces between the nanoparticles. As the pH changes from the iso-electric point, the absolute value of the zeta potential of the nanoparticle surface increases so that agglomeration and collisions between nanoparticles caused by Brownian motion are prevented [10]. Shown in Figure 1 is the measured pH of the distilled water compared with that of the MWCNT–water nanofluid and GA–water mixture. The pH of the distilled water is 7.1 and that of the GA–water mixture for volume concentrations of 0.33%, 0.75% and 1.0% is 6.6, 6.4 and 7.4 respectively.

The MWCNT–water nanofluids follow a similar trend as that of the GA–water mixture except at a concentration of 1.0 vol%, the suspension is not acidic. At volume concentration of 0.33%, 0.75% and 1.0% the pH is 8.0, 7.7 and 7.1 respectively, Xie et al. [11] found that the iso-electric point of their MWCNT suspension was at a pH of 7.3. The 0.33 vol% and 0.75 vol% nanofluid concentrations remained stable for 3 days whereas the 1.0 vol% only remained stable for around 24 h due to the large possibility of its nanoparticles forming agglomerations and settling out of the suspension.

2.1.2. Al$_2$O$_3$–Water Nanofluid

The Al$_2$O$_3$–water nanofluid used in this study is a colloidal dispersion of aluminium oxide, or alumina (Al$_2$O$_3$), at 20% by weight (6 vol%) in liquid water. The nanofluid was purchased from Alfa Aesar®, with Al$_2$O$_3$ particle size of 50 nm specified by the manufacturer. Nanofluids are often described in terms of volume fraction, ϕ, thus the density, weight and volume fractions can be related as

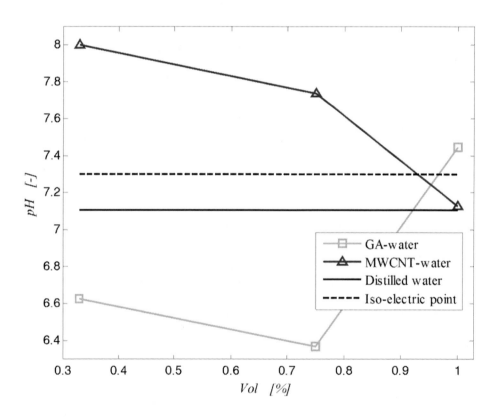

Figure 1. Measured pH of the MWCNT–water nanofluid compared with that of GA–water mixture and distilled water.

$$\rho_{nf} = (1-\phi)\cdot\rho_{bf} + \phi\cdot\rho_p \tag{1}$$

and

$$\phi = \frac{\chi \cdot \rho_{bf}}{\chi \cdot \rho_{bf} + (1 - \chi) \cdot \rho_p} \tag{2}$$

The bulk density of Al_2O_3 is 3965 kg/m^3, as specified by the manufacturer, and compares within ±1% of the value reported in Ref. [12]. The bulk density of the Al_2O_3–water nanofluid is also measured using a digital balance with a readability of 0.01 g. When comparing the measured nanofluid density with the density determined from Eq. (1) and Eq. (2), where ρ_p=3965 kg/m^3 and ρ_{bf}=1000 kg/m^3, the agreement is within ±1.5%.

The specific heat of the nanofluids, $c_{p,nf}$, is determined using the following expression:

$$\rho_{nf} \cdot c_{p,nf} = (1 - \phi) \cdot \rho_{bf} \cdot c_{p,bf} + \phi \cdot \rho_p \cdot c_{p,p} \tag{3}$$

The value of the specific heat for Al_2O_3, $c_{p,p}$, used in this study is 820 J/kg·K, which is adopted from the values reported in Ref. [12]. The specific heat for the base fluid, $c_{p,bf}$, which is liquid water, is 4180 J/kg·K.

2.2. Effective Thermal Conductivity of the Nanofluids

2.2.1. Effective Thermal Conductivity of MWCNT–Water Nanofluid

The thermal conductivity was measured using a KD2 thermal property meter (Decagon Devices), which is based on the transient line heat source method. The KD2 thermal property meter had a measurement uncertainty of ±5% and was calibrated by using distilled water before any set of measurements were taken. The results fall to within 5% of the predicted theory by Popiel and Wojtkowiak [13].

Shown in Figure 2 is the measured thermal conductivity ratio of the 0.33, 0.75 and 1.0 vol% MWCNT–water nanofluids for a temperature range of 20–40°C. The effective thermal conductivity increases with increasing temperature and MWCNT concentration.

For volume concentrations of 0.33%, 0.75% and 1.0%, the thermal conductivities of the MWCNT–water nanofluids were 2%, 3.3% and 8% respectively greater than that of the distilled water.

The results show that the increase in thermal conductivity falls to within the measurement uncertainty of the device. The reason is that the thermal conductivity meter is very sensitive to natural convection. At 20°C the data follows almost a straight line. This is due to that fact that natural convection in the sample is low. The scatter in the data increases with increasing temperature.

Xie et al. [11] reported a 7% increase in thermal conductivity for MWCNT suspended in water at a volume concentration of 1.0%, which compares very well with the measured 8%.

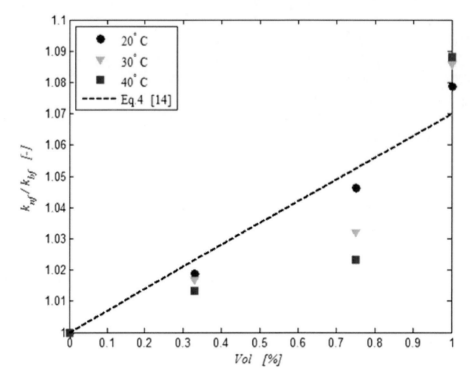

Figure 2. Measured relative thermal conductivity of the MWCNT–water nanofluid as a function of the volume concentration compared to Prasher et al. [14] (Note: At a concentration of 0%, all measurement points are lying on top of each other).

Prasher et al. [14] showed that the thermal conductivity can increase due to agglomeration of the nanoparticles when compared with a well-dispersed system, which is a possible reason for the large increase in the thermal conductivity of the 1.0 vol% MWCNT–water nanofluid. Keblinski et al. [15] also showed that particle clustering, due to Brownian motion, could enhance the thermal conductivity since the particles are much closer together and thus enhance the constant phonon heat transfer.

Prasher et al. [14] showed that the thermal conductivity can be written as follows:

$$\frac{k_{nf}}{k_{bf}} = 1 + C_k \cdot \phi \quad (4)$$

where C_k is a constant depending on the experimental data. In this case, $C_k = 7$ and is shown in Figure 2 as the dotted black line. On average, Eq. (4) predicts the thermal conductivity to within 1.8% for all three volume concentrations.

2.2.2. Effective Thermal Conductivity Al$_2$O$_3$–Water Nanofluid

The thermal conductivity of the Al$_2$O$_3$–water nanofluid was also measured using a Decagon Devices KD2 Pro thermal property analyzer with an accuracy of ±5%. Sample of nanofluid in a vial was inserted in a temperature bath, so that the thermal conductivity of the nanofluid was measured for temperatures between 6 and 55°C.

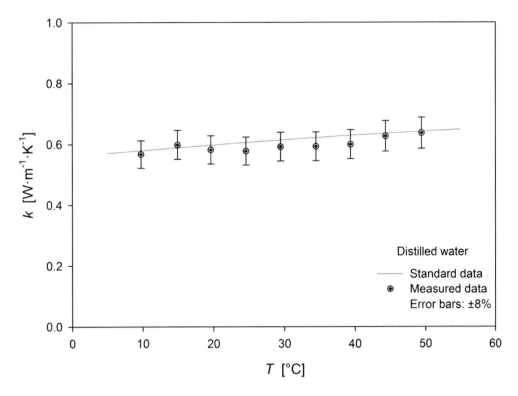

Figure 3. Comparison of measured thermal conductivity of distilled water with standard data from the literature [12] and [16].

Before conducting the measurements for the thermal conductivity of the Al$_2$O$_3$–water nanofluid, the thermal conductivities of distilled water at different temperatures (between 9 and 50°C) were measured.

This process was to verify that the measurement procedure used in this study could correctly measure the thermal conductivity of fluid media. Figure 3 shows the comparison of the measured thermal conductivities of distilled water with the standard data available in the literature [12] and [16].

All the measured thermal conductivities of distilled water are within ±8% agreement with the standard data. The ±8% error bars shown in Figure 3 were determined from the uncertainties due to the accuracy of the instrument and the precision errors from repeated measurements at 95% confidence level. Since the measured thermal conductivities of distilled water compare well with the standard data (within expected deviation), the measurement procedure for thermal conductivity was verified. The same measurement procedure could be implemented for measuring the thermal conductivity of nanofluid.

The thermal conductivities of Al$_2$O$_3$–water (6 vol%) nanofluid were measured at different temperatures, between 7 and 55°C. Figure 4 shows the measured thermal conductivities of the nanofluid with the variation of temperature. The thermal conductivity ratio (k_{nf}/k_{bf}) shows that the thermal conductivity of the nanofluid is about 17% higher than the thermal conductivity of water at 7°C.

The thermal conductivity ratio decreases with increasing temperature, where at 54°C the thermal conductivity of the nanofluid is about 8% higher than the thermal conductivity of water.

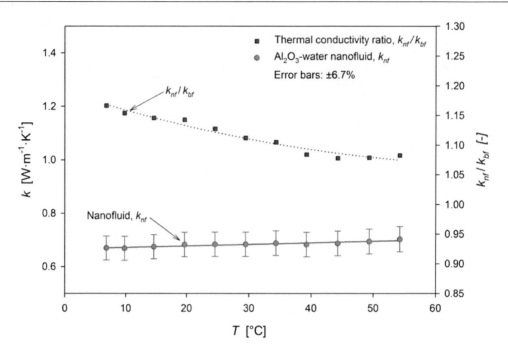

Figure 4. Thermal conductivity of the Al$_2$O$_3$–water (6 vol%) nanofluid with variation of temperature.

2.3. Effective Viscosity of the Nanofluids

2.3.1. Effective Viscosity of MWCNT–Water Nanofluid

When plotting the results on a viscosity versus temperature graph the viscosity of nanofluids increases with increasing MWCNT concentration and decreasing temperature. Figure 5 shows the measured viscosity ratios as a function of the volume concentrations of the MWCNT–water nanofluids. When plotting the results in this fashion the different temperature data points collapse onto one point and the viscosity-ratio is then only dependent on the volume concentration.

The viscosities were measured with a double concentric cylinder viscosity meter. There is a linear increase from a volume concentration of 0% up to 0.75% and then suddenly a steep increase to 1.0%. A possible reason for the large increase in viscosity is that the nanoparticles have agglomerated due to the pH of the solution being too close to the iso-electric point [10].

Einstein, in 1906, developed a correlation for the viscosity of dilute suspensions (<5 vol%) for small and rigid spherical particles [17]:

$$\frac{\mu_{nf}}{\mu_{bf}} = 1 + 2.5 \cdot \phi \tag{5}$$

Eq. (5) is shown as the solid line on Figure 5. Eq. (5) under predicts the data for the 0.33 vol% MWCNT–water nanofluid concentration by 15.9%, for the 0.75 vol% MWCNT–water nanofluid concentration by 29.7% and for the 1.0 vol% MWCNT–water nanofluid concentration by 77.7%.

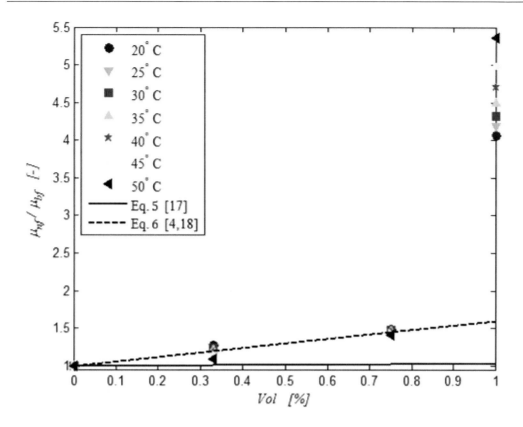

Figure 5. Relative viscosity of the MWCNT–water nanofluid as a function of the volume concentration compared to theory (Note: At a concentration of 0%, all measurement points are lying on top of each other).

The reason for the large under prediction by Eq. (5) is that the correlation was developed for spherical particles. Eq. (5) can be extended to include ellipsoidal particles [18] and [19]:

$$\frac{\mu_{nf}}{\mu_{bf}} = 1 + C_\mu \cdot \phi \tag{6}$$

where C_μ depends on the ratio of the revolution ellipsoid axes and is equal to 2.5 for spherical particles and not tubes with an aspect ratio of 1333. In this case, C_μ equal to 60 correlates the volume concentrations of 0.33% and 0.75% well. At a volume concentration of 0.33%, Eq. (6) over predicts the viscosity by 0.24%, at a volume concentration of 0.75%, the viscosity is under predicted by 0.17% and at a volume concentration of 1.0%, the viscosity is under predicted by 65.1%.

2.3.2. Effective Viscosity of Al$_2$O$_3$–Water Nanofluid

The dynamic viscosity of the Al$_2$O$_3$–water nanofluid was measured by a Brookfield viscometer with an accuracy of ±1%. The nanofluid dynamic viscosity was measured for temperature between 6 and 75°C. The temperature of the nanofluid was maintained using a

Brookfield temperature bath. In addition, the effect of shear rate on the shear stress of the nanofluid is measured to determine whether its rheological behavior is Newtonian or non-Newtonian.

Figure 6 shows the influence of temperature on the viscosity of nanofluid in comparison with the viscosity of water as the base fluid. At 6°C, the nanofluid viscosity is at 12.3 cP. As the temperature increases to 75°C, the nanofluid viscosity decreases to 3.45 cP, representing a 72% decrease in viscosity over the temperature range from 6 to 75°C. For the same temperature range, the nanofluid viscosity is 8.3 to 9.2 times more viscous than water.

The rheological behaviour of the Al$_2$O$_3$–water nanofluid, whether Newtonian or non-Newtonian, is determined by subjecting the mixture to shear rate ranging from 6 to 122 s^{-1}. Figure 7 shows the shear stress results of the nanofluid measured at different temperatures with varying shear rate. The results indicate that the Al2O3–water nanofluid behaves as a Newtonian fluid (within the measurement conditions), with the shear stress being proportional to the shear rate. As the temperature increases from 6 to 75°C, the proportionality constant of the shear stress versus shear rate curve (or viscosity) decreases from 12.3 to 3.45 cP.

2.4. Experimental Setups

2.4.1. Experimental Setup for MWCNT–Water Nanofluid

Shown in Figure 8 was the experimental setup used. An electronically controlled magnetic gear pump (Item 1) was used to pump the working fluid at a stable flow rate. A bypass loop, which was controlled via a gate valve (Item 11), was added to control the flow rate. In order to protect the equipment, a pressure relief valve (Item 2), which was rated at 350 kPa, and a non-return valve (Item 3) were incorporated into the system.

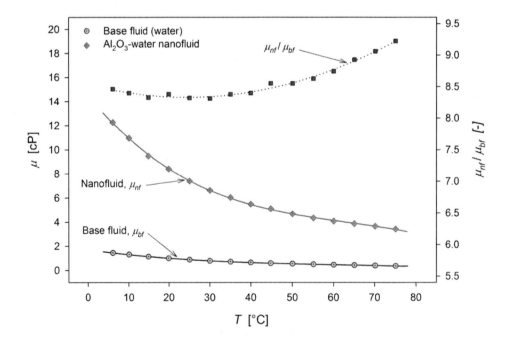

Figure 6. Viscosity of the Al$_2$O$_3$–water nanofluid with variation of temperature.

Convective Heat Transfer of Nanofluids in Tubes 165

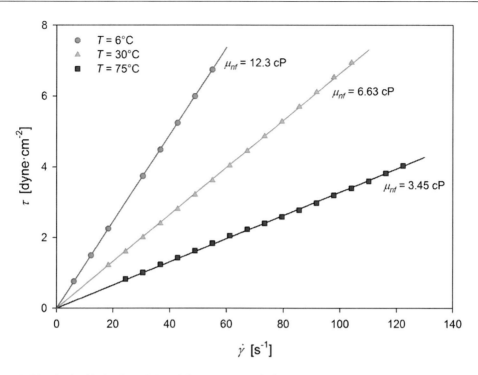

Figure 7. Rheological behavior of the Al$_2$O$_3$–water nanofluids.

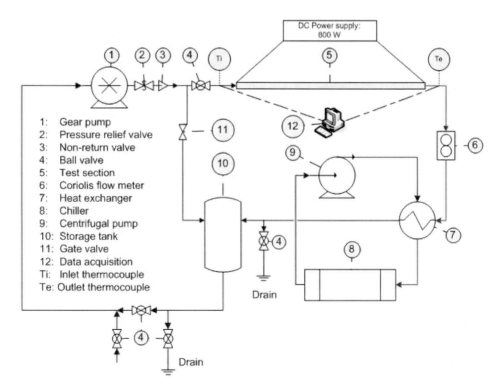

Figure 8. Schematic representation of the experimental setup for MWCNT–water measurements at the University of Pretoria.

The inlet and exit thermocouples (Item Ti and Te respectively) were attached to the system to measure the inlet and outlet temperatures, rather than the test section, since this will avoid any influence that the test section has on the readings. The test section (Item 5) is connected to the system via rubber hosing, so as to prevent axial conduction. The flow rate through the test section was measured via a Coriolis flow meter (Item 6), which was placed after the test section. After the flow meter, the fluid is cooled back to the inlet temperature (≈20°C) via the cooling loop, which consists of a simple tube-in-tube heat exchanger (Item 7), chiller (Item 8) and a centrifugal pump (Item 9), before it is pumped back to the 22 litres storage tank (Item 10). Just after and before the storage there are three ball valves (Item 4) which allow the system to be flushed and cleaned easily. The system and storage tank were kept small to ensure that the volume MWCNT (which is very expensive) used was kept to a minimum.

Shown in Figure 9 is the test section used in the current study. The test section consisted of a 500 mm hydrodynamic entry section, a heat transfer test section and a mixer, and all three sections were separated from each other by acetyl bushes. The acetyl bushes were used to thermally insulate the mixer and inlet sections from the heat transfer test section. The heat transfer test section consisted of a straight copper tube of length 1 m. It had an internal diameter of 5.16 mm and an external diameter of 6.44 mm. The test section was insulated with 60 mm thick insulation. The heat transfer test section was heated via a Constantine wire at 212 W (13,000 W·m^{-2}) with a DC power supply at 200 V and a current of 1.06 A. To measure the wall temperatures, 13 T-type thermocouples were spaced evenly along the test section wall. This was done by drilling a small pilot hole into the test section and securing the thermocouple with a drop of solder. Two T-type thermocouples were inserted into the fluid, before the developing length and after the mixer, in order to measure the inlet and outlet temperatures. The thermocouples were calibrated to an accuracy of 0.1°C. The pressure drops were measured with a pressure transducer with an accuracy of 0.08% at full scale, which was 17 kPa. The pressure range of the experiments was from 155 to 8.2 kPa. The pressure transducer was connected to the pressure taps shown in Figure 9. The pressure tap diameter was 4 mm, which is less than 10% of the tube diameter.

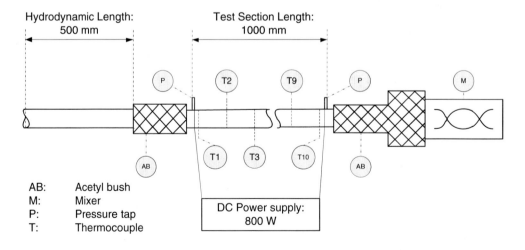

Figure 9. Schematic representation of the test section for MWCNT–water measurements.

The reason for the small diameter is so that the overall flow is not disturbed since a large hole could lead to a localised eddy forming which results in an error in pressure readings [20] and [21]. The flow rate was measured using a Coriolis flow meter, which had an accuracy of 0.05% at full scale.

2.4.2. Experimental Setup for Al_2O_3–Water Nanofluid

The experimental setup used in this study was designed for measuring heat transfer and pressure drop for internal flow in tubes of various diameters. The schematic of the entire experimental setup for the present study is shown in Figure 10.

The working fluid is stored in an 11-litre medium-density polyethylene reservoir. The working fluid is then pumped from the reservoir, by a Liquiflo gear pump, through a shell-and-tube heat exchanger. The purpose of the heat exchanger is to remove heat added to the working fluid during an experiment run. Also, it functions to maintain a consistent inlet temperature during the course of an experiment run.

From the heat exchanger, the working fluid flows through a metering valve, which is used for regulating its flow rate. The working fluid then flows through a Micro Motion Coriolis mass flow meter that is connected to a digital transmitter. The Coriolis mass flow meter has ±0.1% accuracy for 1–50 g/s.

The flow rate signal of the working fluid is conditioned by the digital transmitter for the data acquisition system. From the Coriolis mass flow meter, the working fluid then flows into the test section.

Pressure drop across the test section was measured by three Rosemount pressure transmitters that correspond to different pressure ranges (62, 248, and 2070 kPa, with ±0.15% accuracy). The accuracy of the pressure transmitters were verified using Dwyer digital pressure gages.

The heat transfer test section assembly contains the instruments necessary for measuring the bulk inlet and outlet fluid temperatures, the surface temperatures, and pressure drop across the test section.

The test section assembly was designed to incorporate tubes with various diameters. In this study, AISI 304 stainless steel tubes with inner diameters of 2.97 and 4.45 mm were used as heat transfer test sections. Six small segments for each tube size were cut so that the inner diameters can be verified with a laser scanning microscope (Zeiss LSM 5 Pascal). Analysis from laser scanning microscopy found that, using Student's t-distribution at 95% confidence level, the uncertainty of both inner diameters is about ±1%.

The bulk inlet and outlet fluid temperatures were measured with T-type thermocouples inserted at the upstream and downstream of the test section. The surface temperatures of each stainless tube were measured using T-type thermocouples, with wire diameter of 0.127 mm. The thermocouples are cemented on to the outer surface along the axial locations with thermally conductive epoxy. The epoxy has a thermal conductivity of 1.04 $W \cdot m^{-1} \cdot K^{-1}$ and a high electrical resistivity of 1×10^{15} $\Omega \cdot m$.

Test sections with inner diameters of 2.97 and 4.45 mm and three different heated lengths were used in this study, which gives three different length-to-diameter, L/D, ratios. The axial locations for the placement of the thermocouples on each test section are illustrated in Figure 11.

Two copper connectors are silver-soldered on both ends of each test section, and the length between the copper connectors is the heated length of the test section.

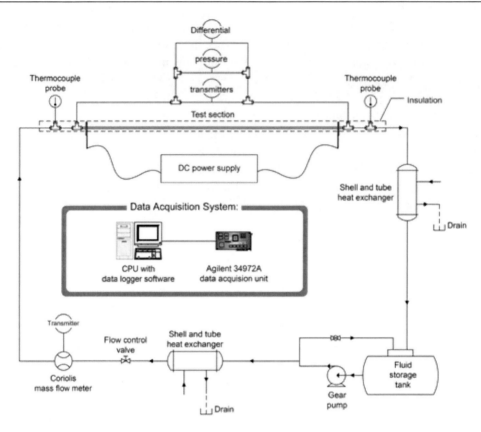

Figure 10. Schematic of the experimental setup for Al$_2$O$_3$–water measurements at the University of North Dakota.

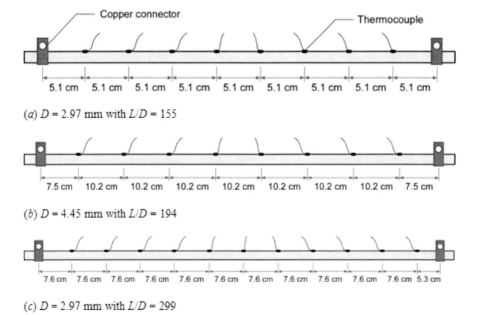

Figure 11. Heat transfer test sections with axial thermocouple locations.

The copper connectors are connected to an Agilent DC power supply, which heats the test section with uniform heat flux.

A remote sense circuit is also connected to the copper connectors with the DC power supply, which uses its remote sense capability to compensate voltage drop due to resistance in the wires.

Output signals from the mass flow meter, pressure transmitters, thermocouples, and DC power supply were recorded by an Agilent data acquisition unit. The convection heat transfer coefficients on the inside surface of the test section were determined from the measured data using the data reduction procedure described in the following section.

2.5. Data Reduction

The local convective heat transfer coefficient, $h(x)$, was calculated as:

$$h(x) = \frac{\dot{q}_{in}}{T_{si}(x) - T_m(x)}$$

$$(7)$$

The heat flux, $\dot{q}_{in}$, was determined from the electrical energy input, $\dot{q}_{in} = V \cdot I$, and the inner surface area, $A_{si} = \pi \cdot D_{si} \cdot L$.

For the MWCNT nanofluids measurements, the electrical energy input remained constant at 212 W throughout the measurements thus resulting in a constant heat flux (≈ 13 kW/m^2).

The inner local surface temperature, $T_{si}(x)$, was determined from the measurements of the outside wall temperature, $T_{so}(x)$, and the heat transfer and resistance through the tube wall:

$$T_{si}(x) = T_{so}(x) - \dot{Q}_{in} \cdot R_w$$

$$(8)$$

where

$$R_w = \frac{\ln(D_{so}/D_{si})}{2 \cdot \pi \cdot k_{Cu} \cdot L}$$

$$(9)$$

The thermal conductivity of the copper was obtained from Abu-Eishah [22] and is given by

$$k_{Cu} = a \cdot T_{Cu}^b \cdot \exp(c T_{Cu} + d/T_{Cu})$$

$$(10)$$

where the constants $a = 82.56648$, $b = 0.262301$, $c = -4.06701 \times 10^{-4}$ and $d = 59.72934$. The mean temperature of the cooper, T_{Cu}, in Eq. (10) is an absolute temperature in the unit of Kelvin. The local heat transfer coefficient was determined from the local mean temperature. The local mean temperature was determined by [23]:

$$T_m(x) = T_i + \frac{\dot{q}_{in} \cdot x \cdot P}{\dot{m} \cdot c_p}$$

(11)

To obtain the average heat transfer coefficient, Eq. (7) was averaged along the length of the tube at different measuring points, n.

This was done by taking the mean of all the local heat transfer coefficients.

$$h_{avg} = \frac{h(x_1) + \cdots + h(x_n)}{n}$$

(12)

The average Nusselt number can then be calculated as follows:

$$Nu_{avg} = \frac{h_{avg} \cdot D}{k}$$

(13)

where k is the thermal conductivity of the working fluid, determined at the bulk fluid temperature, $T_b = (T_i + T_e)/2$. For water the thermal conductivity was determined from the equations developed by Popiel and Wojtkowaik [13].

The testing fluids average Reynolds number and Prandtl number,

$$Re = \frac{4 \cdot \dot{m}}{\pi \cdot D \cdot \mu}$$

(14)

$$Pr = \frac{\mu \cdot c_p}{k}$$

(15)

were calculated based on the viscosity, thermal conductivity and specific heat determined at the bulk fluid temperature, T_b, using the equations developed by Popiel and Wojtkowaik [13] for water and for the MWCNT nanofluid the measured values of its properties were used.

The pressure loss equation is used to calculate the friction factor:

$$f = \frac{\Delta P \cdot D \cdot 2}{L \cdot \rho \cdot V_{avg}^2}$$

(16)

which was then simplified to:

$$f = \frac{\Delta P \cdot \rho \cdot \pi^2 \cdot D^5}{8 \cdot L \cdot \dot{m}^2}$$

(17)

The pressure drop was determined from the pressure drop measurements of the transducer and the mass flow rate was determined from the readings of the Coriolis mass flow meter.

The measured heat transfer of the testing fluid was compared with that of the electrical input energy, q_{in}, which was supplied by the constantan heating wire, by means of an energy balance:

$$EB = \frac{\dot{Q}_{in} - \dot{Q}_{tf}}{(\dot{Q}_{in} + \dot{Q}_{tf})/2} \cdot 100 \qquad (18)$$

and the heat transfer to the testing fluid, Q_{tf}, is

$$\dot{Q}_{tf} = \dot{m} \cdot c_p \cdot (T_e - T_i) \qquad (19)$$

Although energy balances of around 3% were obtained, the input energy was used for the calculations since it was the more accurate of the two.

2.6. Experimental Uncertainty

All experimental uncertainties were calculated within the 95% confidence level using the method of Moffat [24]. Table 1 lists the instruments used in the study with their uncertainties. The uncertainties of the measurements are shown in Table 2 for the low Reynolds number and highest Reynolds number.

Table 1. Ranges and accuracies of instruments

Instrument	Range	Uncertainty
MWCNT–water experimental study		
Thermocouple	−200 to 500°C	0.1°C
Coriolis flow meter	0 to 0.07 kg/s	0.1%
Pressure transducer	0 to 17 kPa	0.16%
Power supply	0 to 320 V	0.33 V
	0 to 12.5 A	0.04 A
Al$_2$O$_3$–water experimental study		
Thermocouple	−200 to 500°C	0.5°C
Coriolis flow meter	0 to 0.05 kg/s	0.1%
Pressure transducer	0 to 2070 kPa	0.15%
Power supply	0 to 40 V	0.08 V
	0 to 38 A	0.152 A

Table 2. Uncertainties of experimental parameters

Measured parameter	Uncertainty
MWCNT–water experimental study	
$\dot{q}_{in}$	3.49%
f	2.0 to 18%
h	1.35%
ΔP	0.3 to 17.5%
Nu	2.45 to 3.19%
Re	1.09 to 2.26%
Al₂O₃–water experimental study	
$\dot{q}_{in}$	3.59%
f	5.5 to 8.5%
h	4.7 to 13.9%
ΔP	0.9 to 7.1%
Nu	6.9 to 14.8%
Re	1.7 to 3.9%

For the MWCNT–water experimental study, the uncertainty of the heat transfer coefficient is between 1.4% for the lowest Reynolds number ($\approx$1000) and up to 2.5% for the highest Reynolds number ($\approx$8000) tested. The friction factor has an uncertainty of 18% for the lowest Reynolds number and goes down to 2% for the highest. The lowest pressure drop recorded is 155 Pa hence the uncertainty of 18% at the low Reynolds number.

For the Al₂O₃–water experimental study, the uncertainty of the heat transfer coefficient is between 4.7% for the lowest Reynolds number ($\approx$50) and up to 13.9% for the highest Reynolds number ($\approx$2000) measured. The friction factor has an uncertainty of 8.5% for the lowest Reynolds number and goes down to 5.5% for the highest. The uncertainty of the tube inside diameter contributed significantly to the uncertainty of the friction factor. This is due to the friction factor being expressed with the inner diameter, D, being raised to the fifth power, see Eq. (17).

2.7. Experimental Procedure

After the start-up of the system, it was necessary to let the system settle for at least two hours in order to reach steady-state conditions. This was due to the thermal inertia of the system being relatively slow before it got to steady-state temperatures and mass flow rates. Once the system was steady, small changes were made in the flow rates in order to achieve the desired mass flow rates for data capturing.

Steady-state conditions were monitored visually, in that there were no observable changes in the temperatures, pressure drops, mass flow rates and energy balances. In the transition region, steady-state conditions were difficult to achieve due to the continuous fluctuations of temperatures, pressure drops and energy balances. But once the fluctuations repeated themselves periodically, measurements were taken. After the change in the mass flow rate from a high flow rate to low flow rate, it took approximately 5–10 min for steady-state conditions to be met. The reason for decreasing the flow rate was to ensure that very little residual heat was stored in the insulation, which has an effect on the next reading. On average, an energy balance of 3% was achieved before the data was captured.

In the experimental study for the MWNCT–water nanofluid, each measured data point consists of 200 readings at a rate of 20 Hz that were captured with a data acquisition system and then averaged for data reduction. In the experimental study for the Al_2O_3–water nanofluid, each measurement run was sampled at a rate of 10 Hz, for duration of 1 minute. The values recorded by the data acquisition system were subsequently analysed using the data reduction procedure discussed previously.

2.8. Validation of Experimental Setup for MWNCT–Water Nanofluid

The friction factors and heat transfer coefficients were validated by taking measurements in the laminar and turbulent flow regimes and comparing these with published data. Adiabatic friction factors were used to validate the experiments since they disregard any influence of heat transfer on the properties. Diabatic friction factors were also measured and compared with published data. Only water was used to validate the experimental set-up as no other data exist for MWCNT–water nanofluids in a similar experimental set-up.

2.8.1. Adiabatic Friction Factors of Water Flow

The friction factor data consisted of 82 data points spanning a Reynolds number range of 1000–8000. Measurements were taken without any heat transfer to eliminate any varying density and viscosity effects.

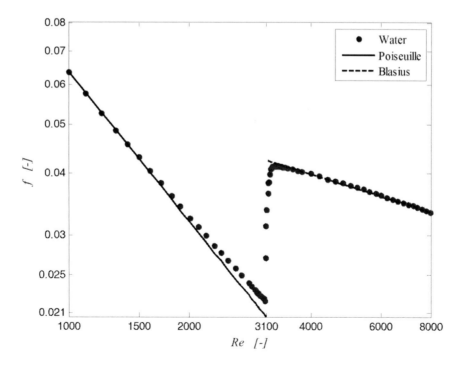

Figure 12. Validation of the water adiabatic friction factor results as a function of Reynolds number.

Figure 12 shows the adiabatic friction factor data as compared with Poiseuille flow ($f = 64/Re$) and the Blasius equation ($f = 0.3164 \cdot Re^{-0.25}$).

Comparing the laminar results with Poiseuille flow, the data is under predicted on average by 3.3% with a maximum deviation of 2.8%. For the turbulent data, the Blasius correlation under predicts the data, on average, by 0.2% with a maximum deviation of 0.5%. Generally, the friction factors in the laminar and turbulent flow regimes compare well with the Poiseuille and Blasius equations.

When considering the transitional flow regime, transition to turbulent flow appears to start at a Reynolds number of approximately 3100 rather than the conventional value of 2300 and the transitional flow range is also very short, around 100 Reynolds number long. The delayed transition and the sharp transition is the result of inlet effects as was shown by Olivier and Meyer [20] and by Meyer and Olivier [21].

2.8.2. Diabatic Friction Factors of Water Flow

Figure 13 shows the experimental data for the diabatic friction factor as a function of the Reynolds number compared with the laminar Poiseuille equation, the Blasius equation and corrected Blasius equation as suggested by Allen and Eckert [25]. Transition starts at a Reynolds number of approximately 2900 and ends at a Reynolds number of approximately 3600. The diabatic friction factors are lower than the adiabatic friction factors in laminar and turbulent flow. Allen and Eckert [25] proposed a viscosity correction factor, $(\mu_b/\mu_w)^{-0.25}$, to be multiplied with the Blasius equation. The turbulent results correlate fairly well with the corrected Blasius equation (on average, the data is under predicted by 0.5% with a maximum deviation of 1.9%). It should be noted that the correction factor is very close to unity (on average, it is 0.96) and the experimental data correlates with the Blasius equation, on average, by 1.9% with a maximum deviation of 2.1%.

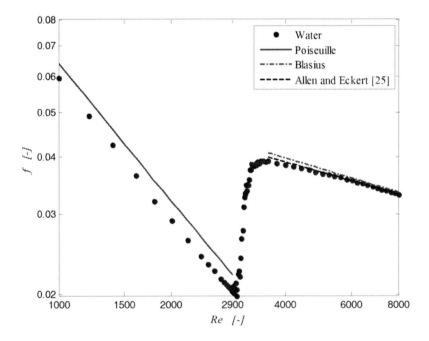

Figure 13. Water diabatic friction factors results as a function of the Reynolds number.

In the laminar flow regime, the same phenomenon is observed as in the turbulent flow regime. Here the experimental data is under predicted by an average of 8.5% (however, it is within the uncertainties of 18% at low Reynolds numbers) with a maximum deviation of 2.2%. According to Shome and Jensen [26] and Tam and Ghajar [27], secondary flow effects increase the friction factor, especially in tubes with uniform heat flux boundary conditions. Tam and Ghajar [27] discovered that by increasing the overall heat flux, the laminar friction factors increased. The reason for this is that the wall-to-bulk temperature difference exists throughout the length of the tube [26].

Metais and Eckert [28] recommend the use of a flow map to distinguish between mixed and forced convection regimes. The map is based on the Reynolds number being a function of the Rayleigh number. Their flow map is for a constant wall temperature boundary condition and hence cannot be used to compare the current data. Ghajar and Tam [29] developed a new flow map for a constant heat flux boundary condition. They used the data for three different inlets to produce a new boundary between laminar, transition and turbulent for both forced and mixed convection. Figure 14 shows the flow map developed by Ghajar and Tam [29] with the current data plotted. From the map, it can be concluded that experimental laminar values for this study are well within the laminar forced convection boundary and secondary flow effects are not present. Hence the reason for the drop in friction factor is the result of the reduction of liquid viscosity in the near-wall region due to heating [26], which seems to be the dominating effect.

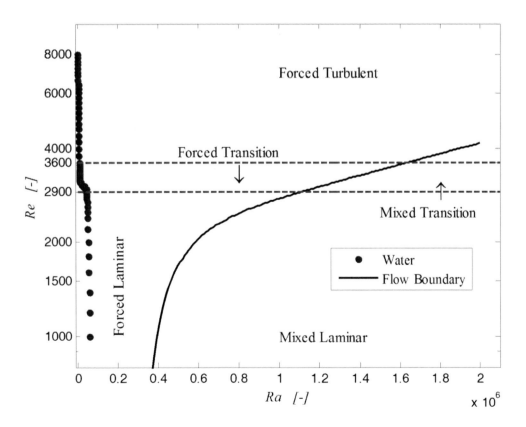

Figure 14. Water laminar-turbulent heat transfer results on the flow regime map of Ghajar and Tam [29].

2.8.3. Nusselt Number of Water Flow

Heat transfer results were compared to the correlations developed by Ghajar and Tam [30] for laminar and turbulent flow, which are shown below:

$$Nu_{lam} = 1.24 \cdot [Gz + 0.025 \cdot (Gr \cdot Pr)^{0.75}]^{1/3} \cdot (\mu_b / \mu_w)^{0.14} \quad (20)$$

$$Nu_{turb} = 0.023 \cdot Re^{0.8} \cdot Pr^{0.385} \cdot (L/D)^{-0.0054} \cdot (\mu_b / \mu_w)^{0.14} \quad (21)$$

For the transitional flow regime, the correlation developed by Ghajar and Tam [30] was modified to account for a developing length inlet condition. The modified correlation is shown below:

$$Nu_{trans} = Nu_{lam} + \left[\exp\left(\frac{Re_{trans} - Re}{36} \right) + Nu_{turb}^{-0.935} \right]^{-0.935} \quad (22)$$

The results are shown in Figure 15. In the laminar flow regime, the results are under predicted on average, by 5.2%, in the turbulent flow regime, on average, by 2.4% and in the transitional flow regime, on average, by 5.3%.

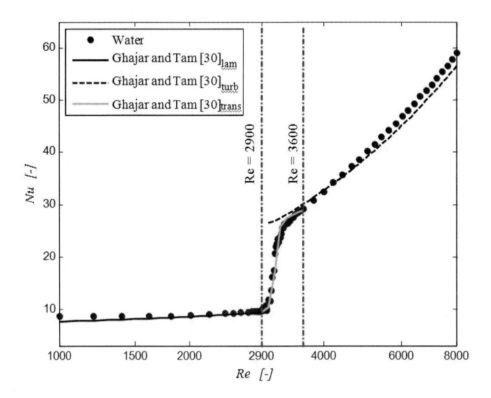

Figure 15. Smooth tube heat transfer results compared with the work of Ghajar and Tam [30] for water.

Convective Heat Transfer of Nanofluids in Tubes 177

Hence there was a good agreement between the equations by Ghajar and Tam [30] and the measurements over the Reynolds number range.

In general, the friction factors and Nusselt numbers in the laminar and turbulent flow regimes compare well with literature. This adds confidence in the measurement technique and further validates the experimental set-up and data reduction methodology.

2.9. Validation of Experimental Setup for Al$_2$O$_3$–Water Nanofluid

Before conducting the measurements for Al$_2$O$_3$–water nanofluid flow, the reliability of the experimental setup and procedures were checked and validated by making several water flow runs. This process is to verify that the experimental setup is capable of measuring the viscous flow and heat transfer parameters that are comparable with established correlations. The validation involved measuring the adiabatic friction factors and the heat transfer coefficients of water flow in the laminar and turbulent regimes.

2.9.1. Adiabatic Friction Factors of Water Flow

The adiabatic friction factor measurements for water flow were conducted for a Reynolds number range of approximately 480 to 8800. Measurements for both inner tube diameters of 2.97 and 4.45 mm were conducted.

The water flow friction factor results that were obtained experimentally are compared with the friction factors calculated from the Blasius equation ($f = 0.3164 \cdot Re^{-0.25}$) for turbulent flow. The Churchill [31] friction factor equation was also used for comparing the water flow results from the laminar to turbulent flow regime:

$$f = 8 \left[\left(\frac{8}{Re} \right)^{12} + \frac{1}{(A_1 + B_1)^{1.5}} \right]^{1/12}$$

(23)

where

$$A_1 = \left\{ 2.457 \cdot \ln \left[\frac{1}{(7/Re)^{0.9} + 0.27 \cdot (\varepsilon/D)} \right] \right\}^{16}$$

(24)

and

$$B_1 = \left(\frac{37530}{Re} \right)^{16}$$

(25)

For smooth tube, the effect of tube roughness (ε) needed for A_1, Eq. (24), is negligible.

When compared with the correlation of Blasius, the measured friction factor results of water flow in the turbulent regime are all within ±5% agreement (see Figure 16). When compared with the Churchill [31] correlation, the measured water flow friction factor results are within ±5% agreement in the laminar and turbulent regimes, and within ±8% agreement in the transition regimes (see Figure 16).

With the exception of the transition region, the measured water flow friction factor results are within ±5% agreement with the correlations considered. The discrepancies between the measured and the calculated results are within the experimental uncertainties. The results for water flow show that the experimental setup and measurement procedures used in this study are able to acquire data that agreed with established friction factor correlations.

2.9.2. Nusselt Number of Water Flow

Heat transfer measurements for water flow are conducted for a Reynolds number range of approximately 480 to 10,000. Measurements for heat transfer test sections of three different L/D ratios were conducted. The measured Nusselt number results for water flow are compared with prediction from the Gnielinski [32] correlation for turbulent flow:

$$Nu = \frac{(f/8) \cdot (Re - 1000) \cdot Pr}{1 + 12.7 \cdot (f/8)^{0.5} \cdot (Pr^{2/3} - 1)} \tag{26}$$

Figure 17 shows that the measured Nusselt number results are within ±15% agreement when compared with the predictions from the Gnielinski [32] correlation. The discrepancies between the measured and the calculated results are within the experimental uncertainties.

For laminar flow with fully-developed velocity profile and developing thermal profile, the following Lienhard and Lienhard [33] correlation is used with uniform surface heat flux boundary condition:

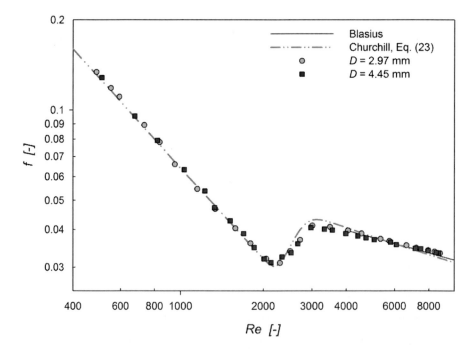

Figure 16. Measured friction factors of water flow in comparison with textbook friction factor equations from laminar to turbulent flow regime.

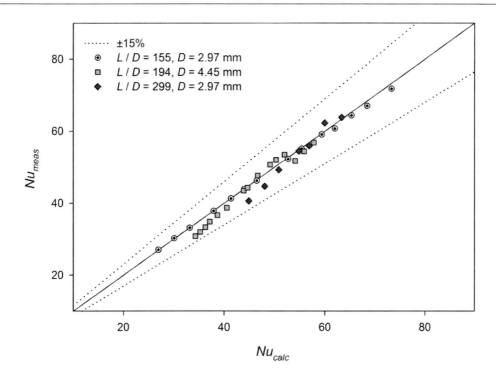

Figure 17. Measured Nusselt number results of water flow in comparison with the calculated results using Gnielinski [32] correlation, Eq. (26).

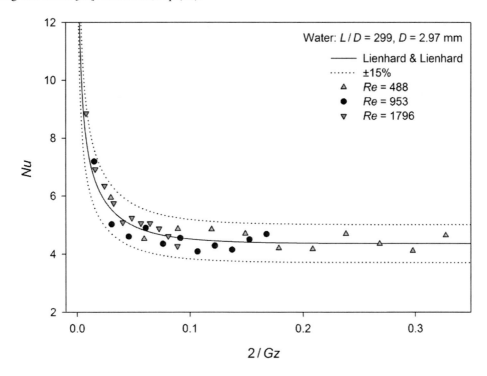

Figure 18. Measured local Nusselt number results of water flow in comparison with the Lienhard and Lienhard [33] correlation, Eq. (27).

$$Nu = \begin{cases} 1.302 \cdot Gz^{1/3} - 1 & \text{for } 2 \times 10^4 \leq Gz \\ 1.302 \cdot Gz^{1/3} - 0.5 & \text{for } 667 \leq Gz \leq 2 \times 10^4 \\ 4.364 + 0.263 \cdot Gz^{0.506} \exp(-41/Gz) & \text{for } 0 \leq Gz \leq 667 \end{cases} \quad (27)$$

where the Graetz number is defined as $Gz = Re \cdot Pr \cdot D/x$.

Figure 18 shows that the measured local Nusselt number results versus the Graetz number for three different Reynolds numbers flow in the tube with a diameter of 2.97 mm. The measured results are all within ±15% agreement with the predictions of the Lienhard and Lienhard [33] equation, Eq. (27).

3. COMPARISON AND DISCUSSION OF RESULTS

The results and comparisons of the MWCNT data and aluminium oxide date are presented in four sections. The average Nusselt numbers as function of Reynolds number, the average heat transfer coefficients as function of average tube velocity, the local developing Nusselt number in the axial tube length, and the average friction factor over the total tube length as function of Reynolds number.

3.1. Average Nusselt Number

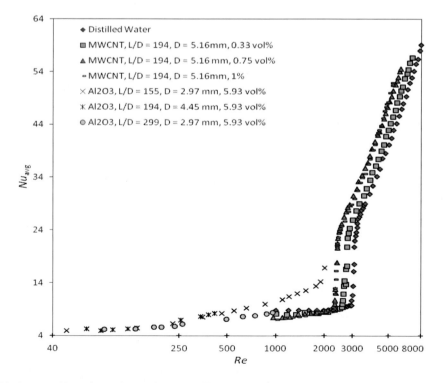

Figure 19. Average Nusselt number as function of Reynolds number and concentration.

The Nusselt numbers as function of Reynolds number for the three different MWCNT nanofluids and three different aluminium oxide nanofluids at different concentrations are shown in Figure 19. The results are also compared to distilled water. The results of the MWCNT are discussed, then the results of the aluminium oxide, where after the results of the two nanofluids are compared.

3.1.1. MWCNT Results

For fully turbulent flow, at a Reynolds number of typically 5000, the Nusselt number if compared to the water of the MWCNT's is enhanced by 9.7% at a concentration of 0.33% , 23.5% at 0.75% and 33% at 1%. It also seems as if the Nusselt number enhancement from a concentration of 0.75% to 1% is small. In general a possible reason for the increase in Nusselt number is that the nanoparticles presented in the base liquid (distilled water) increase the thermal conductivity, delay and disturb the thermal boundary layer and accelerate the energy exchange process in the fluid due to the chaotic movement of the nanoparticles [6, 34, 35, 36].

The transition from laminar to turbulent flow differs and is a function of Reynolds number. The transition for water is at a Reynolds number of approximately 2700. The transition for the 0.33% concentration is at approximately 2400, while the transition for the 0.75% concentration is at approximately 2100, and the transition for 1% is at approximately 2000. Thus the higher the concentrations of the MWCNT the earlier transition occur. The delay is in agreement with the work of Meyer and Olivier [37]. This happens due to the particle-fluid interaction, which damps the instability and reduces the turbulence intensity and Reynolds stress in the flow [6].

The reason for the early transition when comparing the Nusselt number with the Reynolds number is that the viscosities of the MWCNT-water nanofluids are larger than that of water, hence the results of the heat transfer coefficients shift from the position shown in Figure 20 to the position shown in Figure 19 and due to the shift, the MWCNT nanofluids show enhancement in Nusselt numbers.

In the laminar flow regime, at a Reynolds number of 2000, the Nusselt number is reduced by 1.6% when using the 0.33% MWCNT-mixture, enhanced by 2.2% and 2.3% when using the 0.75% and 1% concentrations respectively. The differences are, however, less than the experimental uncertainties reported in Table 2. Thus as a general conclusion for the MWCNT's it was observed that no Nusselt number enhancements occurred in the laminar flow regime.

The concentrations did, however, influence the transition point from laminar to turbulent flow. The higher the concentration, the earlier transition will occur. In the turbulent flow regime the Nusselt number enhancements when compared to water were significant. The enhancements increased with concentration and the maximum enhancements were approximately 33%.

The length of the MWCNT test section was selected to ensure that the flow will be thermally fully developed halfway through the tube length (see Figure 21). The result is that the average Nusselt numbers in Figure 19 is a combination of developing and fully developed values.

Measurements were also not conducted at very low Reynolds numbers to check if the textbook value of 4.36 can be achieved as the fluid outlet temperatures were getting too high and outside the operating range of the pump.

3.1.2. Aluminium Oxide Results

All the Al_2O_3–water nanofluid flows are measured in the laminar region, between the Reynolds number of approximately 50 and 2200. The Prandtl number for the aluminium oxide–water nanofluid is determined to be approximately 35. For the flow to achieve fully-developed thermal boundary layer, the thermal entry length needed can be estimated with $L/D=0.05 \cdot Re \cdot Pr$. Thus, at a Reynolds number of 500, the thermal entry length required for the Al_2O_3–water nanofluid to reached fully-developed thermal boundary layer is approximately $L/D=875$. The textbook value for Nusselt number in fully-developed single-phase flow under constant surface heat flux is 4.36. Figure 19 shows that the average Nusselt number for Al_2O_3–water nanofluid approaching 4.36 at low Reynolds number ($Re<100$).

For the same tube diameter of 2.97 mm, Figure 19 shows that the shorter tube ($L/D=155$) has higher average Nusselt number values than the results from the longer tube ($L/D=299$). The shorter tube is experiencing more significant thermal entrance effect than the longer tube. With a higher level of thermal entrance effect, this means that the average heat transfer coefficient for the shorter tube ($L/D=155$) would be higher than the longer tube ($L/D=299$), even though the diameter is constant at 2.97 mm.

3.1.3. Comparison of MWCNT and Aluminium Oxide

Comparisons are possible between Reynolds numbers of 740 to 2000, which is in the laminar flow regime. In this regime the Nusselt numbers of the MWCNT stay constant while the Nusselt numbers of the aluminium oxide increases. Furthermore the Nusselt numbers of the aluminium oxide are higher than that of the MWCNT. At a Reynolds number of 740, the Nusselt number is 12% higher and at a Reynolds number of 2000 it is 59% higher.

There are three reasons why the Nusselt numbers of the aluminium oxide nanofluids perform better than that of the MWCNT nanofluid. Firstly, and most probably the most important is that the two different sets of results are in different stages of thermal development.

Approximately half of the measurements in the MWCNT set-up show that the flow is fully developed and therefore the average Nusselt numbers are a combination of fully developed and developing flow. The aluminium oxide results, however, indicate that the all the results are for developing flow.

Secondly, the concentration of the aluminium oxide is at least about five times higher than that of the MWCNT concentrations, therefore the increase in Nusselt number is most probably from the increase in thermal conductivity. The thermal conductivity increase of the MWCNT's are at most 1% (Figure 2) while the increase of the thermal conductivity of the aluminium oxide is up to 20% (Figure 4).

Thirdly, in the comparison of the Nusselt number as function of Reynolds number are embedded the tube diameters, viscosities and thermal conductivities. A higher Nusselt number would not necessarily imply a higher heat transfer rate as it is determined by the heat transfer coefficient.

The diameters of the tubes (2.97 and 4.45 mm) in which the aluminium oxide experiments were done were smaller than that of the MWCNT tube (5.16 mm). The bulk densities and viscosities were also different between the two types of nanofluids and also for different concentrations of the same nanofluid. Therefore a fair comparison is sometimes difficult and for that reason many researchers prefer to compare the heat transfer coefficients as function of velocity.

3.2. AVERAGE HEAT TRANSFER COEFFICIENTS

The heat transfer coefficients are compared in Figure 20 as function of the average axial tube velocity for the three MWCNT nanofluids concentrations and three aluminium oxide concentrations. The results for distilled water are also given. As the results are difficult to follow for velocities less than 0.5 m/s an enlarged graph is given in Figure 20b for velocities smaller than 0.5 m/s.

3.2.1. MWCNT Results

As can be expected the heat transfer coefficients increase as the average fluid velocity increase. The transition from laminar to turbulent flow in general occurs at velocities higher than approximately 0.5 m/s. The heat transfer coefficients in the turbulent flow is much higher than in the laminar flow regime as can be expected [23]. The velocity where transition occurs is being influenced by the nanofluid concentration. The higher the concentration the earlier transition occurs.

In general, when compared to water the heat transfer coefficients decrease as the concentrations of the nanofluids are increased. At constant velocity, the heat transfer coefficient on average of the 0.33% concentration is 3.3% lower, at 0.75%, it is 6,6% lower and at 1.0%, it is 12.6% lower than that of the distilled water. This trend was also observed by others [21]. It seems, however, that the heat transfer coefficients converge to the same value as that of water if the velocity increase to values of 2 m/s and higher.

3.2.2. Aluminium Oxide Results

In general the heat transfer coefficients increase as the average fluid velocity increases, as can be expected. For average velocities higher than 0.5 m/s, the distilled water flows are departing from the aluminium nanofluids as the transition occurs for the water. For the aluminium nanofluids it seems as if transition only occurs at a velocity of approximately 3.93 m/s for the 2.97 mm tube. The reason is that at point the Reynolds number is only 2014.

3.2.3. Comparison of MWCNT and Aluminium Oxide

At lower velocities (smaller than 0.5 m/s) and as showed in Figure 20b, the heat transfer coefficients of the aluminium oxide nanofluids are significantly larger than the heat transfer coefficients of the MWCNT nanofluids. When comparing to water the heat transfer coefficients of the MWCNT are up to 20% lower while the aluminium oxide concentrations are up to 30% higher. There are two reasons for the differences.

Firstly, at low velocities of smaller than 0.5 m/s both the nanofluids are in the laminar flow regime. The aluminium oxide nanofluids test sections ensured that the flow was thermally developing from the tube inlet to outlet. The MWCNT test section was relatively longer and ensured that the flow fully developed approximately halfway through the tube. The MWCNT average heat transfer coefficients in Figure 20b, are thus a combination of fully developed heat transfer coefficients (which are lower) and developing heat transfer coefficients (which are higher). The aluminium oxide nanofluids takes much longer to fully developed as the average Prandtl number is approximately 35 in comparison with water and the MWCNT nanofluids with a Prandtl number of approximately 7. Thus, for the same tube diameter and Reynolds number the length to get fully developed flow will be five times longer.

Secondly, the higher heat transfer coefficient values observed in aluminium oxide flow can be attributed to the tube diameters. At a given heat rate, a smaller diameter tube would have higher value of heat flux, thus resulting in higher heat transfer coefficient. This can be observed by comparing the data for aluminium oxide nanofluid flow in the 2.97-mm tube with the data for the other larger tubes. Note that for aluminium oxide nanofluid flow in the 4.45-mm tube, which is comparable with the tube size for MWCNT nanofluids and distilled water (D=5.16 mm), the comparisons between distilled water, aluminium oxide and MWCNT nanofluids have better agreement (perhaps within experimental uncertainty).

For average velocities higher than 0.5 m/s, the MWCNT and distilled water flows are departing the laminar regime for transition and then turbulent flow, hence the drastic increase in the average heat transfer coefficient of the MWCNT nanofluids and of the water. As the aluminium oxide nanofluids are still in the laminar region, they will have much lower average heat transfer coefficient when compared with the values from MWCNT and distilled water flows which are in the turbulent flow regimes.

3.3. Local Heat Transfer Coefficients

The heat transfer coefficients are compared in Figure 21 during developing flow. A Reynolds number of approximately 1000 was chosen for the comparison. The results for distilled water are also given. It is usually very challenging in experiments of this nature to separate the fully developed flow from the developing flow. To get fully developed flow very long tube lengths are usually required (normally longer than what is commercially available) and that is why most research on this topic is done with small diameter tubes.

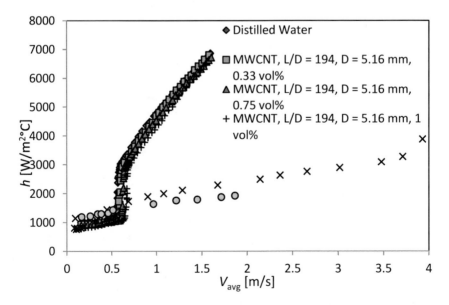

Figure 20a. Average heat transfer coefficient as function of average velocity.

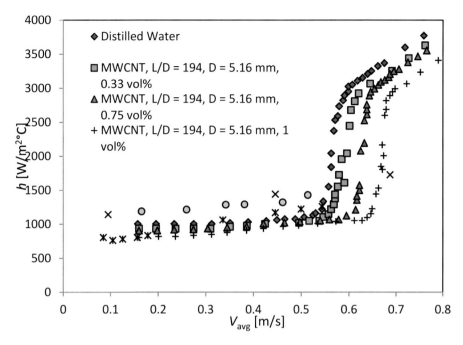

Figure 20b. Average heat transfer coefficients as function of average velocity (enlarged region of Figure 20a) for velocity smaller than 5 m/s.

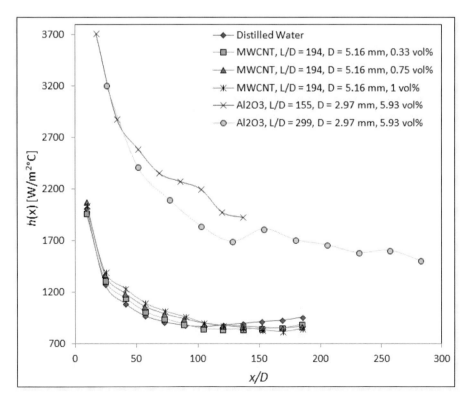

Figure 21. Local heat transfer coefficient as function of axial tube position and concentration. Comparison at $Re = 1000$.

One of the tubes in which the aluminium oxide experiments were conducted had a relative length of (L/D=299) which was longer than the length (L/D=194) in which the MWCNT experiments were done, and one tube was shorter (L/D=155).

3.3.1. MWCNT Results

As expected the heat transfer coefficients decrease as the flow is developing in the tube. It seems as if for most cases and the case of water that the flow is thermally fully developed at approximately x/D=100. This is much earlier than expected (according to L/D=0.05·Re·Pr, it should be at approximately x/D=350. However, it is most probably caused by inlet effects [6, 20, 21].

While the flow is developing the heat transfer coefficients of the MWCNT nanofluids are higher than that of water and increases with concentration. At the 1% concentration the heat transfer coefficient is approximately 10% higher. After the flows of the different concentrations are fully developed the heat transfer coefficients of the water increases downstream and is higher than that of the MWCNT concentrations.

3.3.2. Aluminium Oxide Results

Again, the downward trend of the heat transfer coefficients as the flow is developing is expected. The results showed that the flow in the $L//D$= 155 tube, has not yet fully developed. In the tube with a relative length of L/D=299; there are signs that the flow might be getting closer to be fully developed but the heat transfer coefficients in the range of L/D=150 to L/D=280, are on a downward trend. At Re = 1000, the aluminium oxide nanofluid (Pr≈35) flow in 2.97-mm tube would require L/D≈1750 to reach fully-developed thermal boundary layer.

3.3.3. Comparison of MWCNT and Aluminium Oxide

In general while the flow is developing the heat transfer coefficients of both nanofluid mixtures decreased. The flow of the MWCNT mixtures thermally developed quicker than that of the aluminium oxide mixtures. The reason is that Prandtl numbers of the MWCNT were approximately 7, while the Prandtl numbers of the aluminium oxide mixture were approximately 35. The maximum concentrations of the MWCNT varied from 0.33% to 1%, while the aluminium oxide concentration experiments were at a concentration of 5.93%.

When comparing with the flow of distilled water and MWCNT nanofluids, the higher local heat transfer coefficients for aluminium oxide nanofluid flow is also attributed to its smaller tube size. The higher heat flux for smaller tube means that the heat transfer coefficient would be higher.

3.4. AVERAGE FRICTION FACTOR

The average friction factors are compared in Figure 22 as function of Reynolds number for the two different types of nanofluids as well as for water. Two graphs are given. In Figure 22a a larger range of Reynolds numbers are given while Figure 22b concentrates on a smaller Range from 600 to 4000. In these no results are given for the 1% concentration of MWCNT as it was found that at this concentration it blocked the pressure tap ports.

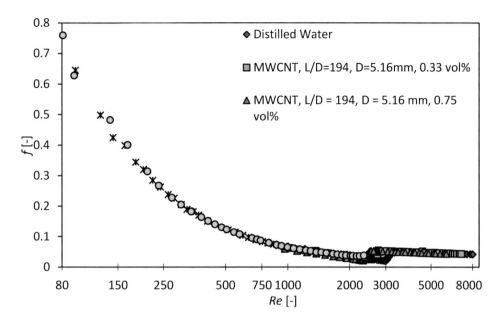

Figure 22a. Average friction factor as function of Reynolds number and concentration.

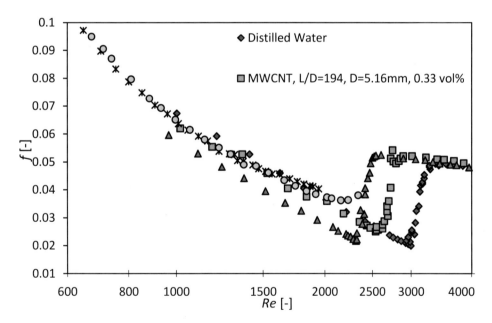

Figure 22b. Average friction factor as function of Reynolds for the smaller Reynolds number range of 600-4000 (with specific reference to the results in Figure 22a.).

3.4.1. MWCNT Results

Figure 22a, shows that the friction factors of the different nanofluid concentrations are very close to each other in the laminar and turbulent flow regimes, but the values are higher than that of water by approximately 17% for the 0.07% concentration and 4% with a 0.33%

concentration. However, significant differences occur in the transitional flow regime (Figure 22b). Transition is a function of concentration as was also observed in Figure 19.

3.4.2. Aluminium Oxide Results

The friction factors of the two different concentrations are very close to each other and follow the trend of the water results well. At a Reynolds number of 1500 the friction factor of the nanofluids are approximately 25% higher than that of water.

3.4.3. Comparison of MWCNT and Aluminium Oxide

Comparisons between the MWCNT nanofluids and aluminium oxide nanofluids were only possible in the laminar flow regime. In general the friction factors of the MWCNT mixtures were lower than that of the aluminium oxide mixtures. This can be expected as the local heat transfer coefficients of the MWCNT as shown in Figure 21 are also lower. The reason for the higher friction factors of the aluminium oxide nanofluids is that the flow throughout the test section was thermally developing while about half of the flow in the MWCNT test-section is fully developed.

The onset of transition flow appears to be affected by the type of nanofluids and the nanoparticle concentration. For distilled water, MWCNT (0.33% vol.) and aluminium oxide nanofluids, the onset of transition appears to be at approximately Reynolds numbers of 2800, 2600, and 2000, respectively.

CONCLUSION

In this chapter the performance of nanofluids were compared during convective heat transfer in smooth tubes. The nanofluids were MWCNT and aluminium oxide nanofluids. Both nanofluids were mixed with water and at different concentrations. Water was also used for comparison purposes. Comparisons were made of the average Nusselt numbers as function of Reynolds number, the average heat transfer coefficients as function of average tube velocity, the local developing heat transfer coefficient in the axial tube length, and the average friction factor over the total tube length as function of Reynolds number.

It was found that when the average Nusselt numbers were compared as function of Reynolds number that the aluminium oxide nanofluids performed better. The most important reasons are that although the tube diameters were almost the same the two nanofluids were at different stages of developing flow for the thermal boundary layer as the Prandtl number of the aluminium oxide nanofluids was approximately five times higher than that of the MWCNT. The concentrations of the aluminium oxide were also 5–15 times higher than that of the MWCNT.

When heat transfer coefficients were compared as function of average axial tube velocity it was found that at velocities lower than 0.5 m/s the heat transfer coefficients of the aluminium oxide were significantly higher than that of the MWCNT. But at velocities higher than 0.5 m/s the heat transfer coefficients of the MWCNT are much larger than that of the aluminium oxide. The main reason is that at a velocity of 0.5 m/s the water and MWCNT nanofluid mixtures flow regime changes from laminar flow to turbulent flow.

When the local heat transfer coefficients were compared as function of axial tube length the aluminium oxide nanofluids performed better than the MWCNT nanofluids. The reasons are because the thermal boundary layer of the aluminium oxide nanofluids were developing while the MWCNT developed fully approximately half way through the tube. A higher heat flux was also used for the aluminium oxide nanofluids for the tube with the smaller diameter.

Average friction factor comparisons as function of Reynolds number were only possible in the laminar flow regime. In general the friction factors of the MWCNT nanofluids were lower than that of the aluminium oxide. This can be expected as it correlates with the local heat transfer coefficients and the fact that the flow in the MWCNT are thermally fully developed halfway through the tube while the flow with the aluminium oxide nanofluids were far from fully developed.

In general it can be concluded in this chapter that it is challenging to compare the performance of experimental nanofluid data of different laboratories with different experimental set-ups, test-sections, and operating conditions. Designers who are considering using different types of nanofluids in the design of heat transfer equipment should be careful to not only consider Nusselt numbers as function of Reynolds number. A large variety of other factors should also be taken into consideration such as the Prandtl number of the nanofluids, operating range (laminar, transition or turbulent), heat flux, possible inlet effects, and if the flow will be developed or fully developed.

ACKNOWLEDGMENT

The first author would like to thank my former Master's degree student Kersten Grote who provided the MWCNT data and data for the graphs. Thanks also to my master's degree student Marilize Everts who assisted with compiling some of the graphs and editing of this manuscript. The second author would like to thank Sanjib Tiwari and Matthew Cox for their contributions in conducting the measurements.

NOMENCLATURE

A_s	surface area, m^2
C_k	constant for thermal conductivity that depends on the experimental data
C_μ	constant for viscosity that depends on the ratio of the revolution ellipsoid axes
c_p	specific heat at constant pressure, $J{\cdot}kg^{-1}{\cdot}K^{-1}$
D	diameter of tube, m
EB	energy balance
Gz	Graetz number
$h(x)$	local convective heat transfer coefficient, $W{\cdot}m^{-2}{\cdot}K^{-1}$
I	electrical current, A
k	thermal conductivity, $W{\cdot}m^{-1}{\cdot}K^{-1}$

L	length of tube, m
$\dot{m}$	mass flow rate, $kg \cdot s^{-1}$
Nu	Nusselt number
P	perimeter, m
ΔP	pressure drop in tube, Pa
Pr	Prandtl number
$\dot{Q}$	heat transfer rate, W
$\dot{q}$	heat flux, $W \cdot m^{-2}$
R	thermal resistance, $K \cdot W^{-1}$
Re	Reynolds number
$T(x)$	local temperature, °C
V	electrical voltage, V
x	axial distance, m

Greek symbols

$\dot{\gamma}$	shear rate, s^{-1}
μ	viscosity, Pa·s or cP
ρ	density, $kg \cdot m^{-3}$
τ	shear stress, $dyne \cdot cm^{-2}$
ϕ	volume concentration
χ	weight fraction

Subscripts

avg	average
b	bulk
bf	base fluid
$calc$	calculated
Cu	copper
e	exit
i	inlet
in	input
lam	laminar
m	mean
$meas$	measured
nf	nanofluid
p	nanoparticle
si	inner surface
so	outer surface
tf	testing fluid
$trans$	transition
$turb$	turbulent
w	wall

REFERENCES

[1] Tang, C. C., Tiwari, S., Cox, M. W. In Proceedings of the ASME 2013 Heat Transfer Summer Conference, Paper HT2013-17550, Minneapolis, USA, 14-19 July 2013.

[2] Ahuja, A. S. *J. Applied Physics*, 1985, 46, 3408-3416.

[3] Kakaç, S., Pramuanjaroenkij, A. *Int. J. Heat Mass Transfer,* 2009, 52, 3187-3196.

[4] Prasher, R., Bhattacharya P., Phelan, P. E. *Physical Review Letters*, 2005, 94, 025901, (4 pages).

[5] Wang, X.-Q., Mujumdar A. S., *Int. J. Thermal Science*, 2007, 46, 1-19.

[6] Meyer, J. P., McKrell T. J., Grote, K. *Int. J. Heat Mass Transfer*, 2013, 58, 597-609.

[7] Tang, C. C., Tiwari, S., Cox, M. W. In *Proceedings of the ASME 2013 Heat Transfer Summer Conference*, Paper HT2013-17509, Minneapolis, USA, 14-19 July 2013.

[8] Elnajjar E., Haik, Y., Hamdan M. O., Khashan S. *Heat Mass Transfer*, 2013, 49, 1681-1687.

[9] Garg, P., Alvarado, J. L., Marsh, C., Carlson, T. A., Kessler, D. A., Annamalai, K. *Int. J. Heat Mass Transfer*, 2009, 52, 5090-5101.z`1.

[10] Lee, S. W., Park, S. D., Kang, S., Bang, I. C., Kim, J. H. *Int. J. Heat Mass Transfer*, 2011, 54, 433-438.

[11] Xie, H., Lee, H., Youn, W., Choi, M. *J. Applied Physics*, 2003, 94, 4967-4971.

[12] CRC Handbook of Chemistry and Physics, 87th ed., CRC Press: Boca Raton, FL, 2006.

[13] Popiel, C. O., Wojtkowiak, J. *Heat Transfer Engineering*, 1998, 19, 87-101.

[14] Prasher, R., Song, D., Wang, J. L., Phelan, P. *Appl. Phys. Lett.*, 2006, 89, 133108-133110.

[15] Keblinski, P., Phillpot, S. R., Choi, S. U. S., Eastman, J. A. *Int. J. Heat Mass Transfer*, 2002, 45, 855-863.

[16] Sengers, J. V., Watson, J. T. R. *J. Physical and Chemical Reference Data*, 1986, 15, 1291-1314.

[17] Hiemenz, P. Principles of Colloid and Surface Chemistry, 2nd ed., Vol. 9, Marcel Dekker, Inc.: New York, 1986.

[18] Ferrouillat, S., Bontemps, A., Ribeiro, J.-P., Gruss, J.-A., Soriano, O. *Int. J. Heat and Fluid Flow*, 2011, 32, 424-439.

[19] Prasher, R., Phelan, P. E., Bhattacharya, P. *Nano Lett.*, 2006, 6, 1529-1534.

[20] Olivier, J. A., Meyer, J. P. *HVAC&R Research*, 2010, 16, 471-496.

[21] Meyer, J. P., Olivier, J. A. *Int. J. Heat Mass Transfer*, 2011, 54, 1587-1597.

[22] Abu-Eishah, S. I. *Int. J. Thermophysics*, 2001, 22, 1855-1868.

[23] Çengel, Y. A., Ghajar, A. J. *Heat and Mass Transfer: Fundamentals and Applications*, 4th ed., McGraw-Hill: New York, 2010.

[24] Moffat, R. J. *Exp. Thermal and Fluid Science*, 1988, 1, 3-17.

[25] Allen, R. W., Eckert, E. R. G. *J. Heat Transfer*, 1964, 86, 301-310.

[26] Shome, B., Jensen, M. K. *Int. J. Heat Mass Transfer*, 1995, 38, 1945-1956.

[27] Tam, L.-M., Ghajar, A. J. *Exp. Thermal and Fluid Science*, 1997, 15, 52-64.

[28] Metais, B., Eckert, E. R. G. *J. Heat Transfer*, 1964, 86, 295-296.

[29] Ghajar, A. J., Tam, L.-M. *Exp. Thermal and Fluid Science*, 1995, 10, 287-297.

[30] Ghajar, A. J., Tam, L.-M. *Exp. Thermal and Fluid Science*, 1994, 8, 79-90.

[31] Churchill, S. W. *Chemical Engineering*, 1977, 84, 91-92.

[32] Gnielinski, V. *Int. Chemical Engineering*, 1976, 16, 359-368.
[33] Lienhard I. V., J. H., Lienhard V., J. H. A Heat Transfer Textbook, 4th ed., Phlogiston Press: Cambridge, MA, 2012.
[34] Wen, D., Ding, Y. *Int. J. Heat Mass Transfer,* 2004, 47, 5181-5188.
[35] Duangthongsuk, W., Wongwises, S. *Int. J. Heat Mass Transfer*, 2010, 53, 334-344.
[36] Tam, L.-M., Ghajar, A. J. *Exp Thermal and Fluid Science*, 1998, 16, 187-194.
[37] Meyer, J. P., Olivier, J. A. *Int. J. Heat Mass Transfer*, 2011, 54, 1598-1607.

In: Nanofluids: Synthesis, Properties and Applications   ISBN: 978-1-63321-677-8
Editors: S.M. Sohel Murshed, C.A. Nieto de Castro   © 2014 Nova Science Publishers, Inc.

Chapter 8

POOL BOILING HEAT TRANSFER OF NANOFLUIDS

Ehsan Ebrahimnia-Bajestan[1], Omid Mahian[2],, Ahmet Selim Dalkilic[3] and Somchai Wongwises[4]*

[1]Department of Energy, Institute of Science and High Technology and
Environmental Sciences, Graduate University of Advanced Technology,
Kerman, Iran
[2]Department of Mechanical Engineering, Ferdowsi University of Mashhad,
Mashhad, Iran
[3]Heat and Thermodynamics Division, Department of Mechanical Engineering,
Mechanical Engineering Faculty, Yildiz Technical University, Yildiz, Besiktas,
Istanbul, Turkey
[4]Fluid Mechanics, Thermal Engineering and Multiphase Flow Research Lab.
(FUTURE), Department of Mechanical Engineering, Faculty of Engineering,
King Mongkut's University of Technology Thonburi, Bangmod,
Bangkok, Thailand

ABSTRACT

Pool boiling as a heat transfer phenomenon that deals with phase change of working fluid has many industrial applications. Furthermore, pool boiling of nanofluids as an advanced type of working fluids has attracted considerable interest. In this chapter, a literature review of pool boiling of nanofluids is provided, focusing on the effects of nanofluid concentration, nanoparticles type, contact surface material, and surface roughness on the heat transfer rate during the process. The study shows that the sedimentation of nanoparticles on the heating surface is a key factor in the critical heat flux (CHF) enhancement. It is also concluded that the available data are not enough to identify the mechanisms of boiling heat transfer characteristics of nanofluids. Therefore, it is suggested to perform more experimental works on bubble dynamics, bubble effective parameters, heating surface characteristics, properties of nanoparticles and nanofluids as well as the operating conditions to recognize the pool boiling heat transfer of nanofluids.

---

* E-mail: omid.mahian@gmail.com; Tel: +985118763304; Fax.: +985118816840.

1. INTRODUCTION

Pool boiling is a heat transfer phenomenon regarding the phase change of the working fluid. Pool boiling has many industrial applications including thermal and nuclear power generation in steam plants, refrigeration systems, heat transmission, and so on. Two different techniques can be utilized as the solution to increase the heat transfer rate in pool boiling. The first one is the improvement of the contact surface and the second approach is the improvement of working fluid. Nanofluids as an advanced group of suspensions containing ultrafine solid particles (with the size of the order of nanometer) could be a possible option to use as the working fluid in the systems in which pool boiling happens. In the studies on the pool boiling, heat transfer coefficient and CHF are the main characteristics that have been investigated by the researchers. The pool boiling is limited by the quantity of CHF. CHF is the highest heat flux in which the working fluid has a high performance. When the heated surface reaches CHF, it is coated with a vapor film that separates the hot surface and the working fluid, and consequently the heat transfer decreases. With the decrease of heat transfer, the wall temperature increases. If the wall temperature reaches a higher temperature than that of allowed temperature for the special material, the heater surface may destroy. Therefore, increasing the CHF point is a goal to avoid the destruction of thermal systems in which pool boiling happens [1]. In this chapter, first a literature review of the studies on pool boiling is conducted. Next, by considering some experimental works conducted in the pool boiling field, the effects of nanofluid concentration, nanoparticles type, contact surface material, and surface roughness on the heat transfer rate are discussed.

2. LITERATURE REVIEW

Since a decade ago, a substantial amount of work has been performed on the boiling heat transfer characteristics of nanofluids that is considered in several review papers [2-8]. However, in this study, the works conducted in the field of pool boiling are classified based on the nanoparticle type used in each study. Firstly, the works in which only a single type of nanoparticles is used are reviewed. Next, the comparative studies in which several types of nanoparticles are utilized have been summarized.

2.1. Pool Boiling Using Al$_2$O$_3$ Nanoparticles

A pioneering study on the issue of pool boiling characteristics of nanofluids was reported by Das et al. [9] using Al$_2$O$_3$/water nanofluid under atmospheric conditions. They concluded that adding nanoparticles reduces the boiling performance due to the change of surface characteristics.

Bang and Chang [10] examined the pool boiling of Al$_2$O$_3$/ water nanofluids (concentrations up to 4%) on a plain surface in both horizontal and vertical positions. They concluded that particle loading results in a decrease in the heat transfer rate of pool boiling. They also found that using nanofluid with the concentration of 4% leads to 32% and 13%

enhancement in the CHF compared to the pure water for the horizontal and vertical positions of the surface, respectively.

An experimental investigation was performed on the nucleate pool boiling of Al_2O_3/water nanofluids by Wen and Ding [11], where the results were inconsistent with those obtained by Bang and Chang [10], Das et al. [9], and You et al. [12]. It was observed that the boiling heat transfer coefficient enhancement increases with nanoparticles concentration. They discussed on the some effective parameters on the boiling heat transfer of nanofluids, such as thermophysical properties of nanofluids and boiling surface. In another study, Wen [13] claimed that structural disjoining pressure, which affects the number of active nucleate sites and bubble dynamics can be responsible for the scattered experimental boiling heat transfer data of nanofluids.

Kedzierski [14] experimentally studied the effects of adding Al_2O_3 nanoparticles with the average size of 10 nm to R134a/polyolester mixtures on the pool boiling on a roughened horizontal flat surface. Polyolester as the lubricant was mixed with the R134a at three mass fractions, including 0.5%, 1%, and 2%. The results showed that nanoparticle loading leads to heat transfer enhancement so that this enhancement has an inverse relation with the heat flux. For heat fluxes less than 40 kW/ m^2, the heat flux enhanced on average about 105%, 49%, and 155% for the 0.5%, the 1% and the 2% mass fractions, respectively. A semi-empirical model was developed to predict the pool boiling enhancement of the mixture of R134a/polyolester and Al_2O_3 nanoparticles. According to the model, the boiling performance is maximized where the volume fraction of nanoparticles and the mass fraction of lubricant are close to one and the particle has the minimum size.

Narayan et al. [15] considered the pool boiling of Al_2O_3/water in a tube with a length of 170 mm, diameter of 33mm, and roughness of 48nm. They considered the effects of the inclination angle of the tube (0°, 45°, 90°), particle size (47 nm and 150 nm), and concentration of nanofluids (0.25%, 1% and 2%) on the pool boiling performance. They found that the pool boiling heat transfer is maximized for the horizontal position of the tube and bigger size of particles.

Harish et al. [16] studied the interactions between the Al_2O_3 nanoparticles and the heater surface in pool boiling of Al_2O_3/water nanofluids. They concluded that the boiling heat transfer can be enhanced or may be deteriorated depends on a parameter called as surface particle interaction parameter which is the ratio of roughness of the heater surface to the nanoparticle size. The bigger surface particle interaction parameter leads to a greater heat transfer enhancement ratio.

Raveshi et al. [17] investigated the effects of adding Al_2O_3 nanoparticles with the size of 20-30nm to a mixture of water and ethylene glycol (50:50) on pool boiling in a cylindrical vessel made of Pyrex glass with the length of 300 mm, inner diameter of 98mm, and outer diameter of 100 mm. They tested six volume concentrations of the nanofluids including 0.05%, 0.1%, 0.25%, 0.5%, 0.75%, and 1% and found that the concentration of 0.75% is the optimal one in which the heat transfer coefficient enhances up to 64%.

Tang et al. [18] studied the nucleate pool boiling of a mixture of δ-Alumina nanoparticles and the refrigerant of R141b at concentrations of 0.001 %, 0.01% and 0.1% by volume with and without the surfactant of sodium dodecyl benzene sulphonate (SDBS) in a stainless steel vessel with an inner diameter of 170-mm and a height of 300 mm. The results showed that for 0.1% of concentration, using the nanofluid without surfactant deteriorates the pool boiling heat transfer because of the sedimentation of nanoparticles on the heater surface. On the other hand, they concluded that at very low volume concentrations, i.e. 0.001%, adding the SDBS may reduce the enhancement in the boiling heat transfer characteristics, because using the SDBS results in a decrease in the effective thermal conductivity and the other effective parameters of the nanofluid which can enhance the heat transfer rate.

Shahmoradi et al. [19] evaluated the pool boiling of Al_2O_3/water nanofluid on a flat plate heater. In the tests, they considered the concentrations of 0.001%, 0.002%, 0.02%, 0.05%, and 0.1%. The results showed somewhat increase in the boiling heat transfer coefficient for the concentration of 0.001% compared to pure water. However, in other concentrations a decrease was observed so that the value of reduction was more considerable for heat fluxes greater than 600 kW/m^2. The reductions in heat transfer are 13%, 22%, 33% and 40% for the concentrations of 0.002%, 0.02%, 0.05%, and 0.1%, respectively. It was also found that the CHF increases with increasing the volume fraction of particles. They concluded that during the boiling of nanofluids, the nanoparticles deposit on the heated surface and form a porous layer which increases the wettability of the surface and consequently the CHF enhancement increases.

Soltani et al. [20] inquired about the pool boiling of non-Newtonian nanofluids in a cylindrical vessel. The nanofluids were synthesized by γ-Al_2O_3 nanoparticles and the mixture of carboxy methyl cellulose (CMC) and water. Different concentrations of CMC non-Newtonian fluids and γ-Al_2O_3/CMC non-Newtonian nanofluids were tested under nucleate pool boiling heat transfer conditions. They found that the increase of volume fractions of γ-Al_2O_3 particles in CMC solutions leads to the decrease of surface temperature as well as the increase of heat transfer coefficient. They attributed the improvement in the heat transfer coefficient to the viscosity of the solution, the collisions between the particles and heater surface and bubbles, and the thickness of the boundary layer.

Xu et al. [21] studied the effects of pulse heating and the addition of Al_2O_3 particles to pure water on the pool boiling heat transfer on a micro-heater surface (with the size of 50 × 20 μm). Nanofluids were provided in the concentrations of 0.1%, 0.2%, 0.5% and 1.0% by weight. The bubble dynamics found during the tests can be classified into three groups as follows.

- The first type of bubble dynamics is belonged to the boiling of pure water. Large bubbles are detected on the surface of micro heater and they remain on the surface during the stage of pulse off. The rate of increase in temperatures of the micro heater surface is high in the begging of a new pulse cycle. During the pulse duration stage the increase in the temperatures is kept.
- The second type of bubble dynamics is observed in the two concentrations of 0.1% and 0.2%. In this type of bubble dynamics, the micro heater temperature increases sharply at the beginning of a new pulse cycle and next the temperature decreases during the pulse duration. The vibration of bubbles on the surface is diminished when the pulse is removed.
- The third type of bubble dynamics happens in the concentrations of 0.5% and 1.0%. The behavior of bubbles in the third type is very similar to the first type, only the surface has lower temperatures in this type compared to the first type.

Jung et al. [22] investigated the pool boiling of the solutions of Al_2O_3 nanoparticles suspended in a mixture of $H_2O/LiBr$ on a plate copper heater with the size of 10×10 mm^2 where polyvinyl alcohol is used to provide a more stable nanofluid. The concentrations of nanofluids were between 0 and 0.1% by volume, and the weight fractions of LiBr are 3, 7 and 10%. They found that using the volume fractions of 0.1% and 10% of LiBr increases the CHF by about 49%. They also found the surfactant has no significant effect on the pool boiling.

In another work close to Ref. [22], Jung et al. [23] focused on the effect of adding polyvinyl alcohol as a surfactant on the pool boiling of alumina/water nanofluids. They observed that at low concentrations (up to 0.001 vol.%), the CHF enhancement (the ratio of the CHF of nanofluid to that of water) for the nanofluid with the surfactant is higher than that of the nanofluid without surfactant, while in higher concentrations the trend is opposite.

Jeong et al. [24] examined the effects of wettability of a heater surface (strips made of stainless steel and the size of $30 \times 30 \times 3$mm) on the pool boiling of Tri-sodium phosphate solutions in the concentrations of 0.01, 0.05, 0.1, 0.3, 0.5, 0.8% by weight and aluminum oxide nanofluids with the concentrations of 0.5, 1, 2, 4% by volume. They found that with increasing the concentration the contact angle decreases and consequently the CHF increases.

Ahn and Kim [25] studied the pool boiling of Al2O3/ water nanofluid with concentration of 0.001% on a copper surface with the diameter of 10mm. Also, pool boiling of water was investigated in two cases, first, when the heater surface is coated with nanoparticles, second, without coating the surface. They found that the CHF for water increases from 1532 to 1900 kW/m^2 after coating the surface of heater by a layer of nanoparticles due to the improvement of surface wettability. They also reported that the water temperature will increase instantly and film boiling is developed once the CHF is reached. However, for the case of nanofluid and water boiling along with the nanoparticle coating, the nucleate boiling is kept only for a couple of seconds.

2.2. Pool Boiling Using CNT Nanoparticles

Peng et al. [26] performed experiments to study the heat transfer characteristics of nucleate pool boiling using various carbon nanotubes having an outer diameter from 15nm to 80nm and the lengths from 1.5mm to 10mm. According to their results, CNTs cause significant enhancement as high as 61% on the heat transfer coefficient in comparison with R113-oil mixture without CNTs. The increase of CNTs outside diameter reduces enhancement factor which is affected by CNTs mass fraction positively. They emphasized on the importance of high thermal conductivity and small diameter CNTs for refrigerant-oil mixtures. Finally, they proposed an empirical correlation to determine the nucleate pool boiling heat transfer coefficient of refrigerant-oil mixtures with CNTs.

Peng et al. [27] studied the pool boiling of various carbon nanotubes with different diameters and lengths suspended in R11, R141b and n-pentane. They found that CNTs dimensions affect the migration ratios that are influenced by dynamic viscosity and density of refrigerant. They also investigated the effects of oil concentration, heat flux, and initial liquid-level height on the migration ratio of CNTs. Their model was found to be in good agreement with the experiments of refrigerant-based nanofluid pool boiling.

Kathiravan et al. [28] investigated the pool boiling of multi-walled carbon nanotubes (CNTs) suspension in pure water and water containing some additives over a stainless steel flat plate heater. They found that CNTs enhance the heat transfer coefficient due to their higher thermal conductivity and large specific surface area. They also discussed on the effects of properties of the base liquid and the dispersed phase, the relation between bubbles and CNTs, CNTs sizes, concentration and morphology, as well as the presence of surfactants on the pool boiling.

They found that a decrease in concentration of CNTs increases the onset nucleate boiling delay and the addition of surfactant is also effective on the delay of onset nucleate boiling. The decrease results in a high heat transfer coefficient of water due to changes in surface tension and surface wettability.

Liu et al. [29] used the carbon nanotubes (CNTs) in deionized water during the pool boiling on a flat copper surface under the atmospheric and sub-atmospheric pressures. They showed the importance of pressure on the improvement of heat transfer coefficient and CHF, using CNT suspensions. They stated that low pressures cause high heat transfer coefficient and CHF.

Xue et al. [30] studied the boiling heat transfer characteristics of CNTs nanofluid, water, and the gum Arabic solution. They found that CNTs display higher CHF than the gum Arabic solution. They also observed that using gum Arabic solution leads to higher CHF compared to water.

Park et al. [31] conducted pool boiling experiments on a smooth square flat copper heater to calculate heat transfer coefficient and CHF of water using various concentrations of multi-walled CNTs. They found that CHF increases with increasing CNTs concentrations. They took surface images by SEM and showed that CNTs deposit on the surface during bubble formation and departure. When CNT concentration decreases, the nucleate boiling heat transfer coefficient (HTC) increases because a thin film of CNTs is formed on the surface and, hence, reduces the contact angle.

The mentioned result was not in agreement with the work of Park and Jung [32], where they found an increase in heat transfer coefficient with the same multi-walled CNTs dispersed in water. The authors [31] concluded that this inconsistency is due to the different methods used to disperse CNTs in water.

2.3. Pool Boiling Using ZnO Nanoparticles

Mourgues et al. [1] studied the pool boiling of water and ZnO/water nanofluid with the volume concentration of 0.01% on a disk in horizontal and vertical positions. They found that:

- CHF in the vertical position is about 19% higher than that of CHF in the horizontal position.
- The enhancement in CHF is not due to using the nanofluid, but it is caused by the sedimentation of nanoparticles on the disk surface.
- When the ZnO nanofluid is used and there is an initial nanoparticle deposition on the surface, a minimum increase as high as 54% is observed compared to pure water.
- At the same heat flux, where the surface is coated with a layer of nanoparticles, the measurements of the wall temperature show a decrease in the temperature.

Kole and Dey [33] provided stable ZnO/ EG nanofluids with concentrations up to 2.6% without using surfactant and only the use of long time sanction (more than 60h) and tested the pool boiling behavior of the suspension on the surface of a copper cylinder and compared the results with the base fluid(EG).

They concluded that:

- Using the nanofluid with the volume concentration of 1.6% leads to an increase in heat transfer coefficient by 22% compared to pure EG.
- For the concentrations higher than 1.6%, the heat transfer coefficient decreases due to the sedimentation of particles on the cavities of surface.

White et al. [34] conducted experiments on the pool boiling of ZnO/ propylene glycol nanofluids. They used an electrophoretic deposition method to create a layer of nanoparticles on the surface in which the boiling happens. Under atmospheric pressure, they observed 200% enhancement in HTC.

This work suggests that the electrophoretic deposition approach is a good way to enhance the boiling heat transfer.

Sharma et al. [35] examined the pool boiling of ZnO/water nanofluids on metallic plate heaters where the applied heat flux increases with time. The amount of heat flux varies from zero to CHF in the intervals of 1, 10 and 100s. They concluded that using nanofluids under very fast applied heat fluxes (i.e. 1s) has no effect on the CHF, although for the two others intervals results in an enhancement in the CHF.

2.4. Pool Boiling Using SiO_2 Nanoparticles

A study on the pool boiling of nanofluids on a 0.4mm diameter horizontal NiCr wire was carried out by Vassallo et al. [36]. They applied the silica particles with nano (15 and 50nm) and micro (3μm) sizes dispersed in water at volume concentration of 0.5%. The results showed remarkable increases in the CHF for all mixtures compared to the base fluid, where the maximum enhancement was obtained by using the particles with the size of 50nm; about 160%.

Bolukbasi and Ciloglu [37] tested the pool boiling of SiO_2/ water nanofluids on a cylindrical rod where the volume concentration varies from 0 to 0.1%. They found that in the nucleate boiling region the heat transfer coefficient of nanofluid is slightly lower than that of pure water.

Similar to other works, the CHF enhancement in nanofluids was attributed to the sedimentation of nanoparticles on the heater surface.

Yang and Liu [38] used functionalized silica/ water nanofluids to test their pool boiling behavior under pressures lower than the atmosphere. The use of functionalized nanofluid results in an increase in the heat transfer coefficient compared to pure water. Also, in contrast the common nanofluid, using a functionalized nanofluid has no effect on the CHF, because there is no sedimentation for the functionalized nanofluid.

Milanova and Kumar [39] demonstrated that employing the strong electrolyte enhances the silica nanofluid CHF up to three times in comparison with conventional fluids.

2.5. Pool Boiling Using Cu Nanoparticles

Peng et al. [40] examined the effects of particle size on the pool boiling characteristics of Cu nanoparticles in three sizes of 20, 50 and 80nm suspended in the mixture of R113 and ester oil VG68 in a transparent glass chamber with the size of $150 \times 150 \times 200mm$.

They found that using the nanoparticles with a smaller size results in a higher heat transfer rate so that when the particle size changes from 80nm to 20nm the heat transfer coefficient increases about 24%. They suggested a general correlation to predict the heat transfer coefficient in which the effects of particle size are involved.

In another work, Peng et al. [41] focused on the effects of three different surfactants (with concentrations between 0 and 5000ppm) on the pool boiling heat transfer coefficient of Cu nanoparticles suspended in the refrigerant R113. Sodium Dodecyl Sulfate (SDS), Cetyltrimethyl Ammonium Bromide (CTAB) and Sorbitan Monooleate (Span-80) were used as the surfactants to disperse the nanoparticles.

The results showed that the surfactant increases the heat transfer coefficient, but there is an optimal concentration in which the heat transfer is maximized. For instance, the concentrations of 2000, 500, and 1000 ppm are respectively the optimum concentrations of SDS, CTAB and Span-80 surfactants in which the heat transfer is the maximum. They concluded that using a surfactant with a smaller density provides greater enhancements in the heat transfer during the nucleate pool boiling.

A correlation was proposed to predict the heat transfer coefficient by considering the surfactant effect.

Kathiravan et al.[42] investigated the pool boiling of Cu nanoparticles with weight fractions of 0.25%, 0.5% and 1.0% suspended in two base fluids including water and a mixture of water and sodium lauryl sulphate anionic surfactant (9 wt.%) on a flat plate heater. It is found that CHF of nanofluid increases with increasing the particle load. By increasing the particle weight fraction from 0 to 1% the enhancement in CHF becomes 48% and 106%, for pure water and the mixture of water and surfactant, respectively. On the effect of surfactant on the CHF, the results showed that adding surfactant reduces the CHF due to the decrease in the surface tension.

2.6. Pool Boiling Using CuO Nanoparticles

Kedzierskia and Gong [43] presented a paper on the pool boiling of CuO particles with volume fraction of 1% suspended in a mixture of R134a/polyoester on a horizontal plate. Plaster was applied in three mass fractions, including 0.5,1 and 2%. They found that the heat transfer enhancement ratio due to adding nanoparticles to the mixture increases with a decrease in the mass fractions of polyoester. They concluded that the enhancement in the thermal conductivity is not the main reason for the enhancement of the pool boiling heat transfer.

Zeinali [44] tested the pool boiling of CuO nanoparticles solved in a mixture of ethylene glycol and water in a cubic vessel. The solid volume fractions were 0.1%, 0.2% and 0.5%. The author found that the heat transfer coefficient can be enhanced by 55% when the particle loading is 0.5%.

2.7. Pool Boiling Using TiO$_2$ Nanoparticles

The effects of boiling time on the boiling heat transfer characteristics of nanofluids examined in detail by Okawa et al. [45] using TiO$_2$/water nanofluids at particle volume concentrations of the 0.000094%-0.047%. It was observed that the CHF enhances and the contact angle decreases with an increase in the boiling time. In addition, the boiling time required to achieve the maximum CHF enhancement dramatically reduces with the adding nanoparticles. It is also reported that firstly, the boiling heat transfer coefficient is deteriorated, and then is gradually recovered and reaches an asymptotic value. The authors asserted that the deposition of nanoparticles on the heater surface is one of the main reasons for the CHF improvement.

Furthermore; the experimental results indicated that the colloidal dispersion of nanoparticles affects the CHF more than the nucleate boiling heat transfer.

Since several researchers confirmed that the deposited nanoparticles on heater surface have a significant effect on boiling heat transfer in nanofluids, Phan et al. [46] investigated the effects of the surface coating of nanoparticles during nucleate boiling in TiO$_2$/water nanofluids.

The results indicated that during the nucleate boiling, the deposition of nanoparticles on the heated surface forms a porous layer whose thickness increases with boiling duration and nanoparticles concentration. It was claimed that the increase in fluid-surface adhesion energy due to the formation of this layer, produces a longer residence time of bubbles at the heated surface, which in turn decreases the bubble emission frequency. Therefore; the higher adhesion energy was assumed to be the main reason for the observed considerable degradation of nucleate boiling heat transfer in nanofluids.

2.8. Pool Boiling Using SiC Nanoparticles

Song et al. [47] investigated the potential of SiC nanoparticles as additives to the water to enhance the pool boiling (volume concentrations of 0.0001%, 0.001%, 0.01%) under atmospheric pressure. They used two test sections (different in aspect ratio) in the experiments to verify the trend of CHF. They found that:

- Using the volume concentration of 0.01% leads to 105% enhancement in the CHF.
- A comparison among the concentrations shows that in the concentration of 0.001%, the CHF enhancement is minimized.

2.9. Pool Boiling Using Fe$_3$O$_4$ Nanoparticles

Sheikhbahai et al. [48] studied the effects of electrical field on the nucleate boiling and CHF of the suspensions of Fe$_3$O$_4$ nanoparticles and a mixture of ethylene glycol and water (50:50 vol.%) on a horizontal thin Ni-Cr wire at atmospheric pressure. They concluded that the CHF is enhanced up to 100% when the particle loading is 0.1vol.%. It was also found that the electrical field has no significant effect on the layer formed by nanoparticles on the heater surface and consequently CHF remains nearly constant. However, in general the heat transfer coefficient increased due to applying the electrical field.

2.10. Pool Boiling Using Diamond Nanoparticles

Peng et al. [49] conducted experiments on the pool boiling of diamond nanoparticles suspended in a refrigerant i.e. R113/ oil on a copper plate. They concluded that the heat transfer coefficient increases with increasing the nanoparticles loading and the weight fraction of the refrigerant (R113). In comparison with the CuO nanoparticles/oil suspension, they found the heat transfer during pool boiling of diamond particles is higher than that of CuO particles.

They suggested a correlation to predict the heat transfer coefficient for nucleate pool boiling of the suspended nanoparticles in refrigerant/oil mixture.

2.11. Comparative Studies on Pool Boiling Using Several Types of Nanoparticles

In 2005, Kim et al. [50] represented a possible mechanism for boiling CHF enhancement in nanofluids through the investigation of the water base nanofluids contained nanoparticle types of alumina (110 to 210 nm), zirconia (110 to 250 nm) and silica (20 to 40 nm) in volume fraction of 0.001%, 0.01% and 0.1%. They asserted that the surface wettability improvement caused by nanoparticle deposition is responsible for the significant CHF enhancement of the nanofluids up to 52%, 75% and 80% with alumina, zirconia and silica nanoparticles, respectively.

Kim et al. [51] investigated the CHF of two water based nanofluids containing TiO_2 and Al_2O_3 nanoparticles. The pool boiling is studied on the two different heated wires including a NiCr wire with diameter of 0.2mm and a Ti wire with diameter of 0.25mm.

They stated that the SEM images of the heater surfaces reveal that the reason behind the increase of CHF compared to pure water is the coated layer of the heated surface of nanoparticles.

In other work, Kim et al. [52] evaluated the pool boiling of TiO_2/water and Al_2O_3/water nanofluids with concentrations less than 0.1% on a cylindrical Ni-Cr wire (0.2 mm diameter) under atmospheric pressure. Their results showed that the behavior of CHF depends on the concentration for both nanofluids. In some concentrations, TiO_2 particles show higher CHF compared to Al_2O_3 particles and in some other concentrations Al_2O_3 particles have higher CHF.

Lee et al. [53] compared the pool boiling of Fe_3O_4, TiO_2 and Al_2O_3 nanoparticles suspended in water on a heated NiCr wire. The results indicated that using Fe_3O_4 and TiO_2 nanoparticles leads to the maximum and the minimum CHF, respectively. By the way, all the nanofluids have higher CHF compared to that of the pure water. On the advantages of Fe_3O_4 nanoparticles, besides higher CHF, the use of Fe_3O_4 nanoparticles is more economical because when these particles are influenced by an external magnetic field the local concentration of particles increases in the high heat flux area, which results in very low bulk concentration with small amount of the nanoparticles.

Peng et al. [54] studied the migration of four types of nanoparticles including Cu (at three sizes of 20, 50 and 80 nm), Al and Al_2O_3 (20nm), and CuO (40nm) mixed with three different refrigerants include R113, R141b and n-pentane during pool boiling. They concluded that:

- The migration ratio of nanoparticles has an inverse relation with the density of the nanoparticles. The results showed that for the three particles of Al and Al_2O_3 and Cu with the size of 20nm the migration ratio for Al is the highest and for Cu is the minimal.
- A decrease in particle size leads to an increase in the migration ratio.
  Truong et al. [55] investigated the modification possibility of sandblasted plate heaters using three water based nanofluids with diamond, zinc oxide and alumina nanoparticles to enhance the CHF. They found that diamond coated plates have lowest CHF compared the two other plates.

Kwark et al. [56] studied the pool boiling characteristics of three water based nanofluids with Al_2O_3, CuO and diamond nanoparticles at low concentrations ($\leqslant$1g/l) on a small flat heater (1cm2) under atmospheric pressure. They concluded that the type of nanoparticles has no significant effect on the CHF.

Pham et al. [57] benefitted from nano-fluids as advanced coolant for in-vessel retention system to certify the protection from accidents for nuclear reactors. Their experimental study includes a heated surface made of stainless steel foil having varied angles from the horizontal position, where the maximum CHF enhancement of nanofluids occurred at horizontal orientation of the heater. They applied three water based nanofluids including 0.05 vol.% Al_2O_3, 0.05 vol.% CNT plus 10 vol. % boric acid, and 0.05% Al_2O_3 plus 0.05% CNT, and observed maximum CHF enhancements of 33%, 108% and 122%, respectively. The results indicated that the nanoparticles deposited layers are not the only effective parameters on the CHF enhancement of the nanofluids. The heater surface roughness found to be the main factor, contributing to CHF enhancement, while the contact angle has no considerable effect on the CHF enhancement mechanism of nanofluids.

Golubovic et al. [58] studied experimentally the pool boiling in a square cavity using two different nanofluids including Al_2O_3 /water (with two different sizes of particles, i.e. 22.6 and 46nm) and BiO_2 /water (with the particle size of 39nm) nanofluids. The aim was to evaluate the effects of particle loading, size and type of nanoparticles on CHF at saturated conditions. They concluded that the size of nanoparticles has no significant effect on the peak heat flux of nanofluids while the type of nanoparticles is important so that using Al_2O_3 particles rather than BiO_2 provides a higher peak heat flux as high as 17%. In other words, the CHF increases by 50% when Al_2O_3 nanoparticles are added to pure water, but this value is 33% in the case of BiO_2 particles. They revealed that the main reason for the increase of CHF during pool boiling of nanofluids is the decrease of static surface contact angle.

Cieslinski and Kaczmarczyk [59] carried out tests on pool boiling of Al_2O_3/water and Cu/water nanofluids with weight concentrations of 0.01%, 0.1%, and 1% on horizontal smooth copper and stainless steel pipes with the outer diameter of 10mm and wall thickness of 0.6mm. The results showed that the type of nanoparticles has no considerable effect on the heat transfer coefficient during pool boiling of the two nanofluids on smooth copper tube.

An experimental investigation [60] was conducted on the pool boiling heat transfer characteristics of the water base nanofluids containing three nanoparticle types of graphene-oxide, SiO_2 and Al_2O_3 at 10^{-4} vol. %, considering the effect of the heater orientation. It was concluded that among the mentioned nanoparticles, the graphene-oxide nanofluid shows highest CHF enhancement of about 40% at vertical orientation and about 200% at the horizontal orientation of the heater. It was stated that the forming of the thin layer of nanoparticles on the heater surface is responsible for the CHF enhancement in nanofluids.

The effects of nanoparticle types (CuO and ZnO) as well as the surfactant type were experimentally investigated by Shoghl and Bahrami [61]. The results indicated that the ZnO/water and CuO/water nanofluids deteriorate the boiling performance, by making a smoother surface using the nanofluids. On the other hand, adding the surfactant of sodium dodecyl sulfate (SDS) to the both nanofluids improved the boiling performance, remarkably.

Pool boiling phenomena for a heated, indium-tin-oxide surface in water-based nanofluids with diamond (0.01vol.%) and silica (0.1vol.%) nanoparticles was examined in details by Gerardi et al. [62]. The average heat transfer coefficient and critical heat flux values, as well as the bubble departure diameter and frequency, growth, also wait times, and nucleation site density were measured. A reduction in the nucleate boiling heat transfer (about 50%) and an enhancement in the CHF (about 100%) were found employing both nanofluids. The decrease in bubble departure frequency and nucleation site density were introduced as the reason of deterioration in the nucleate boiling heat transfer coefficient of nanofluids. It was concluded that the deposited nanoparticle is responsible for the increase in the heater surface wettability which leads to the observed decreases in bubble departure frequency and nucleation site density.

For the first time, Aminfar et al. [63] numerically investigated the pool boiling heat transfer of nanofluids using two-phase and three-phase mixture model and control volume technique. The predicted results were in good agreement with experimental data of SiO_2/water and Al_2O_3/water nanofluids. It was observed that at high wall heat flux values the convection heat transfer becomes more important due to the increase in excess temperatures.

3. EXPERIMENTAL STUDY, CORRELATION AND MODELLING

A series of experimental and modeling studies have been carried out by the present authors (S. Wongwises and his colleagues) [64-68] to explore the effects of nanoparticle concentration and nanoparticle type, heating surface material and surface roughness on pool boiling characteristics of nanofluids. The studies as sample ones are summarized as follows.

3.1. Experimental Study

The experimental studies by the authors of this study were performed using two kinds of apparatuses:

In one setup [64], the study was carried out on nucleate pool boiling heat transfer with a horizontal circular plate heating surface. The experimental apparatus has consisted of three main sections: test section, heating device for the test section, and pressure control in the pressure vessel. The test section has been located in the twin-layered cylindrical pressure vessel which has been made of stainless steel. The inner cylinder of the pressure vessel contains nanofluid, while the outer cylinder has been filled with hot water heated by a preheater (1.5kW) to maintain the nanofluid temperature values in saturation temperature [64]. The boiling test section which has consisted of a heater (1.2kW) inserted inside a copper bar which transfers heat from the heater to the test section installed at the upper end of the copper bar. The copper bar and the test section have been covered with a calcium silicate insulator with a thickness of 50mm to prevent lateral heat loss. The test section area has contained four holes for installing sheath thermocouples with an average diameter of 1 mm and a length of 16mm. Thermocouples have been placed at intervals of 10mm. The distance between the uppermost sheath thermocouple and the heating surface is 20mm. Pressure inside the inner cylindrical tank has been controlled by adjusting the water flow rate in the cooling coil to control the condensation rate of vapor at the outer surface of the cooling coil.

In another setup [65], the experiments were conducted using a cylindrical copper tube as the boiling surface. It has consisted of three main parts: a pressure vessel, a condenser and a boiling test section. The boiling surface is a cylindrical copper hollow sleeve (diameter D=28.5mm, length L=90mm). A resistance cartridge heater has been inserted into the copper sleeve to generate heat flux, where the amount of power supply can be adjusted by an electrical transformer. Four T-type thermocouples have been inserted beneath the boiling surface via the thermocouple grooves through the small holes which have been soldered with lead-tin solder.

When the produced vapor contacts with the coil condenser, the formed liquid returns to the bottom of the vessel for re-evaporation. A pressure gauge has been mounted on top of the vessel to monitor the pressure throughout the experiment.

3.1.1. Effects of Nanoparticles Concentration

The nucleate pool boiling of TiO_2-water nanofluids was examined in details at atmospheric pressure and for various nanoparticles concentrations of 0.00005, 0.0001, 0.0005, 0.005, and 0.01 vol.% [64].

Two different heating surfaces made of copper and aluminium with different roughness values of 0.2 and 4μm were employed in the experiments. Considering Figures 1 and 2, in the case of copper heating surface with the surface roughness value of 4 μm there is an optimum volume concentration of 0.0001 vol.%, in which the maximum heat transfer coefficient enhancement of about 15% is achieved. At volume concentrations more than 0.0001 vol.%, the heat transfer coefficient was found to be less than that of the base fluid at every level of surface roughness. This behavior can be attributed to the reduction of nucleation sites due to the sedimentation of nanoparticles on the heating surface. Furthermore, since the thermal conductivity of TiO_2 nanoparticles is lower than that of copper, the deposition of nanoparticles on the copper heating surface may decrease the heat transfer between the heating surface and the nanofluid.

On the other hand, in the case of aluminium heating surface, the heat transfer coefficient was detected to be less than that of base fluid at every nanoparticles concentration and surface roughness. This is due to the increase of surface thermal resistance and nucleation sites reduction, which corresponds to the cover of heating surface with the nanoparticles.

In another study [65], TiO_2 nanoparticles were dispersed in R141b refrigerant at 0.01%, 0.03% and 0.05% nanoparticle volume concentrations to investigate the nucleate pool boiling heat transfer using a cylindrical copper heating surface. It was observed that, as the nanoparticle concentration increases the boiling heat transfer coefficient tends to decrease, especially at high heat fluxes.

In the next study [66], pool boiling characteristics of Al_2O_3-water nanofluid with nanoparticle concentrations of 0.00005, 0.0001, 0.005 and 0.03 vol.% were investigated using a copper heating surface with surface roughness of 3.14 μm. The observed negative effect of the employed nanofluid on the pool boiling heat transfer coefficient in comparison with water as the base fluid can be explained by reduction of the nucleation sites caused by covering or clogging up tiny pores on the heating surface with nanoparticles. It should be noted that, the nucleate bubbles created in the tiny pores, ascend into the fluid and disturb it which in turn cause a better heat transfer performance. As mentioned, the decrease in boiling heat transfer coefficient can also be attributed to the increase of thermal resistance by nanoparticle sedimentation.

3.1.2. Effects of Surface Roughness and Surface Material

The investigation revealed that the boiling heat transfer behavior of nanofluids at various nanoparticles concentrations tends to be similar for different heater surface roughness values of 4μm and 0.2μm [64]. However, it was found that in general, the roughness of 4μm gives a higher boiling heat transfer coefficient than the roughness of 0.2μm, which is noticeable at high heat flux levels. In addition, the better boiling heat transfer characteristics at higher roughness surface can be attributed to the fact that the high roughness surface has more nucleation sites and more cavities, which can contain more gas. Furthermore, the results indicated that, at the same roughness, the aluminium surface gives a higher heat transfer coefficient than the copper surface, whereas this difference was apparent at high heat flux levels [64].

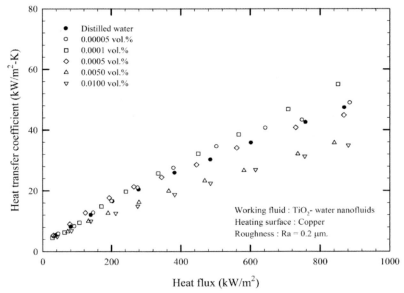

Reprinted from Ref. [64], Copyright, (2010), with permission from Elsevier.

Figure 1. Nucleate pool boiling heat transfer of TiO_2–water nanofluids for copper heating surface with roughness 0.2 μm at 1 atm.

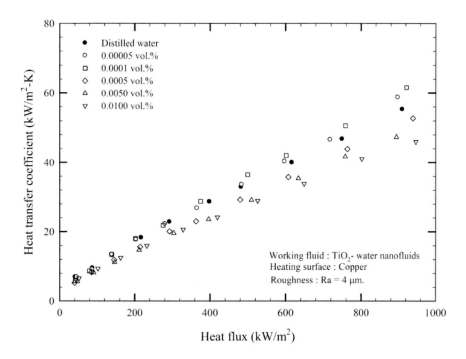

Reprinted from Ref. [64], Copyright, (2010), with permission from Elsevier.

Figure 2. Nucleate pool boiling heat transfer of TiO$_2$–water nanofluids for copper heating surface with roughness 4 μm at 1 atm.

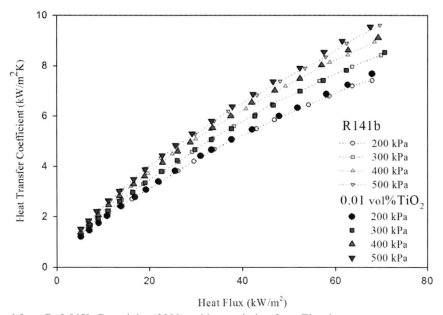

Reprinted from Ref. [65], Copyright, (2009), with permission from Elsevier.

Figure 3. Variation of heat transfer coefficient with pressure for the boiling of 0.01 vol% particle concentration.

3.1.3. Effects of Pressure

The experiments [65, 66] expressed that the boiling heat transfer coefficient of nanofluids increases with pressure, which can be obviously seen at higher heat fluxes. Moreover, the results demonstrated that as the nanoparticles concentration increases, the effectiveness of pressure on the boiling heat transfer coefficient reduces (shown in Figure 3).

3.2. Correlation and Modeling

Beside the aforementioned experimental studies, some attempts were made to determine the effects of important parameters on the boiling heat transfer coefficient of nanofluids and the development of empirical correlation [67, 68].

3.2.1. Correlation Development

The attempts lead to a new correlation [67]which included many relevant variables, i.e., viscosity μ_{nf}, thermal conductivity k_{nf}, specific heat $C_{p,nf}$, and density of nanofluid ρ_{nf}, as well as nanoparticles concentration ϕ, heat flux q'', heating surface roughness ε, surface tension σ, latent heat of vaporization h_{fg}, gravitational acceleration g, and density of water vapor ρ_v. The mentioned correlation for predicting the pool boiling heat transfer coefficient of nanofluids $h_{b,nf}$, is of the form

$$h_{b,nf} = f\left(\mu_{nf}, q'', k_{nf}, \varepsilon, C_{p,nf}, h_{fg}, (\phi + m), \frac{\sigma}{g\,(\rho_{nf}-\rho_v)}\right) \tag{1}$$

Using the Buckingham Pi theorem, a set of dimensionless terms was established as follows:

$$Nu_{nf} = a\left(Pr_{nf}\right)^{b}\left(\frac{q''\varepsilon}{\mu_{nf}\,h_{fg}}\right)^{c}\left(\frac{\varepsilon^2 g\,(\rho_{nf}-\rho_v)}{\sigma}\right)^{d}\left(\frac{L_c}{\varepsilon\,(\phi+m)^{e}}\right) \tag{2}$$

Where

$$Nu_{nf} = \frac{h_{b,nf}L_c}{k_{nf}} \tag{3}$$

$$Pr_{nf} = \frac{??_{nf}C_{p,nf}}{k_{nf}} \tag{4}$$

and L_c represents the characteristic length and a, b, c, d, e, and m are the constant coefficients.

The coefficients in Eq. (2) for TiO_2-water Nanofluidare obtained from the curve-fits applied to the measured data of the other authors' work [64] which is shown in Table 1 for the heating surface materials of copper and aluminum.

Pool Boiling Heat Transfer of Nanofluids

Table 1. The coefficients for the correlation

a	b	c	d	e	m
Coefficients for copper surface					
28.85	0.59	0.70	0.16	0.12	0.001
Coefficients for aluminum surface					
46.63	0.55	0.74	0.15	0.05	0.00001

Table 2. Error rates for the correlation development study [68]

Optimization methods	Proposed Correlations	Error Analysis		
		R^2	Prop. (%)	Square law
NM	$h = \left(\dfrac{1}{45.49}\right)\left[\dfrac{C_{p,l}q}{h_{fg}}\right]\left[\dfrac{q}{\mu_l h_{fg}}\left(\dfrac{\sigma}{g(\rho_l - \rho_v)}\right)^{1/2}\right]^{-0.2445}\left[\dfrac{Cp_l\mu_l}{k_l}\right]^{-1.179}$	0.929	11.954	13.571
NLS	$h = \left(\dfrac{1}{7741.22}\right)\left[\dfrac{C_{p,l}q}{h_{fg}}\right]\left[\dfrac{q}{\mu_l h_{fg}}\left(\dfrac{\sigma}{g(\rho_l - \rho_v)}\right)^{1/2}\right]^{-0.248}\left[\dfrac{Cp_l\mu_l}{k_l}\right]^{-1.989}$	0.938	11.103	11.873
NM	$h = \dfrac{1.1132 P_c^{0.24}}{M^{0.312}T_c^{0.417}}(0.478R_a)^{0.03\left(1-\frac{P}{P_c}\right)}\dfrac{\left(\dfrac{P}{P_c}\right)^{0.318}}{\left[1-0.61\left(\dfrac{P}{P_c}\right)\right]^{-0.012}}q^{0.737}$	0.925	12.673	14.192
NLS	$h = \dfrac{0.544 P_c^{0.411}}{M^{0.817}T_c^{0.747}}(0.716R_a)^{0.03\left(1-\frac{P}{P_c}\right)}\dfrac{\left(\dfrac{P}{P_c}\right)^{0.058}}{\left[1-0.629\left(\dfrac{P}{P_c}\right)\right]^{-0.849}}q^{0.737}$	0.925	12.673	14.192
NM-NLS	$h = 0.311 q^{0.738}$	0.92	12.459	15.306

3.2.2. Modelling by Artificial Neural Network

In another study [68] a comprehensive investigation based on the artificial neural network (ANN) analysis was performed to determine the important effective parameters on the boiling heat transfer coefficient and also to report reliable empirical correlations as shown in Table 2 and Figure 4.

Several parameters such as nanoparticles concentration, thermal conductivity and size, as well as thermophysical properties of the base fluid, surface roughness, and wall superheating, were considered as input variables, while the heat flux and boiling heat transfer coefficient were assumed as the output variables of the ANNs analysis.

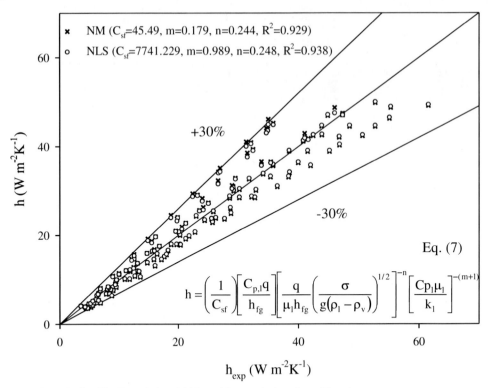

Reprinted from Ref. [68], Copyright, (2012), with permission from Elsevier.

Figure 4. Comparisons of experimental convective heat transfer coefficients with the most predictive proposed correlation obtained from optimization methods using 150 numbers of experimental data points.

A number of ANN methods such as multi-layer perceptron (MLP), generalized regression neural network (GRNN) and radial basis networks (RBF) were employed to model the nucleate pool boiling heat transfer characteristics of TiO_2 nanofluids. The mean relative error criteria with the use of unknown test sets were applied to determine the performances of the ANNs. It was found that the performance of the method of MLP with 10-20-1 architecture, GRNN with the spread coefficient 0.7 and RBFs with the spread coefficient of 1000 and a hidden layer neuron number of 80 are in good agreement, which predicts the experimental pool boiling heat transfer coefficient with deviations within the range of ±5% for all tested conditions.

Furthermore, the neural network analysis revealed that among all effective parameters the pool boiling heat flux of nanofluids depends on the nanofluid dynamic viscosity, heating surface roughness and wall superheating.

CONCLUSION

A comprehensive review of some important aspects of the pool boiling heat transfer of nanofluids considering the employed nanoparticle type along with representative results of present authors' investigation was expressed and analyzed in this chapter.

With reference to a substantial number of works in this area, considerable enhancement in boiling CHF has been observed due to use of nanofluids, while conflicting and inconsistent data have been reported regarding the effects of nanofluids on boiling heat transfer coefficient.

Since the heating surface characteristics have significant effects on boiling heat transfer, it is believed that the deposition of nanoparticles on the heating surface is the major factor in the CHF enhancement of nanofluids, through the mechanisms of modifying the surface area, the surface wettability and the bubble nucleation. Besides, several researchers have stated that nanoparticles deposition is just one of the possible mechanisms of CHF enhancement. They have proposed that the effects of nanoparticles on the physical properties of the base fluid, the growth and detachment of bubbles from heating surface as well as the motion of bubbles inside the base fluid are the other probable reasons of CHF enhancement.

Even though there are several investigations on boiling heat transfer of nanofluids, the available data are still limited and scattered to identify the governing mechanisms of boiling heat transfer characteristics of nanofluids clearly. One reason is that no comprehensive theory exists to explain the energy transfer processes in nanofluids. More organized experimental studies should be carried out on bubble dynamics, bubble effective parameters, heating surface characteristics, properties of nanoparticles and nanofluids as well as the operating conditions to understand the boiling heat transfer mechanism of nanofluids.

ACKNOWLEDGMENT

The first, second and third authors wish to thank the Department of Mechanical Engineering, Faculty of Engineering, King Mongkut's University of Technology Thonburi for providing them facilities during doing their research in Thailand. The fourth author would like to thank the National Science and Technology Development Agency, the Thailand Research Fund and the National Research University Project for the supporting.

REFERENCES

[1] Mourgues, A.; Hourtané, V.; Mullera, T.; Caron-Charles, M. *International Journal of Heat and Mass Transfer,* 2013, 57, 595–607.

[2] Bertsch, S. S.; Groll, E. A.; Garimella, S. V. *Nanoscale and Microscale Thermophysical Engineering,* 2008,12, 187–227.

[3] Taylor, R. A.; Phelan, P. E. *International Journal of Heat and Mass Transfer,* 2009, 52, 5339–5347.

[4] Sohel Murshed, S. M.; Nieto de Castro, C. A.; Lourenc͎ M. J. V.; Lopes, M. L. M.; Santos, F. J. V. *Renewable and Sustainable Energy Reviews*, 2011, 15, 2342–2354.

[5] Barber, J.; Brutin, D.; Tadrist, L. *Nanoscale Research Letters,* 2011, 6, 280.

[6] Kim, H. *Nanoscale Research Letters,* 2011, 6, 415.

[7] Wu, J. M.; Zhao, J. *Progress in Nuclear Energy,* 2013, 66,13-24.

[8] Vafaei, S.; Borca-Tasciuc, T. *Chemical Engineering Research and Design,* In Press.

[9] Das, S. K.; Putra, N.; Roetzel, W. *International Journal of Heat and Mass Transfer*, 2003, 46, 851–862.

[10] Bang, I. C.; Chang,S. H. *International Journal of Heat and Mass Transfer*, 2005, 48, 2407-2419.

[11] Wen, D.; Ding, Y. *Journal of Nanoparticle Research*, 2005, 7, 265–274.

[12] You S. M.; Kim J. H.; Kim K. H. *Applied physics letters*, 2003, 83, 3374–3376.

[13] Wen, D. *J. Nanoparticle Research*, 2008, 10,1129–1140.

[14] Kedzierski,M. A. *International journal of refrigeration*, 2011, 34,498-508.

[15] Prakash Narayan, G.; Anoop, K. B.; Sateesh, G.; Das, S. K. *International Journal of Multiphase Flow*, 2008, 34,145–160.

[16] Harish, G.; Emlin, V.; Sajith, V. *International Journal of Thermal Sciences*, 2011, 50, 2318-2327.

[17] Raveshi, M. R.; Keshavarz, A.; Mojarrad, M. S.; Amiri, S. *Experimental Thermal and Fluid Science*, 2013, 44, 805–814.

[18] Tang, X.; Zhao, Y. H.; Diao,Y. *Experimental Thermal and Fluid Science*, 2014,52, 88-96.

[19] Shahmoradi, Z.; Etesami, N.; Nasr Esfahany, M. *International Communications in Heat and Mass Transfer*, 2013,47,113–120.

[20] Soltani, S.; Etemad, S. Gh.; Thibault, J.; *International Communications in Heat and Mass Transfer,*2010, 37, 29–33.

[21] Xu, L.; Xu, J.; Wang, B.; Zhang, W.; *International Journal of Heat and Mass Transfer*, 2011,54,3309–3322.

[22] Jung, J. Y.; Kim, E. S.; Nam, Y.; Kang, Y. T. *International journal of refrigeration*, 2013, 36,1056-0161.

[23] Jung, J. Y.; Kim, E. S.; Kang, Y. T. *International Journal of Heat and Mass Transfer*, 2012, 55,1941–1946.

[24] Jeong, Y. H.; Chang, W. J.; Chang, S. H. *International Journal of Heat and Mass Transfer*, 2008, 51,3025–3031.

[25] Ahn, H. S.; Kim, M. H. *International Journal of Heat and Mass Transfer*, 2013, 62, 718-728.

[26] Peng, H.; Ding, G.; Hu, H.; Jiang, W. *International Journal of Thermal Sciences*, 2010, 49, 2428-2438.

[27] Peng, H.; Ding, G.; Hu, H. *Nanoscale Research Letters*, 2011, 6, 219.

[28] Kathiravan, R.; Kumar, R. Gupta, A.; Chandr R.; Jain, P. K. *International Journal of Heat and Mass Transfer*, 2011,54, 1289-129.

[29] Liu, Z. H.; Yang, X. F.; Xiong, J. G. *International Journal of Thermal Sciences*, 2010,49, 1156-1164.

[30] Xue, H. S.; Fan, J. R.; Hong, R. H.; Hu, Y. C. *Applied Physics Letters*, 2007, 90, 184107.

[31] Park, K. J.; Jung, D.; Shim, S. E. *International Journal of Multiphase Flow*, 2009, 35, 525-532.

[32] Park, K. J.; Jung, D. *International Journal of Heat and Mass Transfer*, 2007, 50, 4499–4502.

[33] Kole, M.; Dey, T. K. *Applied Thermal Engineering*, 2012, 37, 112-119.

[34] White, S. B.; Shih, A. J.; Pipe, K. P. *International Journal of Heat and Mass Transfer*, 2011, 54,4370-4375.

[35] Sharma, V. I.; Buongiorno, J.; McKrell, J. T.; Hu, L. W. *International Journal of Heat and Mass Transfer,* 2013, 61, 425-431.

[36] Vassallo, P.; Kumar, R.; D'Amico S. *International Journal of Heat and Mass Transfer,* 2004, 47, 407–411.

[37] Bolukbasi,A.; Ciloglu, D. *International Journal of Thermal Sciences,* 2011, 50, 1013-1021.

[38] Yang, X. F.; Liu, Z. H. *International Journal of Thermal Sciences,* 2011, 50,2402-2412.

[39] Milanova, D.; Kumar, R. *Applied physics letters,* 87, 233107, 2005.

[40] Peng, H.; Ding, G.; Hu, H.; Jiang, W. *International Journal of Heat and Mass Transfer,* 2011, 54,1839-1850.

[41] Peng, H.; Ding, G.; Hu, H. *Experimental Thermal and Fluid Science,* 2011, 35, 960-970.

[42] Kathiravan, R.; Kumar, R.; Gupta, A. *International Journal of Heat and Mass Transfer,* 2010, 53, 1673-1681.

[43] Kedzierski, M. A.; Gong, M. *International journal of refrigeration,* 2009, 32,79-799.

[44] Zeinali, S. H. *International Communications in Heat and Mass Transfer,* 2011,38, 1470-1473.

[45] Okawa, T.; Takamura, M.; Kamiya, T. *International Journal of Heat and Mass Transfer,* 2012, 55, 2719–2725.

[46] Phan, H. T.; Caney, N.; Marty, P.; Colasson, S.; Gavillet, J. *Nanoscale and Microscale Thermophysical Engineering,* 2010, 14, 229–244.

[47] Song, S. L.; Lee, J. H.; Chang, S. H. *Experimental Thermal and Fluid Science,* 2014, 52, 12–18.

[48] Sheikhbahai, M.; Nasr Esfahany, M.; Etesami, N. *International Journal of Thermal Sciences,* 2012, 62,149-153.

[49] Peng, H.; Ding, G.; Hu, H.; Jiang, W.; Zhuang, D.; Wang, K. *International journal of refrigeration,* 2010, 33, 347-358.

[50] Kim, S. J.; Bang, I. C.; Buongiorno, J.; Hub, L. W. *International Journal of Heat and Mass Transfer,* 2007, 50, 4105–4116.

[51] Kim, H.; Kim, J.; Kim, M. H. *International Journal of Heat and Mass Transfer,* 2006, 49, 5070-5074.

[52] Kim, H.; Kim, J.; Kim, M. H. *International Journal of Multiphase Flow,* 2007, 33,691–706.

[53] Lee, J. H.; Lee, T.; Jeong, Y. H. *International Journal of Heat and Mass Transfer,* 2012, 55,2656–2663.

[54] Peng, H.; Ding, G.; Hu, H. *International Journal of Refrigeration,* 2011, 34, 1823-1832.

[55] Truong, B.; Hu, L. W.; Buongiorno, J.; McKrell, T. *International Journal of Heat and Mass Transfer,* 2010, 53,85-94.

[56] Kwark, S. M.; Kumar, R.; Moreno, G.; Yoo, J.; You, S. M. *International Journal of Heat and Mass Transfer,* 2010, 53,972-981.

[57] Pham, Q. T.; Kim, T. I.; Lee, S. S.; Chang, S. H. *Applied Thermal Engineering,* 2012, 35, 157-165.

[58] Golubovic, M. N.; Madhawa Hettiarachchi, H. D.; Worek, W. M.; Minkowycz, W. *J. Applied Thermal Engineering,* 2009, 29, 1281–1288.

[59] Cieslinski, J. T.; Kaczmarczyk, T. Z. *Nanoscale Research Letters,* 2011, 6, 220.

[60] Park, S. D.; Lee, S. W.; Kang, S.; Kim, S. M.; Bang, I. C. *Nuclear Engineering and Design,* 2012, 252, 184– 191.

[61] Shoghl, S. N.; Bahrami, M. *International Communications in Heat and Mass Transfer,* 201345 () 122–129.

[62] Gerardi, C.; Buongiorno, J.O.; Hu, L.-W.; McKrell, T. *Nanoscale Research Letters,* 2011, 6, 232.

[63] Aminfar, H.; Mohammad pourfard, M.; Sahraro, M. *Computers and Fluids,* 2012, 66, 29–38.

[64] Suriyawong, A.; Wongwises, S. *Experimental Thermal and Fluid Science,* 2010, 34, 992-999.

[65] Trisaksri, V.; Wongwises, S. *International Journal of Heat and Mass Transfer,* 2009, 52, 1582-1588.

[66] Duangthongsuk, W.; Yiamsawasda, T.; Dalkilic, A. S.; Wongwises, S. *Current Nanoscience,* 2013, 9, 56-60.

[67] Suriyawong, A.; Dalkilic, A. S.; Wongwises, S. *Journal of ASTM International,* 2012, 9, Paper ID JAI104409,1-12.

[68] Balcilar, M.; Dalkilic, A. S.; Suriyawong, A.; Yiamsawas, T.; Wongwises, S. *International Communications in Heat and Mass Transfer,* 2012, 39, 424-431.

In: Nanofluids: Synthesis, Properties and Applications ISBN: 978-1-63321-677-8
Editors: S.M. Sohel Murshed, C.A. Nieto de Castro © 2014 Nova Science Publishers, Inc.

Chapter 9

NANOFLUIDS IN DROPLET-BASED MICROFLUIDICS

S. M. Sohel Murshed[1], and Nam-Trung Nguyen[2]*

[1]Centro de Ciências Moleculares e Materiais,
Faculdade de Ciências, Universidade de Lisboa,
Lisboa, Portugal
[2]Queensland Micro- and Nanotechnology Centre,
Griffith University, Australia

ABSTRACT

This chapter reports two sets of experimental investigations on the droplet formation and size manipulation of nanofluids in the microfluidic T-junction and flow focusing geometries which are commonly used in droplet-based microfluidics. In addition to studying the temperature dependence of droplet formation at both geometries, effect of other factors such as presence of nanoparticles in aqueous fluid, depth of microchannel and flow rate on the droplet formation and size manipulation are investigated. Nanofluids used in these microfluidics devices were prepared by dispersing spherical and cylindrical shaped TiO_2 nanoparticles in deionized water. In order to identify the effect of temperature on the droplet formation, all microfluidic devices were fabricated with integrated microheaters.

Temperature dependence of interfacial tension and viscosity of these nanofluids are also characterized as these properties play major role in droplet generation and flow in microfluidic systems. These nanofluids were found to exhibit substantially smaller oil-based interfacial tension and insignificantly higher viscosity than those of their base fluid and the viscosity of all fluids decreases with increasing temperature.

Results of droplet formation in microfluidic T-junction revealed that nanofluids generate different droplet size than the base fluid and the temperature has significant effect on the droplet formation process. Nanofluids showed different characteristics in droplet formation and size control with the temperature. Except nanofluids containing cylindrical-shaped nanoparticles, the droplet size was found to increase with increasing temperature. For instance, addition of spherical-shaped TiO_2 nanoparticles in deionized water resulted in much smaller droplet size compared to the cylindrically shaped

---

* E-mail: smmurshed@fc.ul.pt.

nanoparticles. This indicates that shape of dispersed nanoparticle influences the droplet formation as well as the droplet size. The depth of the channel also influences the droplet formation.

From the investigations of nanofluids in flow focusing geometry three different droplet breakup regimes and their transition capillary numbers as well as temperatures were identified. The heat generated by the integrated microheater changed the droplet formation process as increasing temperature significantly enlarged the size of the droplets. Results demonstrated that these nanofluids possess similar characteristics in droplet formation at different temperatures and any small change in the flow rate has little impact on the size of the droplets formed in this geometry.

1. INTRODUCTION

Both nanofluids and droplet-based microfluidics are emerging and very rapidly growing technologies in multidisciplinary research fields which can offer immense benefits and applications in numerous important fields. After several decades of efforts by researchers and engineers to develop fluids that have better heat transfer (cooling and heating) performance compared to available conventional fluids and can be applied for cooling modern miniaturized systems, the concept of "nanofluid" was only brought forward by Steve Choi at Argonne National Laboratory of USA in 1995 [1]. Nanofluids are innovative class of fluids and are engineered by dispersing nanometer-sized particles in conventional fluids. These new fluids have attracted great interest from the researchers worldwide due to their enhanced thermophysical properties and heat transfer performance as well as potential applications in various fields such as microelectronics, transportation, manufacturing, medical, and so on [2-11]. Although researchers are mainly focusing on thermophysical properties measurements and applications of nanofluids in thermal management systems and cooling technologies, application of nanofluids in microfluidic technologies has not yet received any attention. However, having superior properties and nano-sized dispersed particles, nanofluids are very suitable for the microfluidic systems and it can diversify the applications of both nanofluids and microfluidic technologies.

On the other hand, droplet-based microfluidics have also shown great potential in numerous applications such as chemical and biochemical analysis (reaction platforms for protein crystallization, cell encapsulation, polymerase chain reaction and DNA analysis), high throughput screening, and synthesis of nanoparticles [12-16]. Thus researchers have shown growing interest in formation and manipulation of microdroplets in microfluidic systems [17-23]. In addition, droplet-based microfluidic techniques are essential for large scale integration of devices for biological or chemical analyses as the microdroplets can ensure both physical and chemical isolation to its contents. The flow of two or more immiscible liquids into a microfluidic device and the formation of emulsion droplets as well as gas bubbles have been adequately studied [24-28]. Also considerable research efforts have been devoted on the droplet formation devices and techniques particularly droplet patterns variation in microchannels [29]. The microfluidic devices which are commonly used for droplet formation and manipulation are T-junction and flow focusing geometries [23-25, 27, 30].

In microfluidic systems, T-junction is one of the most frequently used configurations for droplet formation and manipulation [24, 31-33]. The wide usage of this geometry is due to the ease of droplet formation and uniformity of the formed droplets. Several studies employed different techniques to control the formation of droplets in microfluidic devices. Thorsen et al. [24] used pressure-controlled flow in a T-junction microchannel to generate water droplets using different types of oils as carrier fluids. Their results showed that the droplet size increases with increasing pressure of the dispersed phase. Whereas a decreased in droplet size with increasing flow rate of the continuous phase was observed in a study by Nisisako et al. [31]. The droplet formation at a T- junction geometry through shear force between two immiscible fluids can be controlled by various means such as pressure, flow rates of fluids, properties particularly viscosities and the interfacial tensions of fluids and the channel dimensions particularly depth [30, 34-36]. Thermally mediated droplets breakup in microchannels was also reported in the literature [37].

Microfluidic flow focusing devices can produce monodisperse microdroplets or bubbles at a high frequency. Thus numerous studies on the droplet formation and manipulation through this device have been conducted. For examples, droplet formation and manipulation in flow focusing devices have been achieved by various techniques such as using shear gradients [38], introducing heating effect [23], changing surfactant concentration [21], changing viscosity of the liquids [39], flow rate ratios, and flow rates [25, 40, 41]. Geometrical techniques have also been used to generate droplets of various sizes and size distributions [35]. Anna et al. [25] studied the formation of water droplet in a flow focusing device and showed that the droplets size can be adjusted by controlling the oil flow rate. Xu and Nakajima [42] produced highly monodisperse soybean oil droplets (<1% polydispersity) in 1 wt.% sodium dodecyl sulfate (SDS) solution by using a simple microfluidic flow focusing geometry. Based on the concept of an earlier study on thermally mediated breakup and switching of droplets in T-junction bifurcation [37], Nguyen et al. [23] demonstrated that the droplet formation in a flow focusing device can also be controlled by applying the heat at the droplet formation location. By using different flow rate ratios of two immiscible liquids, the existence of four different droplets formation regimes such as geometry-controlled or squeezing, dripping, thread formation, and jetting was reported by Anna and Mayer [21]. Each regime was identified by a critical capillary number whose value changed with the concentration of the surfactant added [21].

Water is most commonly used as aqueous liquid (also known as dispersed phase) to generate droplets through different droplet formation devices such as T-junction microchannel and flow focusing device. On the other side, nanofluids which exhibit superior thermophysical and interfacial properties [2-9, 43] such as thermal conductivity, thermal diffusivity, viscosity and interfacial tension as compared to their base fluids can also be used for droplet generation and manipulation in those microfluidic devices. Nevertheless, nanofluids containing small concentration of nanoparticles in base fluids are suitable for the application in microfluidic systems. This is mainly possible because nanoparticles can be well-dispersed in base fluids and they are also orders of magnitude smaller than the microfluidic devices or channels. As noted before both nanofluids and droplet-based microfluidics have attracted great interest from the researchers of various disciplines [9-11, 22, 29]. However except research conducted by this group [44, 45] and a recent numerical study [46] no other investigation on the droplet formation and manipulation of nanofluids in droplet-based microfluidic systems has been reported in the literature. In order to exploit the potential of nanofluids in diverse fields, it is worthy to comprehensively study the application of nanofluids in droplet formation and manipulation in microfluidics.

Interfacial tension and viscosity of liquid play an important role in continuous flow- and droplet-based microfluidics. In microfluidics, the size of droplet can be controlled with the interfacial tension and dynamic viscosity, which are directly related to the droplet size at a fixed shear condition. The temperature-dependent interfacial tension and viscosity of the liquids are also crucial in thermally mediated droplet formation [23, 47] as well as breakup [37]. These properties need to be characterized in order to identify their contributions in droplet formation.

In this chapter experimental investigations on the formation and manipulation of droplets of deionized water and deionized water-based TiO_2 nanofluids in both the T-junction microchannel and flow focusing geometries at different temperatures are presented. The temperature dependence of interfacial tension and viscosity of nanofluids are characterized. The effects of various important factors on the droplet formation are investigated. For example, the influences of temperature and channel depth on the droplet formation process and size control are studied. The effects of heating on the droplet formation regimes at constant flow rates of both aqueous and carrier nanofluids are investigated and the transition temperature is identified for flow focusing geometry. The effects of nanoparticles shape and nanofluids flow rate on the droplet formation are also examined. All these investigations are to assess the influence and applicability of nanofluids in droplet formation and manipulation in droplet-based microfluidic systems.

2. SAMPLE PREPARATION AND PROPERTIES CHARACTERIZATION

2.1. Preparation of Sample Nanofluids

In order to use nanofluids in droplet formation and manipulation in microfluidic devices, two types of sample nanofluids (NF) were prepared by dispersing 0.1 volume% of purchased titanium dioxide (TiO_2) nanoparticles of two different shapes in deionized water (DIW). As provided by the supplier the spherical-shaped TiO_2 nanoparticles had average diameter of 15 nm and the cylindrical shaped nanoparticles had size of 10 (diameter) × 40 (length) nm. To ensure proper dispersion of nanoparticles in base fluid the sample nanofluids was homogenized by using an ultrasonic dismembrator (Fisher Scientific Model 500) and a magnetic stirrer. While mineral oil with 2% w/w Span 80 surfactant (Sigma S6760) was used as the carrier fluid in microchannel, deionized water and nanofluids with 0.05% w/w fluorescence dye (Sigma F6377) were used as the aqueous fluids for the droplet formation.

2.2. Properties Characterization of Nanofluids

As mentioned before the interfacial tension and viscosity of the liquids play crucial role in droplet formation and droplet breakup. The influence of temperature on these properties can considerably change the droplet formation and breakup in any microfluidic devices. Thus the viscosities and interfacial tensions of nanofluids, their base fluid (i.e., DIW) as well as carrier fluid (oil) were measured at various temperatures [43]. While the interfacial tensions of these fluids were measured with a FTA 200 video system (First Ten Angstroms, Taiwan), viscosities of all liquids were measured with a Lowshear Rheometer (LS 40, Mettler Toledo, Switzerland) at different temperatures. These equipment was calibrated by measuring the properties of base fluid. Based on the deviation between standard and measured values of the properties of deionized water at room temperature, the present measurement error was estimated to be within ±1%. All measurements were performed at atmospheric pressure.

The normalized viscosities and interfacial tensions of aqueous fluids (DIW and NF) and carrier fluid as a function of temperature are presented in Figure 1. The viscosities and interfacial tensions of these fluids are normalized by their corresponding nominal values at 25°C i.e. $\eta^* = \eta(T)/\eta(T = 25^0 C)$ and $\gamma^* = \gamma(T)/\gamma(T = 25^0 C)$. While viscosities of DIW, mineral oil with 2 w/w % surfactant (Span 80), and TiO_2/DIW-nanofluid at 25°C were respectively 0.90 mPa.s, 26.4 mPa.s and 0.93 mPa.s, the interfacial tensions of DIW/oil and NF/oil systems at 25°C are 27.35 mN/m and 15.9 mN/m, respectively. As anticipated the viscosities of sample fluids were found to decrease significantly with increasing temperature. However, at room temperature and at 30°C nanofluid showed slightly higher viscosity compared to base DIW. Figure 1 also demonstrates that the interfacial tension of this nanofluid in mineral oil is significantly smaller than that of the base fluid in the same oil. Nanofluid showed nearly linear trend of decreasing interfacial tension with increasing temperature. This result was in good agreement with the calculated results from the classical Girifalco and Good [48] model which was shown in our previous study [43]. The possible reason for such smaller interfacial tension of nanofluid is that nanoparticles can easily experience Brownian motion and interact with the liquid molecules resulting to a reduced cohesive energy at the interface. Moreover, temperature intensifies the Brownian motion and it is known that the lower the cohesive energy the smaller the interfacial tension. Furthermore, nanoparticles can be adsorbed at the interfaces (liquid-liquid) and function in similar ways to surfactants to reduce the interfacial tension [49, 50].

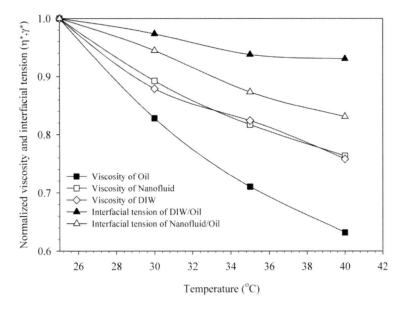

Figure 1. Temperature-dependent viscosity and interfacial tension of the sample fluids (normalized by their nominal values at 25°C).

3. Theoretical Basis

The dimensionless capillary number (Ca) is the most important parameter used to characterize the droplet formation regime in microfluidic devices through the relative importance of viscous stress and interfacial tension or capillary pressure exerted on the droplet. It also identifies the dominant parameter in droplet formation. This capillary number for T-junction systems is commonly defined by

$$Ca = \eta_c u_c / \gamma \tag{1}$$

where η_c and u_c are the viscosity and flow velocity of the carrier fluid, respectively and γ is the interfacial tension of carrier fluid/aqueous fluid. The flow velocity of the carrier fluid can be determined from $u = Q/(hw)$ where Q is the flow rate, h and w are the depth and width of the channel, respectively (e.g., Figure 2).

On the other hand, the capillary number expression for the flow focusing geometry has slight different form and is given as [21]:

$$Ca = \frac{\eta_c aG}{\gamma} \tag{2}$$

where a is the half-width of the microchannel of the aqueous fluid, and G is the effective shear or elongation rate which can be determined from a similar approach adopted by Anna and Mayer [21] as

$$G = \frac{Q_c}{h}[\frac{1}{wz} - \frac{1}{b^2}] \tag{3}$$

where Q_c is the total flow rate of carrier fluid and h, b, w, and z are the depth, width of the oil flow channel, width of the orifice entrance, and the distance from the end of the aqueous channel to the orifice entrance, respectively (e.g., Figure 3). Thus using the above shear rate given by Eq. (3) into Eq. (2), the capillary number for the flow focusing geometry has the form

$$Ca = \frac{\eta_c aQ_c}{\gamma h}[\frac{1}{wz} - \frac{1}{b^2}] \tag{4}$$

Using two empirical correlations for temperature dependence of viscosity and interfacial tension of the aqueous fluid in oil [23] in Eq.(4), the capillary number as a function of temperature can be determined from following expression:

$$Ca(T) = \frac{\eta_0 a Q_c}{\gamma_o h} [\frac{1}{wz} - \frac{1}{b^2}]\exp(-0.02\Delta T) \tag{5}$$

where η_o (23.8×10^{-3} Pa·s for mineral oil with Span 80 [23]) and γ_0 (3.65×10^{-3} N/m for DIW/oil [23] and 15.9 mN/m for nanofluid/oil systems [44]) are respectively viscosity of career fluid and interfacial tension between career fluid and aqueous fluid at room temperature.

In addition to changing flow rate [21], Eq.(5) reveals that the temperature can be used to control the capillary number and consequently the droplet formation regimes. An increase in temperature reduces the capillary number through the influence of viscosity over interfacial tension and thus could move the formation process of the droplets across different regimes.

When the diameter of generated droplets are larger than the channel depth, the droplets will have the form of a disc and for such non-spherical shape of droplets, the concept of equivalent diameter with respect to same volume of spherical droplet needs to be adopted. Based on a correlation used by Nie et al. [39], the equivalent diameter of a spherical droplet with the same volume can be determined from

$$D_{eq} = 2\sqrt[3]{\frac{1}{16}[2D^3 - (D-h)^2(2D+h)]} \tag{6}$$

where h is the depth of the channel and D is the measured diameter of the disk.

4. MICROFLUIDIC EXPERIMENTAL DETAILS

4.1. Design and Fabrication of Microfluidic Devices

As mentioned previously droplets in microchannels are usually generated by two configurations, which are T-junction and flow focusing through a small orifice. Both the T-junction and flow focusing devices with integrated microheaters and heating sensors were designed and fabricated for conducting experiments on temperature dependence of droplet formation and size manipulation. While the microheater provides localized heating, the sensor detects the induced temperature. Both the T-junction and flow focusing devices were fabricated using micromachining of glass and polydimethylsiloxane (PDMS). One of the reasons to used PDMS is that it yields better surface finishing as compared to laser machined polymethyl-methacrylate (PMMA) devices [30, 37]. Photolithography and lift-off technique were used to pattern the microheater and temperature sensor which were made of thin-film platinum. While the microchannel network was fabricated in PDMS using soft lithography, the master mold was fabricated by photolithography of the thick-film resist SU-8 using a transparency mask. The glass wafer with the patterned microheater and microsensor was subsequently coated with a thin PDMS layer before being bonded to the PDMS part with the microchannel network. Bonding is achieved using oxygen plasma treatment on both PDMS surfaces (Plasma-Therm Inc. Florida, USA) at 70W for 50s. The alignment was performed manually after the oxygen plasma treatment. The bonded layers were then thermally treated at

150 °C for two hours to reinforce the bonding strength and to ensure a leak-proof bonding interface. The heat treatment also ensured the hydrophobic recovery of the PDMS surface which was essential for the formation of water-in-oil droplets. Three different T-junction devices were fabricated. All channels of one T-junction device were 90 μm deep and the widths of the carrier and injection channels were 150 μm and 50 μm, respectively. The widths of the carrier channel and the injection channel of other two T-junction devices were 100 μm and 50 μm, respectively whereas the depths of channels of these two devices were 300 μm and 30 μm. The area of the fabricated microdevice measures 1×1 cm^2. The schematic of the T-junction is shown in Figure 2.

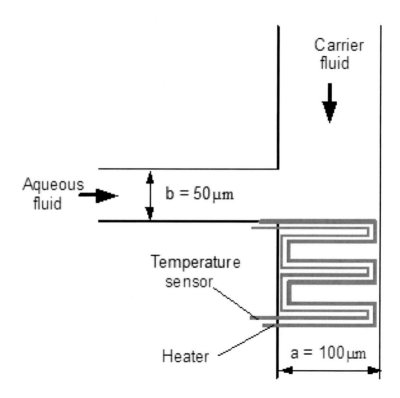

Figure 2. Schematic concept of a T-junction with integrated microheater and temperature sensor.

A microfluidic flow-focusing device (MFFD) with integrated microheater and temperature sensor was designed and fabricated. The MFFD was also fabricated using micromachining of glass and polydimethylsiloxane (PDMS) devices [23, 37]. The techniques and materials used to fabricate the MFFD are the same as discussed above and used for the T-junction device. The footprint of the MFFD is 1cm×1cm. The depth of the flow focusing microchannels was 30 μm and the widths of both the carrier channel and the injection (or aqueous) channel was of 200 μm. Figure 3 depicts the schematic concept and dimensions of the microfluidic device with integrated microheater and temperature sensor used for droplet formation experiments.

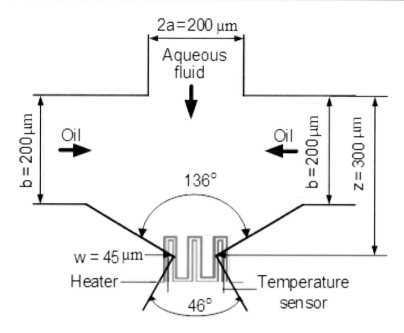

Figure 3. Schematic concept of the flow focusing devices with integrated microheater and temperature sensor.

4.2. Experimental Setup and Procedure

In order to generate droplets at the both microfluidic geometries (T-junction and flow focusing), two precision syringe pumps from KD Scientific Inc., USA were used to drive the oil and the aqueous fluids. The flow rates of both the carrier fluid and aqueous fluids were adjusted to form uniform sized droplets. Prior to conduct experiments the temperature sensor was calibrated and its resistance value was used for the in-situ temperature measurement. The temperature which was adjusted by changing the voltage of the heater was monitored through the resistance of the sensor. An epi-fluorescent inverted microscope with a filter set (Nikon B-2A) was employed to observe the droplets formation and flow. A sensitive CMOS camera (Basler A504K, Basler AG, Germany), which has a maximum frame rate of 500 fps, was used for recording the droplet images. The recorded droplet images were then processed by a customized MATLAB program to obtain the droplet diameter. For droplets of disc or other non-spherical shapes the equivalent diameters were determined using Eq. (6). Details of the experimental procedure can be found elsewhere [44, 45] and is not elaborated here.

5. Results and Discussion

5.1. For Microfluidics T-junction

5.1.1. Droplet Formation

Although the droplet formation process at T-junction has adequately discussed in the literature [19,24, 33,36, 51] and it is briefly discussed here mainly for better understanding by researchers working on nanofluids. In addition, the regimes of droplet formation in experiments are identified and discussed. During the early stage of a droplet formation, the interfacial tension between the carrier fluid and aqueous fluid is large enough to hold the droplet from detachment. At the same time, the extruded aqueous phase volume grows due to its continuous flow which means that the effective drag force increases as the droplet grows. When the droplet growth reaches a stage whereby the drag force is large enough to overcome the interfacial tension force, the droplet detaches and carries away by the flow of carrier fluid to downstream. Thus the force balance between the shear force and the force of interfacial tension determines the droplet size formed inside the microfluidic channel. A recorded image of droplet formation at a T-junction with embedded microheater at the junction and flow of formed droplet is depicted in Figure 4.

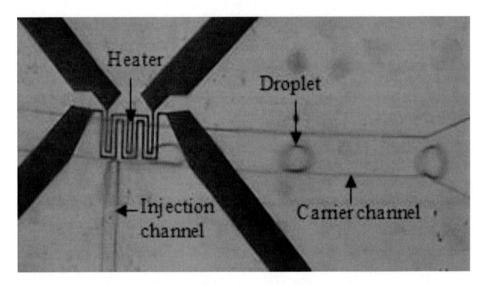

Figure 4. A typical image of T-junction droplet formation and flow through the microchannel.

Figure 5 also presents recorded images of DIW droplet formation process at the T-junction of 30μm channel depth at different stages [45]. At the initial stage of droplet formation (Figure 3(a)), the aqueous phase extrudes into the main channel (channel for carrier fluid) and forms a half disk like extrusion. As the flows proceed the extruded aqueous phase starts to grow and gains more volume. Meanwhile the continuous flow of carrier fluid deforms the extruded aqueous phase. Figure 3(b) depicts the moment right before the detachment of the droplet and form into a droplet. The droplet necking which helps connecting the droplet between the aqueous phase inlets was also observed in the experiments as can be seen from Figure 3(b). Figure 3 (b) also represents unconfined breakup of droplet which is expected because the width of the carrier channel is much larger than that of the injection channel of the T-junction devices used. This unconfined breakup confirms that droplet size is primarily controlled by local shear stress [22]. Figure 3(c) shows a newly formed droplet whose shape is not yet steady.

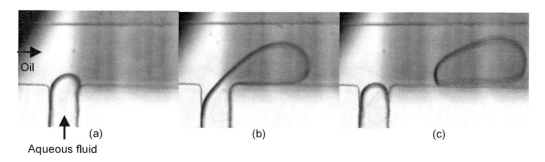

Figure 5. Droplet formation stages at a microfluidic T-junction: (a) forming; (b) before detachment; (c) droplet after detachment.

5.1.2. Effect of Temperature and Nanoparticles on Droplet Size

When all geometrical parameters and flow rates are fixed, the control over droplet size can be achieved through the precise control of the temperature at the T-junction. In addition, dispersion of nanoparticles in the base fluid can also play a role in its droplet formation and size manipulation. In order to investigate the effect of these factors on droplet formation, the droplets of DIW and nanofluids were generated at T-junction microchannels.

As an example of influence of temperature on droplet size, Figure 6 illustrates recorded images of droplets at different temperatures at a 30 μm depth T-junction microchannel [45]. It can clearly be seen that the droplet size increases considerably with increasing temperature. Note that all flow rates were kept constant (6 μl/h for the aqueous fluid and 12 μl/h for the carrier fluid). The heater generates a thermal gradient in the carrier fluid around the T-junction resulting in a lower viscosity and a reduced interfacial tension between aqueous fluid and carrier fluid. The increase of temperature results in a faster decrease in viscosity than the interfacial tension (Figure 1). The extended droplet detachment time leads to larger droplets compared to those formed without any thermal effect.

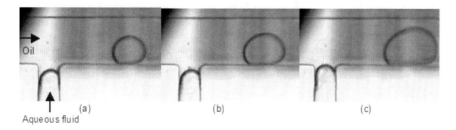

Figure 6. Images of droplet size variation at different temperatures: (a) 25°C, (b) 32°C, and (c) 39°C.

Droplet formation and size of both nanofluid (NF) and its base fluid (DIW) are characterized at different temperatures. For disc or other non-spherical shaped droplets the concept of equivalent diameter was adopted and the size of droplet (diameter) was determined using Eq. (6).

Figure 7 compares the temperature-dependent normalized droplet sizes of DIW and TiO_2 (15 nm)/DIW-based nanofluid at constant flow rates of 6 μl/h for the aqueous fluids (DIW and nanofluid) and 12 μl/h for the carrier fluid (oil). Droplet diameters are normalized by their nominal values at 25°C which are 85 μm for DIW and 68 μm for nanofluid. The droplet sizes of both nanofluid and its base fluid (DIW) were found to increase nonlinearly with increasing temperature. It can be seen (Figure 7) that the droplet sizes of this nanofluid are smaller than that of DIW and the dependency on temperature is less significant compared to the base fluid. The formation of smaller droplets for this nanofluid indicates that in the case of nanofluid the different flow regimes may experience at this depth (30μm) of microchannel. The observed differences between droplets of DIW and nanofluid could be caused by the different temperature-dependent thermophysical and interfacial properties particularly viscosity and interfacial tensions of nanofluid. Furthermore, adsorbed nanoparticles at liquid-liquid interfaces reduce the interfacial tension of nanofluid which can result in the formation of smaller droplets. During flowing through microchannels at such small flow velocity the agglomeration and sedimentation of nanoparticles may take place which may also affect the droplet formation process.

Effect of temperature on the droplet size of DIW and TiO$_2$ (15 nm)/DIW-based nanofluid in a T-junction of 90 μm deep microchannel has been demonstrated in Figure 8. The droplet diameters are normalized by their nominal values at 25°C which are 180 μm for DIW and 223 μm for NF. The flow rates of the carrier fluid (oil) and the aqueous fluids (DIW and nanofluid) were kept at constant values of 300 μl/h and 60 μl/h, respectively. In this case the sizes of droplets of nanofluid are larger than its base fluid and the dependency on temperature is slightly more significant compared to the base fluid (Figure 8). It can be seen that the enhancement in droplet sizes of both fluids fluctuate some extend with increasing temperature and the reasons for such increase in droplet size are not known yet. Since the viscosity of nanofluid does not change significantly with temperature, the droplet formation at this channel size and flow conditions is governed by the interfacial tension, which is much smaller for nanofluid than that of DIW. In this case, the capillary number (*Ca*) for deionized water and nanofluid at room temperature were (calculated using Eq.(1)) 2.8×10^{-3} and 4.1×10^{-3}, respectively and according to Garstecki et al. [51] for such small values of capillary numbers, the droplet formation is in the squeezing regime. This indicates that interfacial tension dominates viscous stress. Nevertheless, it is not yet well-understood how the nanoparticles contribute in changing the droplet size apart from the interfacial tension.

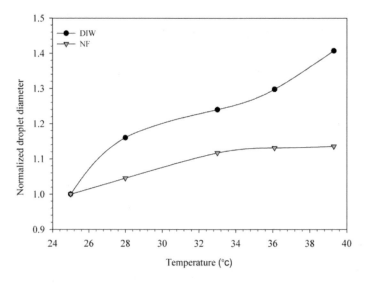

Figure 7. Temperature dependence of droplets size of DIW and NF at T-junction microchannel of 30 μm depth (droplet diameters are normalized by their nominal values at 25°C).

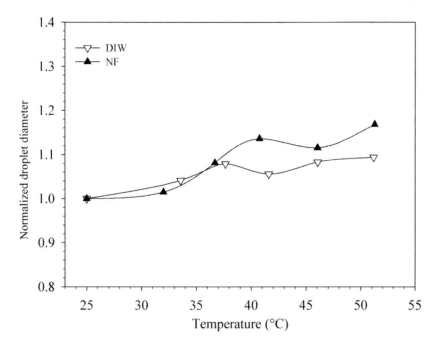

Figure 8. Temperature dependence of droplet size of DIW and NF at T-junction of 90 μm deep microchannel (droplet diameters are normalized by their nominal values at 25°C).

Analyzing data presented in Figures 7 and 8 it can be concluded that the effect of temperature on droplet size depends on various factors such as droplet depth, flow rates as well as aqueous fluids properties particularly viscosity and interfacial tensions. While 30 μm deep T-junction channel showed much lower enhancement in relative droplet size (with corresponding droplet size at room temperature) of this nanofluid compared to that of the base DIW at all temperatures (Figure 7), nanofluid exhibits noticeably larger enhancement of relative droplet size than that of DIW at 90 μm deep T-junction at temperatures above 35°C (Figure 8). Both channel dimensions particularly depth and flow rate are believed to be responsible for such temperature-dependent variation of droplet sizes.

5.1.3. Effect of Shape of Nanoparticle on Droplet Size

The effect of nanoparticle shape on the droplet formation at different temperatures has been studied and the results are presented in Figure 9. Two different shapes (spherical and cylindrical) of TiO_2 nanoparticles were used in order to quantify the effect of nanoparticles shapes on droplet formation. The T-junction device of 30 μm channel depth was used for these experiments. The droplet size of nanofluid containing cylindrical-shaped nanoparticles was found to decrease with increasing temperature particularly at higher temperatures. This is in contrast to the trend observed for deionized water and other nanofluid. The reason for such decreasing of droplet size is unclear at this moment and it requires more investigations. Figure 9 also demonstrates that addition of cylindrical-shaped nanoparticles in deionized water results in much larger droplet size compared to the spherical-shaped nanoparticles. The slip at the interface due to the presence and alignment of cylindrical-shaped nanoparticles could be a possible reason. The slip reduces the shear and allows the droplet to grow. Since no study on the droplet formation of nanofluids at microfluidic T-junction is available in the literature, no comparison of the present results can be made.

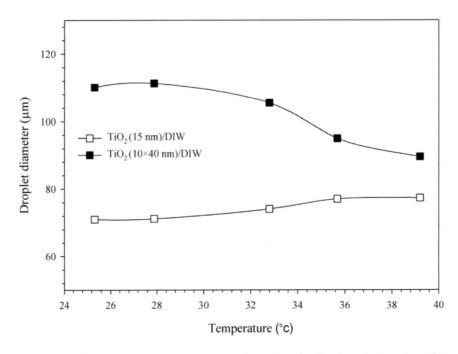

Figure 9. Effect of nanoparticle shape on temperature-dependent droplet size at T-junction of 30μm deep microchannel.

4.2. For Microfluidic Flow Focusing Device

4.2.1. Droplet Formation and Effect if Temperature

In order to explore the potential application of nanofluids in flow focusing devices of droplet-based microfluidics and to understand the effect of dispersed nanoparticles on droplet formation and droplet size, DIW-based TiO_2 (15 nm) nanofluid was used for the droplet formation experiment. While two flow rates of aqueous fluid (NF) were used the flow rate of the carrier fluid (oil) and other aqueous fluid (DIW) were kept constant value of 30 μl/h and 5 μl/h, respectively. Two different flow rates (3 μl/h and 5 μl/h) of nanofluid were used to quantify the effect of flow rate on nanofluid's droplet formation. At these flow rates and room temperature, a strong elongational flow of a continuous stream of the aqueous fluid flowing through a constriction draws a thin filament through the orifice. This filament subsequently breaks into droplets [21]. As the temperature at the constriction can be controlled by the microheater and monitor through the temperature sensor, the droplets are formed at different temperatures within the limit of heater and flow focusing device. Three different droplet formation regimes in a relatively small temperature range from 25 °C to 45 °C were observed.

At room temperature (i.e. 25°C), the droplet formation of DIW clearly followed the dripping regime as can be seen from the recorded images presented in Figure 10. The characteristics of encountering this dripping regime is that the tip of the non-moving water is fixed and the oscillations are localized around the constriction. As a droplet breaks free, immediately starts forming a new droplet as shown in Figure 10(a). In this case, the aqueous phase finger is longer and narrower. Figure 10(b) shows the moment of droplet forming at the constriction and before just breakup. The breakups are observed at a fixed point at the orifice due to the focused velocity gradient created by the nozzle shape geometry [52]. The droplet size distribution is highly monodispersed. The existence of dripping regime was observed within 30°C. It also found that with an increase in temperature beyond 30°C, the droplet formation mode starts to change to another mode. As observed in this experiment, the possible existence of two formation regimes at a constant flow rate was previously proposed in a study by Nguyen et al. [23].

At a temperature of around 35°C the droplet breakup was found to be geometry controlled or squeezing regime as described by Anna et al. [25]. Figure 11 depicts the images of droplet formation and breakup at the squeezing regime (mode A). The tip gradually narrows down as it extends towards the constriction and then penetrates into the constriction and grows into a bulb behind the constriction (Figure 11 (a)) until it breaks off. After breakup, the tip retracts back upstream (Figure 11(b)) and start forming another new droplet. At temperature above 30°C both DIW and nanofluid showed different droplet formation modes in the squeezing regime which were termed as squeezing regime A and regime B. The forming tip is pointed toward the constriction and has a relatively small radius of curvature in the squeezing regime A and the tip grows into a bulb with a large radius of curvature before the constriction and the bulb is then squeezed through the constriction to form the actual droplet behind the constriction for squeezing regime B. Such change of formation mode in a regime is mainly driven by effect of temperature-dependent interfacial tension at two different temperatures (i.e. 35°C and 45°C).

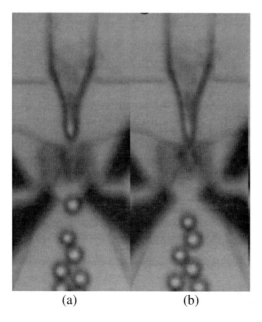

(a)   (b)

Figure 10. Images showing dripping regime encountered at flow focusing device: (a) started forming new droplet after breaking up one and (b) during droplet formation at the constriction.

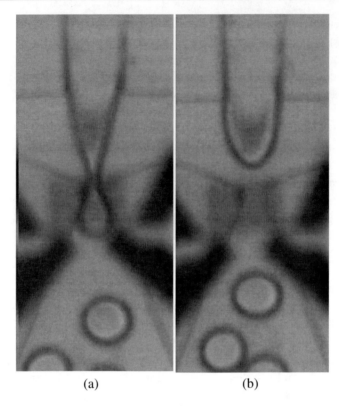

Figure 11. Images of squeezing regime (A) encountered at flow focusing device: (a) during droplet formation (by breaking up) and (b) after releasing droplet.

As discussed before the capillary number (Ca) is commonly used to characterize droplet formation regime. Figure 12 depicts the capillary number as a function of temperature for both the base fluid (DIW) and nanofluid in flow focusing geometry at flow rate ratio (carrier fluid: aqueous fluid: carrier fluid) of 30:5:30 (µl/h). Three different droplet breakup regimes and their transition capillary numbers as well as temperatures were identified (Figure 12). An increase in temperature results in a change (decrease) in the capillary number mainly through the interfacial tension and viscosity of the carrier fluid. The capillary number was calculated from Eq.(5). It can be seen that nanofluid yielded much smaller capillary number than that of the DIW. For DIW, the dripping regime is observed from 25 °C to 30 °C and at the capillary number range of 0.018>Ca>0.016 (Figure 12). The same dripping regime temperature range (as for DIW) was observed for nanofluid suggesting that at low temperature both the nanofluid and DIW droplets break up by dripping. In the temperature range of 30 °C to 42 °C, the squeezing regime A was observed and corresponding range of the capillary number found to be 0.016>Ca>0.013. At a temperature around 42 °C, the formation changes from one mode to another mode of squeezing regime (squeezing regime B) and corresponding range of the capillary number found to be Ca<0.013. It is also noticed that the change of squeezing regime A to regime B for nanofluids started at lower temperature compared to the case of DIW. The present range of capillary number for water in squeezing regime is within the range reported by Zhou et al. [53]. However, since no other study used nanofluid in microfluidic flow focusing geometry, no comparison of droplet formation or breakup regime for nanofluid can be made.

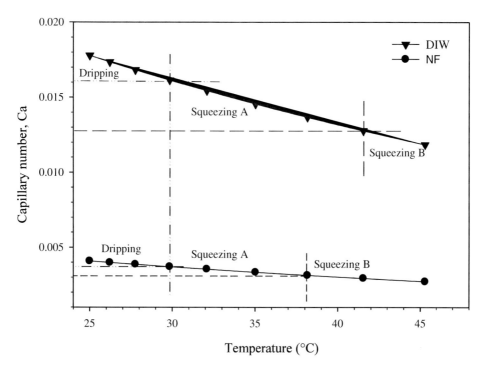

Figure 12. Capillary number and droplet formation regimes as a function of temperature at constant flow rates of fluids.

Like T-junction geometry, temperature plays a major role in changing the droplet size through its formation process at flow focusing devices. The images of droplets of nanofluid at three different temperatures and at constant flow rates of fluids (30 µl/h for oil and 3 µl/h for nanofluid) are depicted in Figures 13 which clearly demonstrate that with an increase in temperature the droplet size also increases significantly.

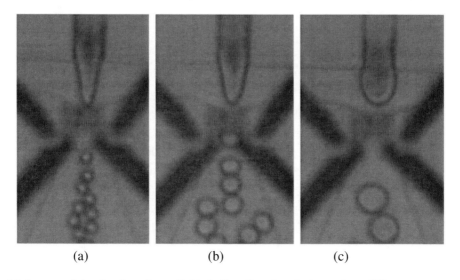

Figure 13. Images of droplet size of nanofluids in flow focusing device at three different temperatures: (a) 25 °C, (b) 35 °C, and (b) 41.5 °C.

4.4.2. Effect of Nanoparticle and Nanofluid Flow Rate on Droplet Size

The different regimes discussed in previous section reflected well in the measured droplet diameters as shown in Figure 14. To investigate the influence of nanoparticle suspension, the results of nanofluid are compared with DIW at constant flow rate ratios of 30:5:30 (µl/h). While at 25°C the sizes of droplets of DIW and nanofluid are about 62 µm and 49 µm, respectively and at 45°C the droplet sizes are found to be 122µm and 125 µm, respectively. The observed differences are mainly due to the differences in the interfacial tension and viscosity of both fluids and the dependence of temperature on these properties as reported in our previous study [43]. Both fluids exhibit similar formation behavior in three different regimes. Dripping regime and squeezing regimes (A and B) are reflected well in the change of droplet diameters as shown in Figure 14.

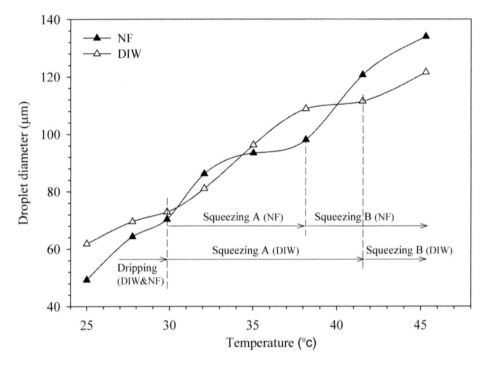

Figure 14. Comparison of droplet size of DIW and DIW-based nanofluid formed at constant flow rates of fluids.

The flow rates of both aqueous and carrier fluids greatly influence the droplet formation process. Thus the effect of different flow rates of nanofluid was investigated. As can be seen from Figure 15, there is a slight decrease in droplet size with increase in NF flow rate from 3 µl/h to 5 µl/h [44]. Although this is not anticipated, at room temperature condition similar decrease in size of water droplet was also observed by Yobas et al. [54] when the water flow rate was increased from 0.83 µl/min to 1.7 µl/min in their study. These results suggest that the droplet size is weakly dependent on the small change in the flow rate of the dispersed phase. In the case of a large change (order magnitude) of flow rate, the dependence of aqueous phase flow rate on the droplet size could be stronger.

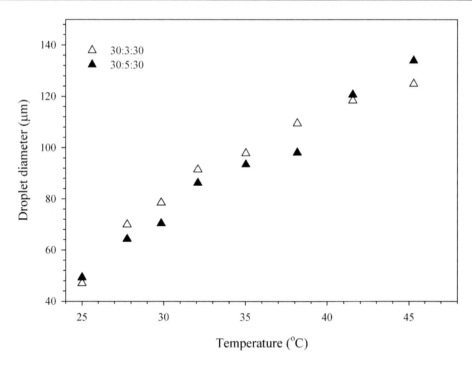

Figure 15. Effect of flow rates of nanofluid on droplet size for 30 μm channel depth.

CONCLUSION

Experimental investigations on the droplet formation and size manipulation of aqueous-TiO$_2$ nanofluids and their base fluid in the T-junction and flow focusing geometries of droplet-based microfluidics are reported. In addition to studying the dependency of temperature on the droplet formation at both geometries, effects of various important factors such as presence of nanoparticles in water, shapes of nanoparticles, microchannels dimension (depth), and fluids flow rates on the droplet formation and size manipulation are comprehensively investigated. All microfluidic devices were fabricated with integrated microheaters in order to investigate the effect of temperature on droplet formation.

As interfacial tension and viscosity of fluids play key role in droplet generation and flow in microfluidic systems, temperature dependence of interfacial tension and viscosity of all sample fluids are also characterized first. Nanofluids were found to exhibit substantially smaller oil-based interfacial tension and insignificantly higher viscosity than those of the base fluid and as usual the viscosity and interfacial tension of nanofluid decrease with increasing temperature. The Brownian motion and adsorption of nanoparticles at the interfaces could be responsible for the reduced interfacial tension of nanofluid.

Results from droplet formation experiments in T-junction showed that the addition of small concentration of nanoparticles in base fluid can significantly alter its droplet formation process, droplet size as well as their dependence on temperature. The temperature was found to have considerable effect on the droplet formation process of both nanofluids and their base fluid. Nanofluids exhibit different characteristics in droplet formation and size control with the temperature. Except nanofluids with cylindrical-shaped TiO$_2$ nanoparticles, the droplet size for nanofluid having spherical-shaped TiO_2 nanoparticles and base fluid were found to increase with increasing temperature. Nanoparticle shape also influences the droplet formation as addition of cylindrical-shaped nanoparticles in deionized water results in much larger droplet size compared to the spherical-shaped nanoparticles. This could be because of the slip at the interface which is generated due to the presence and alignment of cylindrical-shaped nanoparticles. The dimensions of microchannels particularly channel depth can also influence the droplet formation process significantly.

Investigations of droplet formation of both the nanofluid and base fluid (DIW) in flow focusing device demonstrated three different droplet formation regimes and their corresponding transition capillary numbers as well as transition temperatures. At constant flow rates of aqueous and carrier fluids, the droplet formation regimes can be changed by controlling the temperature. The capillary numbers of nanofluid was much smaller than the base fluid. In addition, the change of modes in squeezing regime for nanofluid begins at lower temperature compared to the case of DIW. Like T-junction, the heat generated by an integrated microheater changed the droplet formation process for both fluids as increasing temperature enlarged the size of the droplets significantly. This demonstrates that heating at microfluidic devices can effectively control the droplet formation and size manipulation. Nanofluid was found to possess similar characteristics in droplet formation at different temperatures and any small change in the flow rate of this nanofluid has little impact on the size of the droplets formed in this geometry.

These results demonstrated that the applicability of nanofluids in droplet-based microfluidics is promising and shed light for numerous potential microfluidics-based applications such as drug delivery, cancer or tumor treatments, lab-on-chip based technologies and so on. Nevertheless, it is imperative to conduct more extensive studies employing various types of nanofluids in microfluidic systems in order to elucidate the mechanisms of their microdroplet formation and also to exploit the application of this novel class of fluids in the new fields of microtechnology and biomedical engineering.

REFERENCES

[1]   Choi, S. U. S. In *Developments and applications of non-Newtonian flows*; Siginer, D. A.; Wang, H. P.; Eds.; ASME Publishing: New York,USA,1995; FED-Vol. 231/MD-Vol. 66, pp 99-105.

[2]   Lee, S.; Choi, S. U. S.; Li, S.; Eastman, J. A. *J. Heat Transfer* 1999, 121, 280-289.

[3]   Choi, S. U. S.; Zhang, Z. G.; Keblinski, P. In *Encyclopedia of Nanoscience and Nanotechnology*, Nalwa, H. S.; Ed.; American Scientific Publishers, Los Angeles, USA 2004, Vol.6, pp.757-773.

[4]   Murshed, S. M. S.; Leong, K. C.; Yang, C. *Int. J. Therm. Sci.* 2005, 44, 367-373.

[5]   Murshed, S. M. S.; Leong, K. C.; Yang, C. *J. Phys. D: Appl. Phys.* 2006, 39, 5316-5322.

[6]   Yu, W.; France, D. M.; Routbort, J. L.; Choi, S. U. S. *Heat Transfer Eng.* 2008, 29, 432-460.

[7]   Murshed, S. M. S.; Leong, K. C.; Yang, C. *Int. J. Therm. Sci.* 2008, 47, 560-568.

[8]   Das, S. K.; Choi, S. U. S.; Patel, H. E. *Heat Transf. Eng.* 2006, 27, 3-19.

[9] Murshed, S. M. S.; Leong, K. C.; Yang, C. *Appl. Therm. Eng.* 2008, 28, 2109-2125.

[10] Murshed, S. M. S.; Nieto de Castro, C. A.; Lourenço, M. J. V.; Lopes, M. L. M.; Santos, F. J. V. *Ren. Sust. En. Rev.* 2011, 15, 2342-2354.

[11] Murshed, S. M. S.; Nieto de Castro, C. A. In *Green Solvents I: Properties and Applications in Chemistry*, Ali M.; Inamuddin; Eds.; Springer, London, UK 2012, Ch. 14, pp.397-415.

[12] Burns, M. A.; Johnson, B. N.; Brahmasandra, S. N.; Handique, K.; Webster, J. R.; Krishnan, M.; Sammarco, T. S.; Man, P. M.; Jones, D.; Heldsinger, D.; Mastrangelo, C. H; Burke, D. T. *Science* 1998, 282, 484-487.

[13] Zheng, B.; Roach, L. S.; Ismagilov, R. F. *J. Am. Chem. Soc.* 2003, 125, 11170-11171.

[14] Zheng, B.; Ismagilov, R. F. *Angew. Chem. Int. Ed.* 2005, 44, 2520-2523.

[15] He, M.; Edgar, J. S.; Jeffries, G. D. M.; Lorenz, R. M.; Shelby, J. P.; Chiu, D. T. *Anal. Chem.* 2005, 77, 1539-1544.

[16] Guttenberg, Z.; Müller, H.; Habermüller, H.; Geisbauer, A.; Pipper, J.; Felbel, J.; Kielpinski, M.; Scriba, J.; Wixforth, A. *Lab. Chip.* 2005, 5, 308-317.

[17] Song, H.; Ismagilov, R. F. *J. Am. Chem. Soc.* 2003, 125-47, 14613-14619.

[18] Joanicot, M.; Ajdari. A. *Science* 2005, 309, 887-888.

[19] Song, H.; Chen, D. L.; Ismagilov, R. F. *Angew. Chem. Int. Ed.* 2006, 45, 7336-7356.

[20] Whitesides, G. M. *Nature* 2006, 422, 368-373.

[21] Anna, S. L.; Mayer, H. C. *Phys. Fluid.* 2006, 18, 121512.

[22] Christopher, G. F.; Anna, S. L. *J. Phys. D: Appl. Phys.* 2007, 40, R319-R336.

[23] Nguyen, N. T.; Ting, T. H.; Yap, Y. F.; Wong, T. N.; Chai, J. C. K.; Ong, W. L.; Zhou, J. L.; Tan, S. H.; Yobas, L. *Appl. Phys. Lett.* 2007, 91, 084102.

[24] Thorsen, T.; Roberts, R. W.; Amold, F. H.; Quake, S. R. *Phys. Rev. Lett.* 2001, 86, 4163-4166.

[25] Anna, S. L.; Bontoux, N.; Stone, H. A. *Appl. Phys. Lett.* 2003, 82, 364-366.

[26] Tice, J. D.; Song, H.; Lyon, A. D.; Ismagilov, R. F. *Langmuir* 2003, 19, 9127-9133.

[27] Garstecki, P., Gitlin, I.; Luzio, W. D.; Whitesides, G. M.; Kumacheva, E.; Stone, H. A. *Appl. Phys. Lett.* 2004, 85 2649-2651.

[28] Cubaud, T.; Tatineni, M.; Zhong, X. L.; Ho, C. M. *Phys. Rev. E* 2005, 72, 37302.

[29] Seemann, R.; Brinkmann, M.; Pfohl T.; Herminghaus, S. *Rep. Prog. Phys.* 2012, 75, 016601.

[30] Nguyen, N. T.; Lassemono, S.; Chollet, F. *Sens. Actuat. B* 2006, 117, 431-436.

[31] Nisisako, T.; Torii, T.; Higuchi, T. *Lab. Chip.* 2002, 2, 24-26.

[32] Xu, J. H.; Li, S. W.; Tan, J.; Wang, Y. J.; Luo, G. S. *AIChE J.* 2006, 52, 3005-3010.

[33] Xu, J. H.; Li, S. W.; Chen, G. G.; Luo, G. S. *AIChE J.* 2006, 52, 2254-2259.

[34] Schröder, V.; Behrend, O.; Schubert, H. *J. Colloid. Interf. Sci.* 1998, 202, 334-340.

[35] Link, D. R.; Anna, S. L.; Weitz, D. A.; Stone, H. A. *Phys Rev Lett.* 2004, 92, 054503.

[36] Van der Graaf, S.; Steegmans, M. L. J.; Schroen, C. G. P. H.; Van der Sman, R. G. M.; Boom, R. M. *Langmuir* 2006, 22, 4144-4152.

[37] Ting, T. H.; Yap, Y. F.; Nguyen, N. T.; Wong, T. N.; Chai, J. C. K.; Yobas, L. *Appl. Phys. Lett.* 2006, 89, 2234101.

[38] Ong, W. L.; Hua, J. S.; Zhang, B. L.; Teo, T. Y.; Zhuo, J. L.; Nguyen, N. T.; Ranganathan, R.; Yobas, L. *Sens. Actuat. A* 2007, 138, 203-212.

[39] Nie, Z. H.; Seo, M. S.; Xu, S. Q.; Lewis, P. C., Mok, M.; Kumacheva, E.; Whitesides, G. M.; Garstecki, P.; Stone, H. A. *Microfluid. Nanofluid.* 2008, 5, 585-594.

[40] Ward, T.; Faivre, M.; Abkarian, M.; Stone, H. A. *Electrophoresis* 2005, 26, 3716-3724.

[41] Takeuchi, S.; Garstecki, P.; Weibel, D. B.; Whitesides, G. M. *Adv. Mater.* 2005, 17, 1067-1072.

[42] Xu, Q.; Nakajima, M. *Appl. Phys. Lett.* 2004, 85, 3726-3728.

[43] Murshed, S. M. S.; Tan, S. H.; Nguyen, N. T. *J. Phys. D: Appl. Phys.* 2008, 41, 085502.

[44] Tan, S. H.; Murshed, S. M. S.; Nguyen, N. T.; Wong, T. N.; Yobas, L. *J. Phys. D: Appl. Phys.* 2008, 41, 165501.

[45] Murshed, S. M. S.; Tan, S. H.; Nguyen, N. T.; Wong, T. N.; Yobas, L. *Microfluid. Nanofluid.* 2009, 6, 253–259.

[46] Wang, R. *J. Nanopart. Res.* 2013, 15, 2128.

[47] Tice, J. D.; Lyon, A. D.; Ismagilov, R. F. *Anal. Chim. Acta* 2004, 507, 73-77.

[48] Girifalco, L. A.; Good, R. J. *J. Phys. Chem.* 1957, 61, 904-909.

[49] Binks, B. P. *Curr. Opin. Colloid Interface Sci.* 2002, 7, 21-41.

[50] Bresme, F.; Faraudo, J. *J. Phys: Cond. Matt.* 2007, 19, 375110.

[51] Garstecki. P.; Fuerstman, M. J.; Stone, H. A.; Whitesides, G. M. *Lap. Chip.* 2006, 6, 437-446.

[52] Tan, Y. C.; Cristini, V.; Lee, A. P. *Sens. Actuat. B* 2006, 114, 350-356.

[53] Zhou, C.; Yue, P.; Feng, J. *J. Phys. Fluids* 2006, 18, 092105.

[54] Yobas, L.; Martens, S.; Ong, W. L.; Ranganathan, R. *Lab. Chip.* 2006, 6, 1073-1073.

In: Nanofluids: Synthesis, Properties and Applications   ISBN: 978-1-63321-677-8
Editors: S.M. Sohel Murshed, C.A. Nieto de Castro   © 2014 Nova Science Publishers, Inc.

Chapter 10

NANOFLUID-BASED OPTICAL ENGINEERING: FUNDAMENTALS AND APPLICATIONS

R. A. Taylor[1,2,] and Y. L. Hewakuruppu[1]*
[1]School of Mechanical and Manufacturing Engineering
[2]School of Photovoltaic and Renewable Energy Engineering,
The University of New South Wales, Kensington,
Sydney, Australia

ABSTRACT

Nanofluids are not only valuable for their enhanced thermal properties – a large international research effort has been directed towards developing nanoparticles and nanofluids with tunable *optical* properties. Optical sensors, optical filters, solar absorbers, lasers, cancer therapies, and a whole suite of other applications can benefit from nanofluids with controlled optical properties. While there are many *solid* optical components commercially available, flowing *fluid*-based (both liquids and gases) systems are superior for transient applications. Optical engineering of nanofluids has been made possible by recent advances in fabrication techniques – enabling tight tolerance, highly reliable production of almost any material in a wide variety of shapes and sizes. In the right dose, the addition of well-designed nanoparticles can alter the optical properties of pure fluids (water, oils, and alcohols) from being transparent to bespoke fluids which are highly absorbing, scattering, or a mixture for any portion of the optical spectrum. Metallic nanoparticles, in particular, display the highly selective phenomenon of plasmon resonance which allows them to be utilized to create fluids which can interact strongly with a small band of light. Thus, the development of a new class of nanofluid-based optofluidic devices represents an emerging trans-disciplinary synthesis between nanotechnology, thermofluids, and optics. As a primer to this field and to encourage research activity in this area, this chapter will describe the state-of-the-art and the requisite theory behind this type of technology.

---

[*] E: Robert.Taylor@UNSW.edu.au, T: (+612) 9385 5400; F: (+61 2) 9663 1222.

1. Introduction

Nanoparticle suspensions have mainly been investigated for their interesting thermal properties. The optical properties of nanofluids, on the other hand, have been much less intensively researched [1]. While the thermal properties can realize considerable relative enhancement with the addition of nanoparticles, the optical properties can be changed by several orders of magnitude. This is readily evidenced by the fact that most fluids (both gases and liquids) are essentially transparent to light in the visible spectrum. Through the addition of less than 0.1% by volume of nanoparticles, the absorption and scattering coefficients can be modified from 1×10^{-6} to 1×10^{4} cm^{-1} (e.g. a change of *10 orders of magnitude*) [2–4]. Though they may be smaller than the wavelength of light, their high scattering and absorption cross sections allow nanoparticles to interact with electromagnetic radiation as if they are much larger particles. This fact enables applications which can be engineered to interact in a controllable way with electromagnetic fields/fluxes [5–8]. With few limits on the combinatorial variety of nanoparticles (size, shape, material) and fluids (water, organic solvents, ionic liquids, gases), a nearly inexhaustible research space awaits.

1.1. Motivation for Developing Nanofluid-Based Optics

A wide range of applications rely on optics: imaging, sensing, communications, lighting, photography, solar energy harvesting, etc. [9,10]. Today's optical components are constructed, by and large, using solid substrates. Lenses, reflectors, filters, waveguides, diffusers, and polarizers usually incorporate active materials into *solid* substrates through mixing, coating, doping, etc. This chapter takes a slight departure from this approach and discusses an analogous set of *fluid*-based optical components, sometimes referred to as optofluidics [11,12]. To date, fluids which interact with light have mostly been limited to pure fluids [13–19], but we propose that by incorporating active nanomaterials into fluid 'substrates' could significantly advance this field. Although the word "optical" is commonly used to describe visible light (~400-700 nm), it should be noted that nanofluids can also be engineered to influence other parts the electromagnetic spectrum. The advantage – and challenge – of creating fluid-based optics is that they are dynamic in nature. That is, pumps, fans, and electromagnetic fields can be used to move the particles and fluids in the optical system, enabling rapid changes in the optical properties. This type of sophisticated transient control could prove to be invaluable in many applications, a subset of which is discussed below. Some specific advantages this type of dynamic technology are: i) particles/fluids can be mixed, matched, and changed to suit the application, ii) spatial and temporal resolutions of the optical components can be very high due to the small size of the particles, iii) stacking functions may be possible where the optical component also serves another function (e.g. heat removal, material transport, hydraulic/pneumatic action).

1.2. Applications of Nanofluid-Based Optofluidic Components

At present, these technologies are relatively underdeveloped as only a limited amount of research (discussed below) has gone into developing 'optofluidics' and the optical properties of nanofluids in general. However, as depicted in Figure 1, advances in materials fabrication techniques have opened up the potential to develop nanoparticle-based optofluidic lenses, reflectors, filters, waveguides, diffusers, and polarizers, etc.

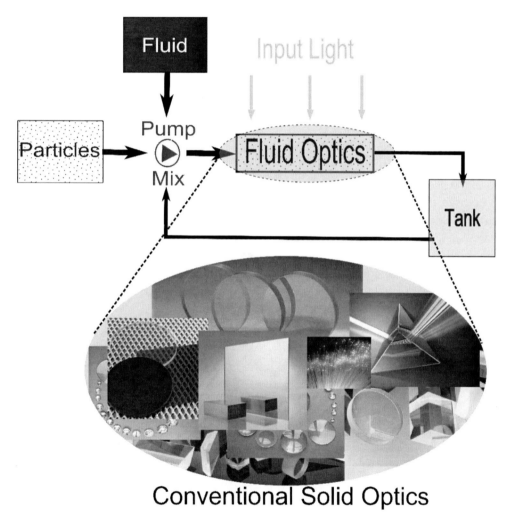

Figure 1. A generalized nanofluid-based configuration as compared to conventional solid-based optical components.

1.2.1. Dynamic Control Methods

As mentioned above, applications where this is best suited are those which require dynamic changes in optical properties. Dynamic control can be achieved simply by flow control via traditional pumps and fans. Since nanoparticles are so small, in most cases they can be incorporated into convention systems without major modification to pumps and plumbing.

Alternatively, in an optical application, the goal may be to achieve optical control of fluid and particle motion. Most notably, a lot of research in recent years has gone into optical tweezers which utilize large gradients in the electromagnetic field to confine small particles [20,21]. Much of the recent work in that area has gone into using near-field optics to get higher optical forces on the scale of nanoparticles and even DNA [22]. In another application, laser heating of nanoparticle dispersions has been shown as a driving force for fluid flow in microchannels [23].

Another method which can be used to create finer control over particle distributions can be achieved if the nanoparticles used are charged or magnetic, wherein motion can be induced with electromagnetic fields. Winslow and Rainbow [24,25] discovered in the 1940s that by incorporating ferromagnetic particles it is possible to move non-magnetic fluids. Since then magnetorheological and ferro- fluids have been applied to a wide variety of engineering systems (sensors, motors, dampers, seals, bearings and brakes) [26–28].

1.2.2. Ferrofluids

Ferrofluids consist of < 100nm sized particles dispersed in fluids [28–34]. Researchers, such as Rosenweig, discussed ferrofluid dispersions of 10nm particles in the 1960s [27,35,36]. Although it is not a major focus for ferrofluids researchers, this class of nanofluids has been used for optical applications. In 1980, Brady et al. patented the ferrofluid optical switch [37]. Mehendran and Philip [38], used the magnetic flux gradients found around structural defects to visualize internal damage with ferrofluids. Bacri et al. [39] experimentally showed that polydisperse (3-30 nm, γ-Fe_2O_3) particles in anionic suspensions exhibit a large degree of control over both static and dynamical birefringence depending on the imposed magnetic field. Ferrofluids coated with silver nanoparticle colloids have been proposed for use in highly controllable deformable mirrors [40].

1.2.3. Volumetric Solar Collectors

By creating optical absorbers and/or filters using nanoparticles it is possible to create volumetric (also known as direct-absorption) solar collectors [2, 41–47]. The basic enhancement found in these systems is related to the fact that energy is absorbed where it is needed (directly in the working fluid) rather than on an outside surface, representing a simplified heat transfer path. With proper design this can allow the highest temperatures to be created inside the solar receiver. These systems have been proposed for use in *low* (< 100 °C), *medium* (100-400 °C), and *high* (> 400 °C) temperature direct-absorption solar collectors. The temperature range determines the base fluid material and/or state of matter that will be used as the energy carrier. In each operational regime there are many solid particle options which can obtain broadband or selective absorption (and scattering) of incoming photons. Metals, semiconductors, insulators, and allotropes of carbon, have all been studied for this purpose in recent literature [4, 41–83]. Generalized methods for engineering these designs will be described in section 2 below.

An obvious and low-cost fluid option for low temperature applications is water, although it may need to be pressurized to prevent liquid-vapor phase change [84]. Water is used as a base fluid in the majority of the reported direct-absorption research and much of this work is dedicated to finding optimal performance with respect to particle size, material, shape, volume fraction, and operational conditions [4,41,46,48,50–52,56–58,85]. Overall, these studies have shown enhancements in the range of 10% are possible in comparison with solid surface absorbers. If higher temperatures are desired, glycols and synthetic heat transfer fluids have also been proposed – e.g. Therminol VP-1^{TM} [86]. However, a unique requirement for solar applications is that the fluids must also be able to withstand UV light. Since UV consists of destructive, high energy photons, it can actually break relatively the weak single carbon bonds (e.g. C-O, C-C, and C-N) which are present in organic liquids. According to Blunden and Chapman [87], UV light at 200 nm represents an energy input of 600 kJ/mol. In the absence of chemical inhibitors, these fluids become brownish in color and foul surfaces after substantial exposure, since single carbon bonds have energies less than 360 kJ/mol [88].

Other, higher temperature, candidates for direct absorption fluid optics are molten salts [43,65,68,71,75,89–91]. Although managing salts in a system is more complex than water/oils, higher temperature systems will allow for collection of higher thermodynamic quality energy. There are a host of binary and ternary salt mixtures available, but nitrate salts (i.e. a 40-60 eutectic mixture of KNO_3-$NaNO_3$) are the most developed. Due to their low cost (< \$0.5/kg) they have been frequently used in solar systems [92]. Recently MIT researchers [43,89] modeled and tested direct absorption molten salts in thermocline tanks. These studies conclude that under optimal conditions, heat losses can be minimized which enables high efficiency collection (and storage).

Table 1. Selected Published Work on Direct Absorption Solar Systems

Author	Type	Development
Ladjevardi et al. [93]	Research Article	Verified model of a low temp. nanofluid collector
Lenert et al. [43]	Research Article	Prototype carbon-coated nanofluid receiver
Lu et al. [50]	Research Article	Prototype Copper Oxide thermosyphon solar collector
Taylor et al. [83]	Research Article	Predicted impact of nanofluids to CSP systems
Sani et al. [72]	Research Article	Characterization of carbon nanohorns for solar collectors
Otanicar et al. [57]	Research Article	Testing of a nanofluid-based flat plate collector prototype
Goldman [94]	Patent	Receiver with semi-transparent / translucent working fluid tubes
Tyagi et al. [95]	Research Article	Flat plate solar thermal collector based on nanofluids
Schunk [96]	Research Article	Rotating zinc oxide particle receiver
Hirsch et al. [97]	Research Article	Carbon particle thermochemical high temperature reactor
Yogev [98]	Patent	Concentrated light focused directly on organic fluid
Parker et al. [99]	Patent	Secondary direct absorber for high temperature
Bohn [66]	Research Article	Experiments with a 'pure' flowing molten salt absorber
Abdelrahman [64]	Research Article	Particle-gas suspensions for ideal absorption
Hunt [100]	Research Article	Particle-gas absorber
Kraus et al. [101]	Patent	Particle-fluid absorber

1.2.4. Bio Applications

Optical properties [102] are also important inside the biological environment – which of course is naturally a fluidic environment. To date, the most significant bio applications utilizing nanofluids are: photo-thermal cancer treatment, medical imaging (e.g. using 1-10nm quantum dots), and lab-on-a-chip optical diagnostics. Laser induced-hyperthermia, which requires dispersed nanoparticles to selectively attach to diseased regions, has seen increasing interest in recent years [103–105]. The key challenge of this treatment is to irradiate the nanoparticles to cause irreparable damage (thermal ablation) to the cancerous cells, but with enough control/selectivity to leave healthy tissues unharmed [106–109]. As demonstrated in

Table 2, nanofluids – though they may not often referred to as such – are finding a myriad of applications in medical applications [103,105,107,110–118]. In many cases these can potentially utilize multiple property enhancements (e.g. enhanced local thermal conductivity is very beneficial locally in hyperthermia treatments). Based on the number of impactful publications, the potential market and the growth potential of 'opto-bio-nanofluid' research may surpass that of conductive/convective heat transfer research.

Table 2. Selected biological optofluidic studies

Author	Particle Type	Development
Huang et al. [104]	Au	Test Tube Photothermal Hyperthermia Experiments: 70-80°C threshold reached easier with high particle loading
Dombrovsky et al. [119]	Au-Silica (Shell-core)	Photothermal Hyperthermia Model: Developed parameterized model
Kikumori et al. [109]	Fe_2O_3	In-Vitro (Mice) Magnetic Hyperthermia: hyperthermia treatments stopped tumor growth
Parveen and Sahoo [120]	Polymer drug carrier	In-Vitro (Mice) Drug Delivery: Promising, long circulation efficacy
Aryal et al. [121]	Hydrophobic- hydrophilic drug conjugate	Test Tube Drug Delivery Experiment: Simple, effective drug delivery mechanism
Kalantar-zadeh et al. [122]	Silica	Tuneable optical waveguide using dielectrophoretically controlled nanoparticles
Schaap et al. [123]	Algae	Algae and other particle detection using a microfluidic photo-detector.
Yang et al. [22]	Dielectric nanoparticles & DNA molecules	Sub- wavelength liquid-core slot waveguides which can trap particles as small as 60nm.

1.2.5. E-Readers

A well-known (but potentially unrecognized) success in colloidal suspension-based optical components is electrophoretic ink [19,124], popularly called 'E-ink'. In this technology, small, charged particles are circulated by electric fields in a microcapsule of clear dielectric fluid. When positively charged, white (scattering) particles are near the outer surface of an e-reader screen, a white pixel is seen. When the negatively charged, black (absorbing) particles are rotated to the top, a dark pixel is seen. By adding particles which preferentially scatter certain wavelengths, color electrophoretic displays have even been demonstrated [124]. Currently, the majority of these systems use particles which are ≥ 100 nm, but if this technology shrinks to achieve higher resolution, it may actually become the biggest commercial application of nanofluids. Sales of Kindle e-readers (and tablets) are now reportedly nearing 4 to 5 billion dollars per year [125].

1.2.6. Other Applications

Although they have not yet incorporated nanoparticles, re-configurable optofluidic components (e.g. liquid lenses) have garnered a lot of interest in recent years [12–16]. These devices usually use the difference in refractive index found in liquids to alter a light path. In the case of liquid lens, the shape of the interface between immiscible fluids (e.g. oil and water) is controlled by applying a voltage to the device [16]. This creates a change in the contact angle at the solid boundary which enables sophisticated, transient focusing and tracking control. Wolf et al. [13] demonstrated an optical splitter and wavelength filter in a creeping flow microfluidic device by controlling diffusion between liquid-liquid interfaces. By adding nanoparticles it is also possible to create liquid filters which can be pumped into and out of a system [2,126,127]. The tunable optical properties of nanoparticles make them ideal for selective absorbing and/or scattering filters. Short-pass, long-pass, and band-pass nanofluid-based filter designs have been reported by the co-authors using one and two nanoparticle component mixtures [2].

Despite the promise for many industries, nanofluid optical components can only truly come to commercial fruition through further research and development. Achieving controlled optical properties requires careful design of the numerous parameters: materials (of both the nanoparticles and the fluids), nanoparticle size, nanoparticle volume fraction, and the configuration. The next section will give a simplified analysis of the fundamentals needed to make advancements in this emerging field.

2. NANOFLUID OPTICAL PROPERTIES

Determining the interaction between light and nanoparticles is of vital importance in the research and development of optical applications of nanofluids. This section discusses theoretical tools and experimental methods which are commonly used to determine the optical properties of nanofluids.

2.1. Theoretical Methods

The theoretical calculation of nanofluid optical properties starts with the interaction of light and a single nanoparticle. This is built upon by then calculating the absorption and scattering coefficients of the nanofluid which consists of a large number of particles and a host fluid. Knowing this information allows one to determine how radiation propagates within a nanofluid sample. The next few sub sections will introduce the mathematical tools used during each of these steps.

2.1.1. Optical Properties of Single Particles

The first logical step is to calculate the interaction between incident light and a single nanoparticle. A choice must be made amongst the various tools available which calculate absorption and scattering of light by nanoparticles with different geometries and dimensions.

Mie theory [128] is one of the most fundamental and commonly used tools to determine the extinction of light by *spherical* particles (e.g. the most common geometry). Mie theory provides not only the exact solution to Maxwell's equations for spheres, but it also for spheroids and infinitely long cylinders [129,130]. It is based on the single (independent) scattering hypothesis where the absorbed and scattered of light of each particle is exactly the same regardless of its relative position in the medium. In other words, each particle is assumed to be in the far-field zone of other particles and neighboring particles do not affect the considered particle. Thus, scattering from individual particles is incoherent. This assumption has been well proven experimentally when aggregation of particles is negligible. Mie theory determines the efficiency factors of extinction (Q_{ext}) and scattering (Q_{sct}) of a particle in a non-participating media (vacuum) as per the derivation given below. The efficiency factor of absorption (Q_{abs}) can be calculated as the difference between extinction and scattering efficiencies:

$$Q_{ext} = \frac{2}{x^2} \sum_{k=1}^{\infty} (2k + 1) Re(a_k + b_k) \tag{1}$$

$$Q_{sct} = \frac{2}{x^2} \sum_{k=1}^{\infty} (2k + 1)(|a_k|^2 + |b_k|^2) \tag{2}$$

$$Q_{abs} = Q_{ext} - Q_{sct} \tag{3}$$

Here a_k and b_k are complex numbers called Mie coefficients, x is the size parameter equal to $2\pi r/\lambda$, r is the radius of the sphere, and λ is the wavelength of the incident light. These 'efficiency factors' represent the ratio between the area of light scattered/absorbed by the particle and the cross sectional area of the particle. Also, the 'extinction' term represents the sum of absorption and scattering. Moreover, Mie theory also calculates the asymmetric factor of scattering which describes the shape of the scattering phase function.

Full mathematical relations of Mie theory can be found in Bohren and Hoffman [129] and Dombrovsky [130,131] along with open source Mie theory codes for homogeneous and two layered spherical particles. The input information for these codes is the size parameter (x), the index of refraction (n), and the index of absorption (k) of the particle. To use the codes for particles embedded in a participating medium these inputs should be modified. In this case, the size parameter should be multiplied and the index of refraction and absorption should be divided by the index of refraction of the surrounding medium. The absorption (C_{abs}) and scattering (C_{sct}) cross sections can then be calculated by multiplying the respective efficiency factors by the cross sectional area of the particle.

As the exact solution of Maxwell equations is limited for spheres, spheroids and infinitely long cylinders, approximate solutions were developed for particles with other geometries. The discrete dipole approximation (DDA) is one such tool which is widely used to determine the optical properties of small particles [132,133]. It provides an approximate numerical solution for the Maxwell equation for arbitrary shaped particles by treating it as a large array of dipoles. Hence the application of DDA is not only limited to spherical particles but is also applicable to particles such as nanorods [134–136], nanocages [137], nanotriangles [138] and other irregular and exotic shapes. Yurkin and Hoekstra [139,140] and Draine and Flatua [141] have reviewed the development of the DDA method in great detail. They also provided freely available DDA codes, which can be downloaded from (*code.google.com/p/ddscat/*). The input information for these DDA codes is the particle geometry and dimensions, the optical constants of the particle substance, the refractive index of the surrounding medium and the orientation of the particle with respect to the direction of the incident light. Similar to Mie theory, DDA provides the efficiency factors of absorption and scattering, when the above information is given. Moreover, it is also possible to calculate the orientation averaged optical properties of irregularly shaped particles. For instance, these DDA codes can include the effects of different cylindrical particles having different geometrical orientations with respect to the direction of the incident radiation within a nanofluid.

Finite difference time domain (FDTD) is another powerful numerical method used for the calculation of nanoparticle optical properties [142–144]. Here the particle and its surroundings are modeled as a spatial grid. The FDTD method then simulates the propagation of an electromagnetic field through this spatial grid in consecutive time steps [132]. In FDTD simulations, the dielectric function of the particle and surrounding substances are modeled as an analytical function of frequency of the incident light. The main advantages of the FDTD method are the applicability for particles with complex geometries and the simulation of the full spectrum in a single run [132].

Other than the above mentioned techniques, methods such as the Boundary element method [132,145], the T-matrix [133,146] approach, and the volume integral equation method have also been employed to calculate optical properties of nanoparticles. In addition, simple methods which ignore the higher order terms of the Mie theory, such as the electrostatic approach, have also been used for nanoparticles [147,148]. The mathematical relations of the electrostatic approach can be found in several texts – e.g. Bohren and Huffman [129]. However, these first order methods are strictly limited to cases of Rayleigh's scattering when the dimensions of the particle are much smaller than the wavelength of the incident light (i.e. $x<<1$) [129].

2.1.2. Size Modification for Nano-Scale Dimensions

It is possible for nanoparticle to have dimensions smaller than the mean free path length of conduction electrons in bulk material. This results in limitations to electron collisions within the particle [129] which alters the dampening constant of conduction electrons and results in dielectric constants which are altered from those of the bulk material. Hence, to accurately calculate the optical properties of metallic nanoparticles, it is necessary to perform a modification of the dielectric constants of the nanoparticle material to incorporate this effect.

The real and imaginary components of the bulk dielectric constant (ε_r and ε_i) can be calculated using the index of refraction (n) and the index of absorption (k) of the bulk material [129]:

$$\varepsilon_r = n^2 - k^2 \tag{4}$$

$$\varepsilon_i = 2nk \tag{5}$$

Both free and interband (bound) electron oscillations contribute to the complex dielectric constant in metals. While free electrons dominate the dielectric constant of material such as silver, the interband electrons dominate in materials such as gold [129]. Due to this phenomena, the model presented by Kreibig and Vollmer [149] includes both free and interband electron terms and is suitable to perform the necessary modifications to the dielectric constant of metallic nanoparticles. According to this method, the dielectric constant modified for size effects (ε_m) is given by [149]:

$$\varepsilon_m = \varepsilon_{bulk} + \omega_p \frac{1}{\omega^2 + i\omega\gamma_{bulk}} - \omega_p \frac{1}{\omega^2 + i\omega\gamma_{eff}} \tag{6}$$

Here ε_{bulk} is the dielectric constant of the bulk material is calculated in Eqs. (4) and (5). The other parameters are the bulk plasmon frequency, ω_p, angular frequency, ω, the relaxation frequency of the bulk material, γ_{bulk}, and the effective relaxation frequency, γ_{eff}, which is given by:

$$\gamma_{eff} = \gamma_{bulk} + \frac{A v_f}{l_{eff}} \tag{7}$$

where v_f is the Fermi velocity and A is a geometrical parameter (generally set to 1 by assuming isotropic scattering) [149]. l_{eff} in Eq. (7) is the effective mean free path. For instance l_{eff} for a spherical particle is equal to the radius. For cylindrical particles (nanorods and nanodisks) this modification is made for both radial and axial dimensions where the radius and length are used as l_{eff}, respectively.

This size modification of the dielectric constant should be performed when the effective mean free path of a given nanoparticle is smaller than the mean free path of the bulk material. The modified complex refractive index can then be calculated from the modified dielectric constant (the modified ε_r and ε_i values) using same the relations in Eqs. (4) and (5).

2.1.3. Optical Properties of the Nanofluid

After determining the properties of a single nanoparticle, it is now possible to calculate the total optical properties of a nanofluid containing a large number of nanoparticles suspended in a basefluid. In general, optical applications of nanofluids only require a very small volume fraction of nanoparticles ($<10E^{-3}$) to be suspended within the nanofluid. Therefore it is reasonable to assume independent scattering which makes it possible to simply add the contribution of each nanoparticle to calculate the optical properties of an ensemble of nanoparticles [130].

Even with self-assembly fabrication, an ensemble of nanoparticles does not consist of particles with uniform dimensions. Polydispersity can be taken into account by assuming a Gaussian distribution of particle dimensions. The mean size and standard deviation of particle dimensions can be determined based on experimental data obtained by electron microscopy, nanoparticle tracking analysis, and/or dynamic light scattering. This information can then be used to determine the probability density function of particle dimensions for each size found in the ensemble. Using this additive method, the following relation calculates the absorption coefficient ($\alpha_{particles}$) of an ensemble of spherical nanoparticles including the effect of polydispersity [130]:

$$\alpha_{particles} = \frac{3f_v}{4\pi} \frac{\int_0^\infty \int_0^\infty C_a P(r) dr}{\int_0^\infty \int_0^\infty r^3 P(r) dr} \tag{8}$$

Here, f_v, is the volume fraction of nanoparticles in the nanofluid and $P(r)$ is the probability density functions of the particle radius. The same procedure is used to account for size variations in other particle types. For instance, the variation in both diameter and length should be considered for cylindrical particles. The scattering coefficient of the sample of particles $\sigma_{particles}$, can also be calculated using the above relation by replacing the C_{abs} with C_{sct}.

Finally it is important to remember that the base fluid also contributes to the optical properties of a nanofluid. The total absorption and scattering coefficients of the nanofluid ($\alpha_{nanofluid}$, $\sigma_{nanofluid}$) can thus be simply calculated as the sum of the contribution by all the particles and the basefluid [147]:

$$\alpha_{nanofluid} = \alpha_{particles} + \alpha_{basefluid} \tag{9}$$

This simple additive method also works in the case of a nanofluid which contains more than one type of nanoparticles, given the independent scattering hypothesis is valid. In this case, the contribution of each type of nanoparticle can be calculated by using separately Eq. (8) for each nanoparticle type. Then similar to Eq. (9), the total absorption/scattering coefficient of the fluid is simply the sum of contributions by each nanoparticle type and the basefluid.

2.1.4. Propagation of Radiation Within a Nanofluid

When the absorption and scattering coefficients of a nanofluid is known it is then possible to determine how radiation propagates within it. Determining the propagation of radiation through a nanofluid enables the calculation of optical performance. For instance, when the intensity of radiation within the medium is known, one can calculate the amount of energy absorbed (scattered/emitted) by a nanofluid in tumor hyperthermia and solar harvesting applications. In addition it is possible to calculate the directional-hemispherical reflectance and transmittance of a nanofluid sample which can be of assistance in designing deformable mirrors and optical filters.

The continuum approach and the resulting radiative transfer equation (RTE) is a suitable tool to describe the propagation of radiation through a medium [130,131].

$$\vec{\Omega}\nabla I_\lambda(\vec{r},\vec{\Omega}) + (\alpha_\lambda + \sigma_\lambda)\, I_\lambda(\vec{r},\vec{\Omega}) = \frac{\sigma_\lambda}{4\pi}\int_0^{4\pi} I_\lambda\left(\vec{r},\vec{\Omega'}\right)\Phi_\lambda\left(\vec{\Omega'},\vec{\Omega}\right)d\vec{\Omega'} + \alpha_\lambda I_{\lambda,BB} \tag{10}$$

Here $I_\lambda(\vec{r},\vec{\Omega})$ is the spectral radiation intensity in the direction $\vec{\Omega}$ at the spatial coordinate $\vec{r}$. Variables α_λ and σ_λ, are the spectral absorption and scattering coefficients, respectively. Φ_λ is the scattering phase function and $\vec{\Omega'}$ represents directions other than $\vec{\Omega}$. Lastly, $I_{\lambda,BB}$ is the spectral black body radiation intensity. The RTE simply states that the change in radiation intensity in a given direction is due to the following:

- The extinction of radiation due to absorption and scattering by the medium (second term on the left hand side)
- Augmentation of the radiation intensity in the considered direction by radiation scattered from other directions (the integral term on the right hand side)
- Augmentation of radiation due to emission of the medium (second term on the right hand side)

Due the complexity of solving the RTE in an absorbing and scattering medium, many approximate solutions have been derived. Amongst these approximations, the methods presented by Dombrovsky and colleagues provide some of the most rapid solutions to the

RTE. First they propose the use of the 'transport approximation' [150] which simplifies the integral term by assuming that the scattering by other directions is isotropic. This removes the complex phase function from the RTE. Secondly the 'modified two flux' approximation [151] provides a simplification for the angular dependence of radiation intensity. These approximations have been shown to give solutions with good accuracy and small computational times as compared to other approaches such as the Monte-Carlo method [130,152–154]. This solution method has been used in tumor hyperthermia studies to calculate the energy absorbed by a nanoparticle embedded tumor when it is illuminated by near infrared light [152,155]. It has also been used to derive analytical solutions for the directional-hemispherical transmittance and reflectance of an absorbing, scattering and refracting medium [151]. For more details about the solutions to the RTE, the reader is referred to [130]. Also note that for a medium with negligible scattering (such as a black nanofluid) and emission (due to low temperatures), the RTE greatly simplifies to the well-known Beer-Lambert law.

2.2. Experimental Determination of Nanofluid Optical Properties

It is also possible to determine the optical properties (absorption and scattering coefficients) of a nanofluid using experimental techniques. Experimental determination of nanofluid properties may be easier in cases when the nanofluid contains nanoparticles of arbitrary shapes (which are hard to model), when the size distribution of nanoparticles is not known, or when mixtures of particles are present. Experimental methods may also be a simple alternative when theoretical determination is computationally infeasible (e.g. DDA and FTDT simulations are time consuming).

Dombrovsky and colleagues [154] presented a method to determine the optical properties of an absorbing and scattering medium using measurements of directional-hemispherical (also known as diffuse) reflectance (R_{dh}) and transmittance (T_{dh}). Steps of this technique are summarized by the flow chart in Figure 2.

First the nanofluid is prepared with the required concentration. Then the R_{dh} and T_{dh} of a nanofluid sample are measured by a spectrophotometer using an integrating sphere. Using the mathematical relations given in [154], it is possible to theoretically calculate R_{dh} and T_{dh} when the absorption and scattering coefficients (α and σ) are known. A guess is made for α and σ and the resulting R_{dh} and T_{dh} values are calculated. If the experimentally measured and theoretically calculated R_{dh} and T_{dh} values match, the guessed α and σ values are the optical properties of the nanofluid. If not, the guess for α and σ are adjusted using an iterative technique until they match.

The inverse adding doubling algorithm is another flexible and fast tool which can calculate the optical properties of turbid media when R_{dh} and T_{dh} measurements are given [156]. It can also calculate the asymmetry factor of scattering when the un-scattered transmittance measurements are available.

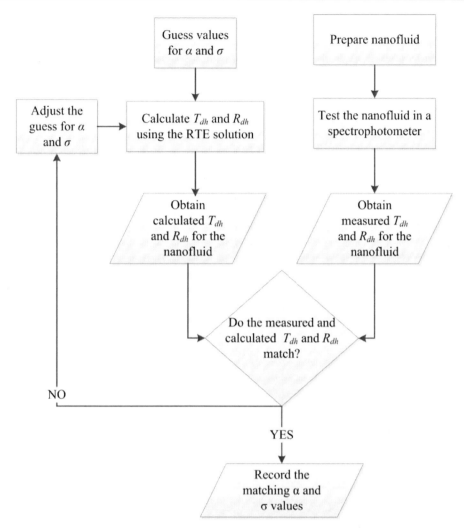

Figure 2. Flow chart describing the experimental nanofluid optical properties determination method.

3. NANOFLUID OPTICAL ENGINEERING

Nanofluid optical engineering sits at the trans-disciplinary intersection of materials science, optics, colloidal science, and fluid mechanics. Good design can result in compact components with an extremely large dynamic range in their optical properties via fluid mobility. At present, much of the fundamental theory needed for engineering a wide range of novel devices has already been developed (as discussed in the previous section). Inventing useful optofluidic lenses, reflectors, filters, waveguides, diffusers, and polarizers is simply an exercise in system design and in tailoring optical properties to the specific application. Thus, engineered solutions await those who can integrate these elements into platforms where nanoparticles, fluid, and light are driven to interact in a controlled manner. While this may seem daunting, the first generation of these is available today – filters, deformable components, and switches. Volumetric solar absorbers represent a facile example since nanofluids need only absorb (and convert to heat) some, or all, of the incident light. The co-authors have proposed and tested several nanofluid recipes and component designs for solar harvesting technology [2–4,57,58,83,85,157,158]. As nanofluids are incorporated into optofluidic and complex microfluidic applications, more advanced applications will result. As was mentioned in section 1, many advanced applications are in the research pipeline. For these, the challenge is apply today's knowledge in a creative way and choose from the infinite combination of particle materials, particle sizes, component configurations, and other parameters to design tomorrow's optical lab-on-a-chip devices which incorporate photodectors/sensor, optoelectronics, and/or optical flow control can be integrated into optofluidic architectures. Of course, a well-engineered design will not only achieve the desired optical performance, but will also prove to be reliable and cost-effective for the job at hand.

3.1. Fabrication Tolerances and Reliability

Tight control over nanofluid optical properties relies on tight control over nanoparticle morphology. Small changes in size and/or non-uniformities in shape or material can drastically shift the optical properties. One dramatic example of this is for plasmonic core-shell nanoparticles where the ratio of the thickness of the shell to the diameter of the core is critical. Comparing particle A (with a 40 nm diameter silica core / 4 nm thick gold shell) to particle B (a 50 nm diameter silica core / 2 nm thick gold shell), will shift the peak absorption wavelength from approximately 0.95 μm to 1.55 μm [126], as is shown in figure 3.

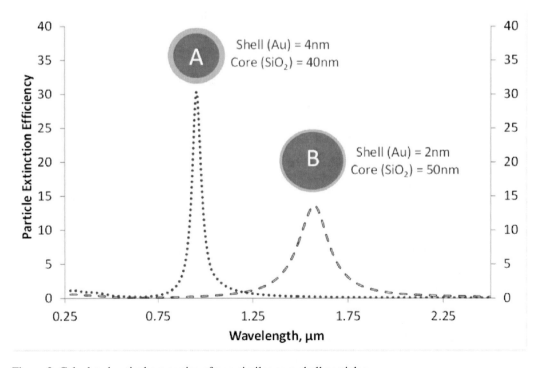

Figure 3. Calculated optical properties of two similar core shell particles.

For an application where this peak cannot tolerate much variation from the nominal optical properties, fabrication tolerances need to be on the order of ± 1 nanometer in shell thickness and < ± 1% in the core/shell radius ratio. Further, even the best fabrication methods result in some distribution of particle sizes. Depending on the allowable tolerance of the desired optical properties, size distributions need to be controlled in the range of < ± 1 nanometer and achieve nearly monodisperse conditions. Averitt et al. noted an 18-20% deviation from the mean size for core/shell nanoparticles during synthesis [159]. It is also worth noting that core/shell nanoparticles require a bit more processing. The most common method for making core/shell particles is to first create the particle cores − the Stöber process is used for silica [160] or for a polymer core, polymerization methods such as RAFT can be used [161]. Next, very small (1-3 nm) seeds of metal are formed in chemical synthesis. Linker agents are used to attach the seeds around the surface of the core. These seeds then serve as growth sites as the metal shell is formed. At present, there are a few of companies which sell core/shell nanoparticle suspensions [162,163].

For reliable operation, nanofluid optical components must maintain their size for time scales on the order of many years. If the base fluid is volatile (e.g. water, alcohol, and refrigerant) the system will need to be vapor tight or will need replacement base fluid over time. In terms of agglomeration, many approaches have been used to obtain stable colloidal suspensions [164–166]. This is critical because particle agglomeration will significantly change the optical properties, as is demonstrated by the small changes in figure 3. Although they do not represent an ideal long term stabilization strategy, surfactants are probably the most commonly used method in the nanofluid literature. For shorter term, low temperature applications where surfactants may be effective, they seem to have little effect on nanofluid optical properties [4]. For longer term stability, surface functionalization is much better, but can have a larger effect on the optical properties. For example, particles functionalized with PVP (polyvinylpyridine) have been shown to significantly suppress (and red shift) plasmon resonance in core/shell particles with thin shells due to dielectric screening [167].

Adequate chemical engineering is very important with regard to stability since *optical* nanofluids likely face even more destabilizing forces than *thermal* nanofluids. Thermal cycling (via absorption of light), exposure to intense light, and other factors can potentially destroy what was otherwise a stable nanofluid. Depending on the intensity of the light which interacts with the nanofluid, normal functionalization not be suitable over time.

3.2. Future Work Required

There are several avenues of future work that need to be pursued before nanoparticle-based optical devices will become commercially viable. As mentioned in the last section, work is needed to ensure that the tight tolerances in fabrication can be repeatedly achieved and preserved over time. While this is readily done on a laboratory scale, it is not clear that the fabrication methods used in the lab are easily scaled to commercial processes. On the topic of stability, accelerated lifetime experiments are needed to determine stability for each specific application. Lastly, the cost of nanomaterials is rather high and quite variable between suppliers. For example, gold nanoparticles cost between 288 USD/gram (nanopowder from Sigma Aldrich) and 32,000 USD/gram (custom made NanoXact Silica/Gold Nanoshells from [162]). Since the average price for gold bullion in 2013 was approximately 45 USD/gram, the majority of the cost can be attributed to nanofabrication. Fortuitously, nanofluid optical application used very small amounts of particles, so high $/gram prices may be manageable in some applications [4]. Nevertheless, these costs will reduce with research [168] and sound engineering optimization should ensure the best value products make it to market.

CONCLUSION

The key aspects of engineering nanoparticle-based optofluidic devices and a review of selected developments in this area were presented in this chapter. The most interesting designs, and the ones we hope that this chapter will help to foster, integrate tunable optical properties with nanofluid flow to achieve dynamic control. In section 2 the fundamental tools to begin designing these systems were given. Since many of these tools are freely available for download, researchers do not need to reinvent the wheel to get started. Section 3 presented some relatively simple methods for experimentally determining the optical properties of nanofluids. Experimental studies and prototype systems are needed in this area, since there can sometimes be a large gap between theoretical predictions and achievable results. Ultimately, this chapter concludes that due to the wide variety of nanofluid options available, nanofluid-based optofluidics represents an exciting, emerging research field with ample room for growth.

NOMENCLATURE

A	geometric parameter
a_k, b_k	Mie coefficients
C	cross section
DDA	discrete dipole approximation
$FTDT$	finite different time domain
f_v	volume fraction
I	radiation intensity
n	refractive index
k	absorption index
l_{eff}	effective mean free path
P	probability density function
Q	efficiency factor
R	reflectance
r	radius
	spatial coordinate
RTE	radiative transfer equation
T	transmittance
v_f	Fermi velocity

Greek symbols

α	absorption coefficient
γ	relaxation frequency
$\varepsilon_r\, \varepsilon_i$	real and imaginary dielectric function
λ	wavelength
μ	cosine of direction
σ	scattering coefficient
ω	angular frequency
ω_p	plasmon frequency
Φ	scattering phase function
$\vec{\Omega}$	unit vector for direction

Subscripts

abs	absorption
basefluid	value for the basefluid
BB	black body
bulk	bulk values
dh	directional-hemispherical
eff	effective
ext	extinction
nanofluid	value for the nanofluid
m	modified
particle	value for the particles
sct	Scattering
λ	spectral

REFERENCES

[1] Taylor, R. A..; Coulombe, S.; Otanicar, T.; Phelan, P.; Gunawan, A.; Lv, W.; Rosengarten, G.; Prasher, R.; Tyagi, H. *J. Appl. Phys.* 2013, *113*, 011301.

[2] Taylor, R. A.; Otanicar, T. P.; Rosengarten, G. *Nat. Light Sci. Appl.* 2012, *1*, 1–7.

[3] Taylor, R. A.; Phelan, P. E.; Adrian, R. J.; Gunawan, A.; Otanicar, T. P. *Int. J. Therm. Sci.* 2012, *56*, 1–11.

[4] Taylor, R. A.; Phelan, P. E.; Otanicar, T. P.; Adrian, R.; Prasher, R. *Nanoscale Res. Lett.* 2011, *6*, 225.

[5] Liz-marza, L. M. *Langmuir* 2006, *22*, 32–41.

[6] Mock, J. J.; Barbic, M.; Smith, D. R.; Schultz, D. A.; Schultz, S. *Chem. Phys.* 2002, *6755*.

[7] Vekas, L. *Nanostructure Mater.* 2004, *49*, 707–721.

[8] Lv, W.; Phelan, P. E.; Swaminathan, R.; Otanicar, T. P.; Taylor, R. A. *J. Sol. Energy Eng.* 2013, *135*, 021005.

[9] Martinu, L.; Poitras, D. *J. Vac. Sci. Technol. A.* 2000, *18*, 2619–2645.

[10] Imenes, A. G.; Mills, D. R. *Sol. Energ. Mat. Sol. C* 2004, *84*, 19–69.

[11] Erickson, D.; Sinton, D.; Psaltis, D. *Nat. Photonics* 2011, *5*, 583-590.

[12] Monat, C.; Domachuk, P.; Grillet, C.; Collins, M.; Eggleton, B. J.; Cronin-Golomb, M.; Mutzenich, S.; Mahmud, T.; Rosengarten, G.; Mitchell, A. *Microfluid. Nanofluid* 2007, *4*, 81–95.

[13] Wolfe, D. B.; Vezenov, D. V.; Mayers, B. T.; Whitesides, G. M.; Conroy, R. S.; Prentiss, M. G. *Appl. Phys. Lett.* 2005, *87*, 181105.

[14] Schmidt, H.; Hawkins, A. R. *Nat. Photonics* 2011, *5*, 598–604.

[15] Monolayer, S.; Gorman, C. B.; Biebuyck, H. A.; Whitesides, G. M. *Langmuir* 1995, *11*, 2242–2246.

[16] Berge, B.; Normale, E.; Lyon, S. De; France, L. In *18th IEEE International Conference on Micro Electro Mechanical Systems:* Miami Beach,USA, January 2005..

[17] Kuiper, S.; Hendriks, B. H. W. *Appl. Phys. Lett.* 2004, *85*, 1128.

[18] Shi, J.; Stratton, Z.; Lin, S.; Huang, H.; Huang, T. J. *Microfluid. Nanofluid* 2009, *9*, 313–318.

[19] Levy, U.; Shamai, R. *Microfluid. Nanofluid.* 2007, *4*, 97–105.

[20] Wang, K.; Schonbrun, E.; Steinvurzel, P.; Crozier, K. *Nat. Commun.* 2011, *2*, 1–6.

[21] Monat, C.; Domachuk, P.; Eggleton, B. J. J. *Nat. Photonics* 2007, *1*, 106–114.

[22] Yang, A. H. J.; Moore, S. D.; Schmidt, B. S.; Klug, M.; Lipson, M.; Erickson, D. *Nature* 2009, *457*, 71–75.

[23] Liu, G. L.; Kim, J.; Lu, Y.; Lee, L. P. *Nat. Mater.* 2006, *5*, 27–32.

[24] Winslow, W. M. *J. Appl. Phys.* 1949, *20*, 1137.

[25] Rainbow, J. *Trans. Am. Inst. Electr. Eng.* 1948, *67*, 1308.

[26] Vékás, L.; Bica, D.; Avdeev, M. V. *China Part* 2007, *5*, 43–49.

[27] Neuringer, J. L.; Rosensweig, R. E. *Phys. Fluids* 1964, *7*, 1927.

[28] Raj, K.; Moskowitz, B.; Casciari, R. *J. Magn. Magn. Mater.* 1995, *149*, 174–180.

[29] Singh, D. K.; Pandey, D. K.; Yadav, R. R. *Ultrasonics* 2009, *49*, 634–637.

[30] Patil, S. *Int. J. Pharma Res. Dev.* 2010, *2*, 25–29.

[31] Büscher, K.; Helm, C. A.; Gross, C.; Glöckl, G.; Romanus, E.; Weitschies, W. *Langmuir* 2004, *20*, 2435–2444.

[32] Li, J.; Dai, D.; Zhao, B.; Lin, Y.; Liu, C. *J. Nanopart Res* 2002, *4*, 261–264.

[33] Ummartyotin, S.; Juntaro, J.; Sain, M.; Manuspiya, H. *Chem. Eng. J.* 2012, *193-194*, 16–20.

[34] Gao, Y.; Huang, J. P.; Liu, Y. M.; Gao, L.; Yu, K. W.; Zhang, X. *Phys. Rev. Lett.* 2010, *104*, 1–4.

[35] Rosenweig, R. E. *Nature* 1966, *210*, 613–614.

[36] Moskowitz, R.; Rosenweig, R. E. *Appl. Phys. Lett.* 1967, *11*, 301.

[37] Brady, M.; Gregor, L. V.; Johnson, M. *Ferrofluid Optical Switches.* US patent, 4,384,761, 1980.

[38] Mahendran, V.; Philip, J. *Appl. Phys. Lett.* 2012, *100*, 073104.

[39] Bacri J. C., Cabuil V., Massart R., Perzynski R., and Salin D., *J. Magn. Magn. Mater.* 1987, 65, 285–288.

[40] Brousseau, D.; Borra, E. F.; Thibault, S. *Opt. Express* 2007, *15*, 18190–18199.

[41] Otanicar, T. P.; Phelan, P. E.; Golden, *J. S. Sol. Energy* 2009, *83*, 969–977.

[42] Veeraragavan, A.; Lenert, A.; Yilbas, B.; Al-Dini, S.; Wang, E. N. *Int. J. Heat Mass Tran.* 2012, *55*, 556–564.

[43] Lenert, A.; Wang, E. N. *Sol. Energy* 2012, *86*, 253–265.

[44] Lenert, A.; Zuniga, Y. S. P.; Wang, E. N. In *Proceedings of the 14th International Heat Transfer Conference;* Washington, USA August 2010.

[45] Otanicar, T. P.; Chowdhury, I.; Prasher, R.; Phelan, P. E. *J. Sol. Energy Eng.* 2011, *133*, 041014.

[46] Mercatelli, L.; Sani, E.; Zaccanti, G.; Martelli, F.; Di Ninni, P.; Barison, S.; Pagura, C.; Agresti, F.; Jafrancesco, D. *Nanoscale Res. Lett.* 2011, *6*, 282.

[47] Sani, E.; Mercatelli, L.; Barison, S.; Pagura, C.; Agresti, F.; Colla, L.; Sansoni, P. *Sol. Energ. Mat. Sol. C* 2011, *95*, 2994–3000.

[48] Han, D.; Meng, Z.; Wu, D.; Zhang, C.; Zhu, H. *Nanoscale Res. Lett.* 2011, *6*, 457.

[49] Yousefi, T.; Shojaeizadeh, E.; Veysi, F.; Zinadini, S. *Sol. Energy* 2012, *86*, 771–779.

[50] Lu, L.; Liu, Z.-H.; Xiao, H.-S. *Sol. Energy* 2011, *85*, 379–387.

[51] Kameya, Y.; Hanamura, K. *Sol. Energy* 2011, *85*, 299–307.

[52] Tyagi, H.; Phelan, P.; Prasher, R. *J. Sol. Energy Eng.* 2009, *131*, 041004.

[53] Natarajan, E.; Sathish, R. *Int. J. Adv. Manuf. Tech.* 2009, *1 [Special]*, 3–7.

[54] Yousefi, T.; Veysi, F.; Shojaeizadeh, E.; Zinadini, S. *Renew. Energ* 2012, *39*, 293–298.

[55] Sani, E.; Mercatelli, L.; Zaccanti, G.; Martellj, F.; Ninni, P. Di; Pagura, C.; Giannini, A.; Francini, D. J. D. F. F. In *Proceedings of the European Conference on Lasers and Electro-Optics*; Munich, Germany, May 2011.

[56] Saidur, R.; Meng, T. C.; Said, Z.; Hasanuzzaman, M.; Kamyar, A. *Int. J. Heat Mass Tran.* 2012, *55*, 5899–5907.

[57] Otanicar, T. P.; Phelan, P. E.; Prasher, R. S.; Rosengarten, G.; Taylor, R. A. *J. Renew. Sustain. Energy* 2010, *2*, 033102.

[58] Otanicar, T. P.; Phelan, P. E.; Taylor, R. A.; Tyagi, H. *J. Sol. Energy Eng.* 2011, *133*, 024501.

[59] Garcia, G.; Buonsanti, R.; Runnerstrom, E. L.; Mendelsberg, R. J.; Llordes, A.; Anders, A.; Richardson, T. J.; Milliron, D. J. *Nano Lett.* 2011, *11*, 4415–4420.

[60] Beydoun, D.; Amal, R.; Low, G.; Mcevoy, S. *J. Nanopart. Res.* 1999, *1*, 439–458.

[61] Otanicar, T. P.; Golden, J. S. *Environ. Sci. Technol.* 2009, *43*, 6082–6087.

[62] Boerema, N.; Morrison, G., Taylor, R.A., Rosengarten, G. *Sol. Energy*; 2012, 86, 2293-2305.

[63] Arancibia-bulnes, C. A.; Bandala, E. R.; Estrada, C. A., *Catalysis Today;* 2002, *76*, 149–159.

[64] Abdelrahman, M.; Fumeaux, P.; Suter, P. *Sol. Energy* 1979, *22*, 45–48.

[65] Jorgensen, G.; Schissel, P.; Burrows, R. *Sol. Energy Mater.* 1986, *14*, 385–394.

[66] Bohn, M. S. *Energy* 1987, *12*, 227–233.

[67] Otanicar, T. P. *Direct Absorption Solar Thermal Collectors Utilizing Liquid-Nanoparticle Suspentions;* PhD thesis; Arizona State University, Arizona, USA, 2009.

[68] Bohn, M. S.; Wang, K. Y. *J. Sol. Energy Eng.* 1988, *110*, 45–51.

[69] Sasse, C.; Ingel, G. *Sol. Energ. Mat. Sol. C* 1993, *31*, 61–73.

[70] Bohn, M. S.; Green, H. J. *Sol. Energy* 1989, *42*, 57–66.

[71] Webb, B. W.; Viskanta, R. *J. Sol. Energy Eng.* 1985, *107*, 113.

[72] Sani, E.; Barison, S.; Pagura, C.; Mercatelli, L.; Sansoni, P.; Fontani, D.; Jafrancesco, D.; Francini, F. *Opt. Express* 2010, *18*, 4613–4616.

[73] Griffin, J. W.; Stahl, K. A.; Pettit, R. B. *Sol. Energy Mater.* 1986, *14*, 395–416.

[74] Liu, Z.; Hou, W.; Pavaskar, P.; Aykol, M.; Cronin, S. B. *Nano Lett.* 2011, *11*, 1111–1116.

[75] Drotning, W. D. *Sol. Energy* 1977, *20*, 313–319.

[76] Miller, F. J.; Koenigsdorff, R. W. *J. Sol. Energy Eng.* 2000, *122*, 23–29.

[77] Hunt, A. J. *Small Particle Heat Exchangers*; 1978.

[78] Karni, J.; Kribus, A.; Rubin, R.; Doron, P. *J. Sol. Energy Eng.* 1998, *120*, 85.

[79] Bertocchi, R.; Kribus, A.; Karni, J. *J. Sol. Energy Eng.* 2004, *126*, 833.

[80] Kumar, S.; Tien, C. L. *J. Heat Transf.* 1990, *112:1*.

[81] Tien, C. L. *J. Heat Transf.* 1988, *110*, 1230–1242.

[82] Prasher, R. *J. Appl. Phys.* 2007, *102*, 1–9.

[83] Taylor, R. A.; Phelan, P. E.; Otanicar, T. P.; Walker, C. A.; Nguyen, M.; Trimble, S.; Prasher, R. *J. Renew. Sustain. Energy* 2011, *3*, 1–8.

[84] Cengel, Y.; Boles, M. *Thermodynamics: An Engineering Approach*; McGraw-Hill Science/Engineering/Math: New York, USA, 2010.

[85] Taylor, R. A.; Phelan, P. E.; Otanicar, T.; Adrian, R. J.; Prasher, R. S. *Appl. Phys. Lett.* 2009, *95*, 161907.

[86] Solutia Therminol VP-1, www.therminol.com/pages/products/vp-1.asp [accessed Oct 19, 2013].

[87] Blunden, S. J.; Chapman, A. H. *Environ. Technol. Lett.* 1982, *3*, 267–277.

[88] Luo, Y.-R. *Comprehensive Handbook of Chemical Bond Energies*; 1st ed.; CRC Press (Taylor & Francis): Boca Raton, USA, 2007.

[89] Codd, D. S. *Concentrated Solar Power on Demand*; Ph.D thesis; Massachusetts Institute of Technology, Boston, USA 2011.

[90] Kurosaki, Y.; Viskanta, R. In *Proceedings of the ASME Winter Annual Meeting,*; San Francisco, USA, December 1978.

[91] Hssatani, M.; Arai, N.; Bando, H. *Heat Transf. Japanese Res.* 1982, *11*, 17–30.

[92] Bradshaw, R. W.; Siegel, N. P. In *Proceedings of the ASME Energey Sustainability*; Jacksonville, USA, August 2008.

[93] Ladjevardi, S. M.; Asnaghi, A.; Izadkhast, P. S.; Kashani, A. H. *Sol. Energy* 2013, *94*, 327–334.

[94] Goldman, A.; Meitav, R.; Yakupov, R.; Krozier, I.; Kokotov, Y.; Gilon, Y. High Temperature Solar Reciever. US Patent, 7690377, 2010.

[95] Tyagi, H.; Phelan, P.; Prasher, R. *J. Sol. Energy Eng.* 2009, *131*, 41004.

[96] Schunk, L. O.; Haeberling, P.; Wepf, S.; Wuillemin, D.; Meier, A.; Steinfeld, A. *J. Sol. Energy Eng.* 2008, *130*, 021009.

[97] Hirsch, D.; Steinfeld, A. *Int. J. Hydrogen Energ.* 2004, *29*, 47–55.

[98] Yogev, A. *Solar Energy System with Direct Absorption of Solar Radiation*, US Patent, 6776154, 2004.

[99] Parker, R. Z.; Langhoff, P. W. *Fluid Absorption Receiver for Solar Radiation,* US Patent, 5241824, 1993.

[100] Hunt, A. J. In *Proceedings of the 13th Intersociety Energy Conversion Engineering Conference*; Boston, USA, 1979.

[101] Kraus, R. A.; Kraus, E. J. *Solar Thermal-Radiation, Absorption and Conversion System*, US Patent, 4055948, 1977.

[102] Kelly, L. K.; Coronado, E.; Zhao, L. L.; Schatz, G.; Coranado, E. *J. Phys Chem. B.* 2003, *107*, 668–677.

[103] Kong, G.; Braun, R. D.; Dewhirst, M. W. *Cancer Res.* 2000, *60*, 4440–4445.

[104] Huang, X.; Jain, P. K.; El-Sayed, I. H.; El-Sayed, M. A. *Photochem. Photobiol.* 2006, *82*, 412–7.

[105] Bergey, E. J.; Levy, L.; Wang, X.; Krebs, L. J.; Lal, M.; Kim, K.; Pakatchi, S.; Liebow, C.; Paras, N. *Biomed. Microdevices* 2002, *4*, 293–299.

[106] Johannsen, M.; Thiesen, B.; Wust, P.; Jordan, A. *Int. J. Hyperther* 2010, *26*, 790–795.

[107] Hergt, R.; Dutz, S.; Müller, R.; Zeisberger, M. *J. Phys-Condens. Mat.* 2006, *18*, 2919–2934.

[108] Salloum, M.; Ma, R. H.; Weeks, D.; Zhu, L. *Int. J. Hyperther* 2008, *24*, 337–345.

[109] Kikumori, T.; Kobayashi, T.; Sawaki, M.; Imai, T. *Breast Cancer Res. Tr.* 2009, *113*, 435–441.

[110] Nie, S.; Xing, Y.; Kim, G. J.; Simons, J. W. *Annu. Rev. Biomed. Eng.* 2007, *9*, 257–288.

[111] Grodzinski, P.; Silver, M.; Molnar, L. K. *Expert Rev. Mol. Diagn.* 2006, *6*, 307–318.

[112] Laconte, L.; Nitin, N.; Bao, G. *Nano Today* 2005, 32–38.

[113] Gu, Y.; Sun, W.; Wang, G.; Fang, N. *Biophys. J.* 2011, *100*.

[114] Hirsch, L. R.; Stafford, R. J.; Bankson, J. A.; Sershen, S. R.; Rivera, B.; Price, R. E.; Hazle, J. D.; Halas, N. J.; West, J. L. *Proc. Natl. Acad. Sci. U. S. A.* 2003, *100*, 13549–13554.

[115] Peer, D.; Karp, J. M.; Hong, S.; Farokhzad, O. C.; Margalit, R.; Langer, R. *Nat. Nanotechnol.* 2007, *2*, 751–760.

[116] Psaltis, D.; Quake, S. R.; Yang, C. *Nature* 2006, *442*, 381–386.

[117] Ferrari, M. *Nat. Rev. Cancer* 2005, *5*, 161–171.

[118] Keblinski, P.; Cahill, D. G.; Bodapati, A.; Sullivan, C. R.; Taton, T. A. *J. Appl. Phys.* 2006, *100*, 054305.

[119] Dombrovsky, L. A.; Timchenko, V.; Jackson, M.; Yeoh, G. H. *Int. J. Heat Mass Tran.* 2011, *54*, 5459–5469.

[120] Parveen, S.; Sahoo, S. K. *Eur. J. Pharmacol.* 2011, *670*, 372–383.

[121] Aryal, S.; Hu, J. C.-M.; Fu, V.; Zhang, L. *J. Mater. Chem.* 2012, *22*, 994–999.

[122] Kalantar-zadeh, K.; Khoshmanesh, K.; Kayani, A. A.; Nahavandi, S.; Mitchell, A. *Appl. Phys. Lett.* 2010, *96*, 101108.

[123] Schaap, A.; Bellouard, Y.; Rohrlack, T. *Biomed. Opt. Express* 2011, *2*, 658–664.

[124] Kim, C. A.; Joung, M. J.; Ahn, S. D.; Kim, G. H.; Kang, S.-Y.; You, I.-K.; Oh, J.; Myoung, H. J.; Baek, K. H.; Suh, K. S. *Synth. Met.* 2005, *151*, 181–185.

[125] Reed, B. Amazon's annual Kindle sales estimated at $4.5 billion http://news.yahoo.com/amazon-annual-kindle-sales-estimated-4-5-billion-011530328.html [accessed Oct 10, 2013].

[126] Taylor, R. A.; Otanicar, T. P.; Hewakerrppu, Y.; Bremond, F.; Rosengarten, G.; Hawkes, E.; Jiang, X.; Coulombe, S. *Appl. Optics.* 2013, *52*, 1413–1422.

[127] Hewakuruppu, Y. L.; Dombrovsky, L. A.; Timchenko, V.; Yeoh, G.; Jiang, X.; Taylor, R. A. *Int. J. Transport Phenomena,* 2013, *13*, 233-244.

[128] Mie, G. *Ann. Phys.* 1908, *25*, 377–445.

[129] Bohren, C. F.; Huffman, D. R. *Absorption and scattering of light by small particles*; Wiley-VCH, Weinheim, Germany, 1998.

[130] Dombrovsky, L. A.; Baillis, D. *Thermal Radiation in Disperse Systems: An Engineering Approach*; Begell House: New York, USA, 2010.

[131] Dombrovsky, L. A. *Radiation Heat Transfer in Disperse Systems*; Begell House: New York, USA, 1996.

[132] Myroshnychenko, V.; Rodríguez-Fernández, J.; Pastoriza-Santos, I.; Funston, A. M.; Novo, C.; Mulvaney, P.; Liz-Marzán, L. M.; García de Abajo, F. J. *Chem. Soc. Rev.* 2008, *37*, 1792–1805.

[133] Rastar, A.; Yazdanshenas, M. E.; Rashidi, A.; Bidoki, S. M. *J. Eng. Fiber. Fabr.* 2013, *8*, 85–96.

[134] He, G. S.; Zhu, J.; Yong, K.T.; Baev, A.; Cai, H.X.; Hu, R.; Cui, Y.; Zhang, X.H.; Prasad, P. N. *J. Phys. Chem. C* 2010, *114*, 2853–2860.

[135] Payne, E. K.; Shuford, K. L.; Park, S.; Schatz, G. C.; Mirkin, C. A. *J. Phys. Chem. B* 2006, *110*, 2150–2154.

[136] Prescott, S. W.; Mulvaney, P. *J. Appl. Phys.* 2006, *99*, 123504.

[137] Chen, J.; Saeki, F.; Wiley, B. J.; Cang, H.; Cobb, M. J.; Li, Z.-Y.; Au, L.; Zhang, H.; Kimmey, M. B.; Li, X.; Xia, Y. *Nano Lett.* 2005, *5*, 473–477.

[138] Félidj, N.; Grand, J.; Laurent, G.; Aubard, J.; Lévi, G.; Hohenau, A.; Galler, N.; Aussenegg, F. R.; Krenn, J. R. *J. Chem. Phys.* 2008, *128*, 094702.

[139] Yurkin, M. A.; Hoekstra, A. G. *J. Quant. Spectrosc. Ra.* 2007, *106*, 558–589.

[140] Yurkin, M. A.; Hoekstra, A. G. *J. Quant. Spectrosc. Ra.* 2011, *112*, 2234–2247.

[141] Draine, B. T.; Flatau, P. J. *J. Opt. Soc. Am. A* 1994, *11*, 1491–1499.

[142] Oubre, C.; Nordlander, P. *J. Phys. Chem. B* 2005, *109*, 10042–10051.

[143] Oubre, C.; Nordlander, P. *J. Phys. Chem. B* 2004, *108*, 17740–17747.

[144] Hao, F.; Nordlander, P. *Chem. Phys. Lett.* 2007, *446*, 115–118.

[145] Myroshnychenko, V.; Carbó-Argibay, E.; Pastoriza-Santos, I.; Pérez-Juste, J.; Liz-Marzán, L. M.; García de Abajo, F. J. *Adv. Mater.* 2008, *20*, 4288–4293.

[146] Mishchenko, M. I.; Videen, G.; Babenko, V. A.; Khlebtsov, N. G.; Wriedt, T. *J. Quant. Spectrosc. Ra.* 2004, *88*, 357–406.

[147] Taylor, R. A.; Phelan, P. E.; Otanicar, T. P.; Adrian, R.; Prasher, R. *Nanoscale Res. Lett.* 2011, *6*, 225.

[148] Soni, S.; Tyagi, H.; Taylor, R. A.; Kumar, A. *Int. J. Hyperther* 2013, *29*, 87–97.

[149] Kreibig, U.; Vollmer, M. *Optical properties of metal clusters*; Springer: New York, USA, 1995.

[150] Dombrovsky, L. A.; Randrianalisoa, J.; Baillis, D.; Pilon, L. *Appl. Optics.* 2005, *44*, 7021–7031.

[151] Dombrovsky, L. A.; Randrianalisoa, J.; Baillis, D. *J. Opt. Soc. Am. A.* 2006, *23*, 91–98.

[152] Dombrovsky, L. A.; Timchenko, V.; Jackson, M.; Yeoh, G. H. *Int. J. Heat Mass Tran.* 2011, *54*, 5459–5469.

[153] Hewakuruppu, Y. L.; Dombrovsky, L. A.; Chen, C.; Timchenko, V.; Jiang, X.; Baek, S.; Taylor, R. A. *Appl. Opt.* 2013, *52*, 6041.

[154] Dombrovsky, L. A.; Randrianalisoa, J.; Baillis, D. *J. Opt. Soc. Am. A.* 2006, *23*, 91–98.

[155] Dombrovsky, L. A.; Timchenko, V.; Jackson, M. *Int. J. Heat Mass Tran.* 2012, *55*, 4688–4700.

[156] Prahl, S. A.; Gemert, M. J. C. Van; Welch, A. J. *Appl. Optics.* 1993, *32*, 559–568.

[157] Khullar, V.; Tyagi, H.; Phelan, P. E.; Otanicar, T. P.; Singh, H.; Taylor, R. A. *J. Nanotechnol. Eng. Med.* 2012, 3 (3), 031003-031003-9.

[158] Otanicar, T. P.; Taylor, R. A.; Telang, C. *J. Renew. Sustain. Energy* 2013, *5*, 033124.

[159] Averitt, R. D.; Westcott, S. L.; Halas, N. J. *J. Opt. Soc. Am. B* 1999, *16*, 1824.

[160] Oldenburg, S. J.; Averitt, R. D.; Westcott, S. L.; Halas, N. J. *Chem. Phys. Lett.* 1998, *288*, 243–247.

[161] Boyer, C.; Stenzel, M. H.; Davis, T. P. *J Polym. Sci. Part A.* 2011, *49*, 551–595.

[162] Nanocomposix Gold Nanoshells, http://nanocomposix.com/products [accessed Oct 10, 2013].

[163] Nanopartz Gold nanoparticles for nano- technology, https://www.nanopartz.com [accessed Oct 10, 2013]

[164] Ghadimi, A.; Saidur, R.; Metselaar, H. S. C. *Int. J. Heat Mass Tran.* 2011, *54*, 4051–4068.

[165] Hwang, Y.; Lee, J.; Lee, C.; Jung, Y.; Cheong, S.; Ku, B.; Jang, S. *Thermochim. Acta* 2007, *455*, 70–74.

[166] Tavares, J.; Coulombe, S. *Powder Technol.* 2011, *210*, 132–142.

[167] Le, F.; Lwin, N.; Halas, N.; Nordlander, P. *Phys. Rev. B* 2007, *76*, 165410.

[168] Taylor, R. A.; Phelan, P. E.; Otanicar, T.; Prasher, R. S.; Phelan, B. E. *Int. Commun. Heat Mass.* 2012, *39*, 1467–1473.

In: Nanofluids: Synthesis, Properties and Applications   ISBN: 978-1-63321-677-8
Editors: S.M. Sohel Murshed, C.A. Nieto de Castro   © 2014 Nova Science Publishers, Inc.

Chapter 11

PROGRESS AND CHALLENGES IN NANOFLUIDS RESEARCH

C. A. Nieto de Castro[*] and S. M. Sohel Murshed

Centro de Ciências Moleculares e Materiais
Faculdade de Ciências, Universidade de Lisboa, Lisboa, Portugal

ABSTRACT

Although extensive research on nanofluids have been conducted since coining this new class of fluids and good progress has also been made in some areas particularly experimental measurements of thermophysical properties, there are lot of tough challenges and unsolved issues remain in the nanofluids research. The success of nanofluids research towards their practical application and commercialization for the socioeconomic benefits depends on the true research progress by resolving the unknown or mysterious issues, identifying and understanding the underlying heat transfer mechanisms and overcoming the challenges of nanofluids. Arguably about two decades have been passed since the innovation of the concept of nanofluids at Argonne National Laboratory of USA and according to the recent records of research publications from all over the world suggested that this novel class of fluids is a hot research topic. It is therefore important and timely to summarize the progresses and addresses the major challenges of nanofluids research and this chapter briefly reviews the development and challenges of research in various areas of nanofluids.

1. BACKGROUND

It has been almost two decades since the term "nanofluids" was proposed by Steve Choi at Argonne National Laboratory in 1995 [1]. However, research on this topic has started to intensify from the beginning of 21st century. In the evolution of the concept of nanofluids credit should also be given to other earlier works related to suspensions of nanoparticles which are in fact nanofluids. For example, Yang and Maa [2] used the suspensions of

---

[*] E-mail: cacastro@fc.ul.pt.

nanoparticles in their boiling heat transfer study as early as 1984 and another comprehensive experimental study on the thermal conductivity and viscosity of several types of nanoparticles-suspensions was performed by Masuda and co-workers [3] in 1993. At present this is one of the hottest research fields which can be evidenced from the exponential growth of research publications on this topic. Although extensive research have been performed in the last decade, except few areas progress on nanofluids research are rather slow. Researchers are still facing enormous challenges to uncover the true mechanisms behind the observed anomalous thermal properties and heat transfer features of these new fluids. After showing some agreements between experimental results and predicted thermal conductivity of several nanofluids using effective medium theory (EMT) based upper and lower bounds models proposed by Hashin and Shtrikman [4] for magnetic permeability of multiphase materials, some researchers [5-6] were convinced that based on aggregation of nanoparticles these upper and lower bounds models can explain the enhancement of thermal conductivity of nanofluids and the controversy regarding the heat conduction mechanisms of nanofluids may be over. However, a subsequent comment on the drawback of such argumentations on mechanisms [7] and continued findings on anomalous enhancement of thermal conductivity of nanofluids justify that variability and controversies in the thermal characteristics still remain [8-10]. Among other drawbacks in nanofluids research include inconsistent and contradictory results, nonsystematic investigations, and so on [8-12].

In order to resolve the uncertainties surrounding the reported thermophysical properties of nanofluids a round-robin exercise named-International Nanofluid Properties Benchmark Exercise (INPBE) was carried out on the measurement of thermal conductivity of nanofluids. Dozens of research groups or organizations worldwide which were involved in the exercise found no anomalous enhancement of thermal conductivity of various nanofluids tested in the exercise [13]. The effective medium theory developed by Maxwell [14] for the thermal conductivity of suspensions of homogenously dispersed particles and its generalized expression of Nan et al. [15] were found to be in good agreement with the experimental data, suggesting that no special nanoscale-based model is necessary for nanofluids [13]. Later a more specific round-robin exercise on the measurement of thermal conductivity of ethylene glycol-based ZnO nanofluids was conducted by five Korean research groups [16] who also compared their results with several EMT based thermal conductivity models. In contrary to the first exercise (INPBE) these groups [16] convincingly demonstrated that the enhancement of thermal conductivity of their nanofluids is considerably higher than the lower and upper bounds predictions using both the Maxwell [14] and Nan et al. [15] models. Thus instead of resolving controversies and uncertainties these round-robin exercises also added some new controversy about the scale of thermal conductivity enhancement and applicability of EMT-based classical theories. Nevertheless, except very limited studies including the first benchmark exercise almost all researchers reported significant enhancement of the thermal conductivity of nanofluids which cannot be predicted by those classical EMT-based models [14,15]. Thus researchers have kept their efforts up on identifying the heat transfer mechanisms behind such enhanced thermal conductivity and on the model development for the prediction of the thermal conductivity of nanofluids.

On the other hand, as compared to studies on thermal conductivity limited research efforts have been devoted on the viscosity of nanofluids. However, there is a growing interest in studying this crucial property of nanofluids and findings on viscosity of nanofluids are more consistent than those of thermal conductivity results. Though it's not desired the viscosity of nanofluids was found to be much higher compared to that of the base fluids and it further increased substantially with increasing the concentration of nanoparticles. The viscosity of nanofluids also cannot be predicted by any existing viscosity models. As a part of the first benchmark exercise (INPBE) the round-robin testing was also carried out on the rheological properties of various nanofluids and the effects of nanoparticle concentration and shape were examined [17]. Results obtained from the exercise revealed that while some nanofluids showed shear-thinning behavior, others are Newtonian. It was also concluded that nanoparticle concentration as well as shape have significant influence with linear-dependent on the viscosity of those nanofluids and the classical Einstein´s model [18] considerably under-predict their viscosity results. The agglomeration of nanoparticles was considered as the main reason for the observed differences between the predictions and measured data.

Research on the other key areas such as convective and boiling heat transfer of nanofluids are also receiving more attention in recent years and the findings in these important areas are even more promising as nanofluids showed substantially higher convective and boiling heat transfer features (e.g., heat transfer coefficient and critical heat flux) compared to those of their base conventional fluids.

Nonetheless, the major developments in research on nanofluids include almost undisputed anomalous enhancement of boiling and convective heat transfer of nanofluids, promising findings from the practical application-based research, exploring more new and highly demanding areas for nanofluids such as solar energy harvesting, nuclear reactor and other advanced cooling technologies. Despite inconsistencies in thermal conductivity data reported by many research groups, nanofluids undoubtedly exhibit considerably enhanced thermal conductivity as compared to their conventional base heat transfer fluids. Numerous experimental measurement techniques have also been deployed to measure the thermophysical properties of nanofluids.

Research showed that nanofluids exhibit much higher thermophysical properties and heat transfer features such as thermal conductivity, viscosity, specific heat and convective as well boiling heat transfer performance as compared to their base fluids [8, 9, 11, 12, 19-21]. With anomalously high thermophysical properties and heat transfer characteristics nanofluids showed great promises as advanced heat transfer fluids and can meet the cooling and heating challenges facing numerous high-tech industries and thermal management systems.

Given the prospect and potentials of nanofluids, it is imperative and timely to provide a systematic overview of research progress and challenges of this new class of emerging fluids.

This chapter therefore, summarizes and addresses the recent research progress in key thermal properties and features of nanofluids. Besides briefing on nanofluids preparation and heat transfer mechanisms, practical applications of nanofluids are also discussed. Furthermore major research challenges in various areas of nanofluids are addressed.

2. RESEARCH PROGRESS

Research progress in various areas of nanofluids including nanofluids preparation, experimental techniques, thermal features, heat transfer mechanisms and model development, practical applications and development of new nanofluids are systematically summarized in this section.

2.1. Nanofluids Preparation

The preparations or productions of nanofluids are widely performed through two routes. As described in Chapter 1, only recently the methodology for the preparation of nanofluids reached a state that researchers and end users can rely on, as the properties of the complex systems are strongly affected by the quality of the suspensions, their thermodynamic and time stability. When property data is compared, even for the same base fluid and the same nanomaterial (sometimes impossible), and gross deviations are found between different laboratories, seldom it is possible to understand completely the preparation of each nanofluid in the bibliography. These two "classical" routes are currently under strong screening and validation, and different methodologies will be developed and reported in the near future, as analyzed below. The first route is the direct fabrication of nanoparticles in host fluids, normally by chemical reaction. The other route is to synthesize or purchase nanoparticles first (from a known producer) and then disperse them in the base fluids. While the former route is known as one-step method, the latter is called two-step method. Recently some other techniques are used to prepare nanofluids in small quantity and most of them are chemical solution or reaction based-route [22]. Some researchers also adopted different methods to prepare nanofluids for their research. For example, while phase transfer method was reported to use in preparing nanofluids by some researchers [23-24], microfluidic continuous flow systems were used to synthesis gold and silver based nanofluids by Boleininger et al. [25] and copper dispersed nanofluids by Wei et al. [26]. With the help of microwave irradiation and ultrasonication precursor transformation was performed to successfully prepare CuO nanofluid in a study by Zhu et al. [27]. Besides short-term stability of prepared nanofluids, the other limitations of these new one-step methods are that they cannot be used to produce large quantity of nanofluids as well as hard to control the size and size distribution of nanoparticles, and the purity of the nanofluid. Whereas other than nanofluids nanoparticles have wide range of application and thus synthesizing nanoparticles (nanopowders) in large quantity received much attention from both the research and commercial sectors. There are two general approaches for synthesizing and fabrication of nanoparticles which are the bottom-up and top-down approaches. The bottom-up approaches begin with atoms or molecules and build up to nanostructures of desired sizes. Among the bottom-up techniques vapor deposition techniques particularly chemical vapor deposition (CVD) is the most widely used method for producing large scale nanoparticles. Nowadays arc discharge and laser ablation (vaporization) methods are also commonly used for nanoparticles production. In recent years, nanoparticles production techniques have been improved considerably. Therefore wide varieties of nanoparticles (various types and shapes) are commercially available and some of them are not very expensive as well. Thus in spite of advantages and disadvantages of both methods, researchers working or dealing with nanofluids mostly used two-step methods. Research advancements in the areas of nanofluids preparation methods, stability mechanisms and other related issues have been well-discussed in the literature [22, 28] as well as in other chapters of this book, namely Chapter 1.

2.2. Nanofluids Thermal Properties and Features

There is an increasing research interest on various thermal properties and features such as thermal conductivity (TC), convective heat transfer (CHT) and boiling heat transfer (BHT) characteristics of nanofluids (NF). This can be evidenced from the growth of annual publications as shown in Figure 1 which is generated from the records of publications searched by topic "nanofluids" and then refined by terms "thermal conductivity", "convection heat transfer" and "boiling" in Web of Science on April 01, 2014. As can be seen the major share of the reported publications during past five years on nanofluids belongs to the thermal conductivity followed by convection and boiling heat transfers.

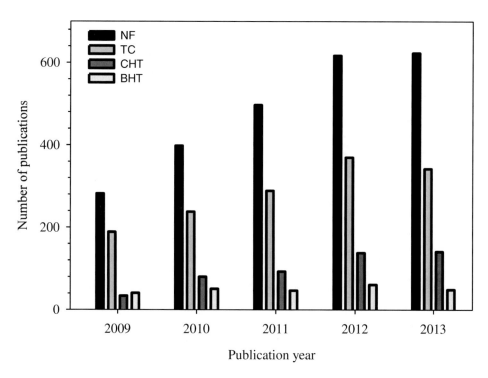

Figure 1. Past five years publication records on various areas of nanofluids (Publications include all types of journal and conference articles, patent, news, letter and others recorded in Web of Science).

2.2.1. Thermal Properties Measurement Methods

Researchers used various types of techniques to measure the thermal properties, particularly thermal conductivity of nanofluids. Although in early stage of nanofluids research, thermal conductivity measurement methods such as transient hot-wire techniques or steady state methods are built in the laboratories, researchers are now relying more on the commercial devices (such as KD2, thermal probes, and 3-omega) to measure the thermal conductivity of nanofluids. While the accuracy of measurements by all these techniques or devices are not the same, some devices need additional procedure, carefulness, and corrections to obtain accurate data from the measurements. The literature survey showed that most researchers used transient hot-wire (THW) techniques to measure the thermal conductivity of the nanofluids [8, 9, 19]. Nevertheless, most of the inconsistencies and scatterings of measured thermal conductivity data appeared due to employing various techniques and devices by different research groups. While thermal conductivity of nanofluids was widely measured and noticeable progress has been made, only a handful of studies measured other properties such as thermal diffusivity and specific heat of nanofluids [21, 29].

2.2.2. Thermal Conductivity

Among the properties of nanofluids, thermal conductivity has been the main focus of research which can be justified from the publication records on this key property of nanofluids (Figure 1). Apart from the controversial and inconsistent findings most of the researchers found anomalously enhanced thermal conductivity of nanofluids which cannot be predicted by the existing classical models. The enhanced thermal conductivity further increases with increasing the concentration of nanoparticles as well as temperature [8, 9, 11, 19]. Figure 2 depicts some representative results on the enhancement of the thermal conductivity of various types of nanofluids at room temperature from the literature. It is seen that although results from various groups are scattered nanofluids clearly exhibit much higher thermal conductivity than their base fluids even when the concentration of nanoparticles are very low and the enhanced further increases with volumetric loading of nanoparticles.

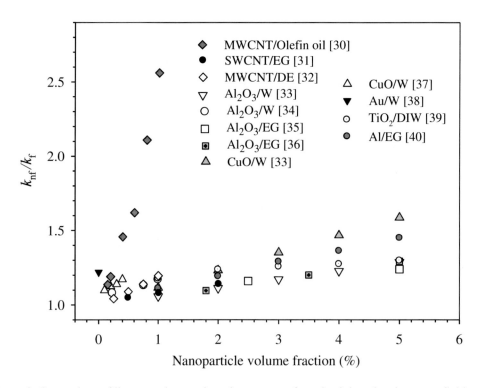

Figure 2. Comparison of literature data on the enhancement of conductivity of various nanofluids as a function of nanoparticle volumetric concentration.

2.2.3. Convective and Boiling Heat Transfer Performance

Although investigations on convective and boiling heat transfer characteristics are very important in order for their various heat transfer applications, considerably less research focus have been given on these areas of nanofluids. A state-of-the-art review on the advances in nanofluids´ boiling and convective heat transfer research was previously reported by the authors [20].

A comparison of some results from representative studies on Nusselt number versus Reynolds number for both laminar and turbulent flow conditions is shown in Figure 3 [20]. It is demonstrated that nanofluids exhibit enhanced heat transfer coefficient compared to their base fluid and this coefficient further increases significantly with increasing the concentration of nanoparticles as well as Reynolds number. The heat transfer coefficient is even more significant at turbulent regime than the laminar regime (Figure 3). This confirms that nanofluids can perform better heat transfer in flow conditions compared to their base fluids. Nonetheless, a clear understanding of the convective heat transfer mechanisms of nanofluids is also not yet achieved. In an attempt to establish a strong explanation of the reported anomalously enhanced convective heat transfer coefficient of nanofluids, Buongiorno [47] considered seven-slip mechanisms and concluded that among those seven only Brownian diffusion and thermophoresis are the two most important particle/fluid slip mechanisms in nanofluids. Besides proposing a new correlation, he also claimed that the enhanced laminar flow convective heat transfer can be attributed to a reduction of viscosity within and consequent thinning of the laminar sublayer.

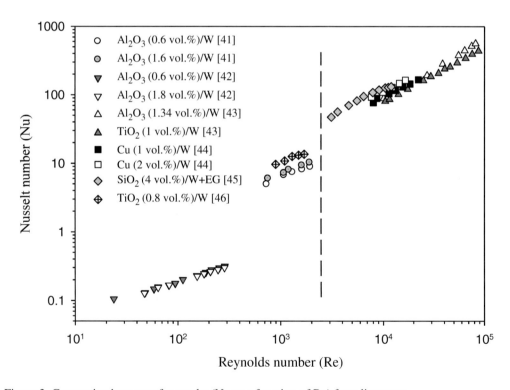

Figure 3. Convective heat transfer results (Nu as a function of Re) from literature.

In order for practical application of nanofluids it is of great importance to study their boiling heat transfer performance and to identify the underlying mechanisms. In boiling heat transfer nanofluids showed very high critical heat flux (CHF) and causing no fouling on the heat transfer surface. Due to significant enhancement in boiling heat transfer performance, there is a growing interest on boiling particularly pool boiling heat transfer investigations of nanofluids and unlike thermal conductivity the significant increase in the boiling critical heat flux of nanofluids is so far undisputed. However, reported data are still limited and scattered to clearly understand the underlying mechanisms and trend of boiling heat transfer performance of nanofluids [20]. The authors recently reviewed available studies and patents on boiling heat transfer and droplet spreading of nanofluids [48]. Figure 4 shows that the enhanced critical heat flux of nanofluids increases substantially with increasing the concentration of nanofluids. Studies showed that various factors such as deposition of nanoparticles on heater surface, heater surface roughness, and nanoparticles concentration are responsible for the enhanced boiling heat transfer of nanofluids [20, 48].

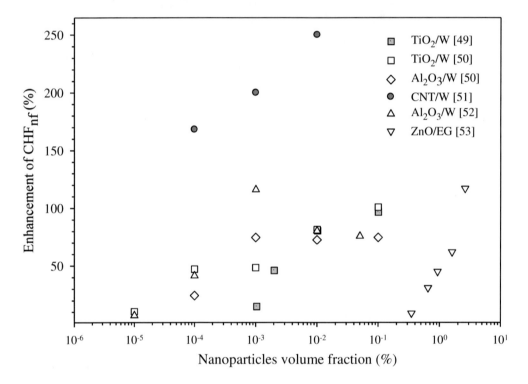

Figure 4. Enhancement of critical heat flux of various nanofluids over their base fluids as a function of nanoparticle concentration.

2.3. Nanofluids Heat Transfer Mechanisms

Most of the studies on thermal conductivity of nanofluids [8-11,19] showed that the observed significantly enhanced thermal conductivity of nanofluids cannot be predicted and explained by using existing effective medium theory (EMT)-based classical models such as Maxwell [14], Bruggeman [54], and Hamilton and Crosser [55] models. This is mainly due to the fact that these models considered thermal conductivities of both liquid and dispersed phases, concentration of dispersed phase and shape factor not any nanoscale mechanisms and activities of nanoparticles in host fluids. Only Hamilton and Crosser model [55] considers the shape of the disperse particles and thus it is particularly used for nanofluids containing cylindrical shape nanoparticles or nanotubes. At the early stage of nanofluids research, several heat transfer mechanisms for nanofluids were proposed and analyzed by Wang et al. [35] and Keblinski et al. [56]. These mechanisms include nanoparticles Brownian motion, interfacial nanolayer at the nanoparticle/fluid interface (nanolayer), nature of heat transport in the nanoparticle, and nanoparticle clustering. However, these mechanisms were mainly proposed for the observed anomalously high thermal conductivity of nanofluids containing spherical shape nanoparticles. Since then many theoretical studies have been carried out to predict the effective thermal conductivity of nanofluids. Several models have also been proposed with consideration of various factors. Based on the emphasized mechanisms the reported theoretical studies can be categorized into two groups. The first group of researchers used the concept of nanolayering at the liquid/nanoparticles interface to develop models and to explain the anomalous enhancement of the thermal conductivity in nanofluids while the other group emphasized the contribution of dynamic mechanisms related to nanoparticle Brownian motion in their model development. Interestingly both of these groups mainly superimposed the contribution their considered nanoscale mechanisms with the thermal conductivity of the effective medium theory-based classical models. Another group of researchers later [5-6, 13] considered that the agglomeration of nanoparticles are mainly responsible for the enhanced thermal conductivity of nanofluids which can even be predicted by the EMT-based models [4, 15]. Nevertheless, although interfacial nanolayer and nanoparticles agglomeration could be the key factors for the thermal conductivity of nanofluids, the effects of other factors such as nanoparticles surface chemistry and interactions can also be significant in enhancing the thermal conductivity of nanofluids. Since there remain controversies about the actual heat transfer mechanisms of nanofluids, it is in urgent need of resolving this issue before nanofluids can be used in practical applications. Detailed review and analysis of various conduction heat transfer mechanisms of nanofluids are explicitly reported in the literature [8-11, 19, 57] and will not be elaborated further.

2.4. Nanofluids in Various Applications

In addition to highly desirable thermal properties and potential benefits, this new class of fluids is believed to have wide range of industrial, electronics and medical applications. Nanofluids can be used for better heat transfer and other performances of systems or technologies in many engineering applications including transportations (engine cooling or vehicle thermal management), solar energy technologies, micro-electromechanical systems (MEMS), electronics and instrumentations, heat exchangers, heating-ventilating and air-conditioning (HVAC), cooling electronics, microfluidics systems, defense, medical and many other applications. Nanofluids have shown great promises in the medical and boimedical fields such as cancer treatment, drug delivery, and control of biofluids motion. Details of the applications of nanofluids have been discussed in the literature [8, 11, 12 19, 57-60]. Thus only some selected applications of these new fluids are briefly discussed here.

2.4.1. Engineering Applications

Nanofluids can be used to improve thermal management systems in many engineering applications including transportation, MEMS and instrumentations, HVAC and so on.

The transportation industry could benefit from the high thermal conductivity offered by nanofluids. Nanofluids would allow for smaller, lighter engines, pumps, radiators, and other components in cooling processes.

Miniaturization has been a major trend in science and technology. Micro-electromechanical systems (MEMS) generate a high heat during operation. Conventional coolants do not work well with high-power MEMS because they do not have enough cooling capability. Since nanofluids can flow in microchannels without clogging, they would be suitable coolants in such applications. They could enhance cooling of MEMS under extreme heat flux conditions. Some efforts have been made to study the cooling performance of nanofluids in commercial cooling systems used for electronics devices and computers and nanofluids were found to be very promising for cooling such devices [61-63].

Nanofluids can improve heat transfer capabilities of current industrial HVAC and refrigeration systems. They can also be used in the buildings heating systems and refrigeration in land and sea units. Kulkarni et al. [64] showed that nanofluids can save pumping power in heat exchangers used to warm to room and buildings particularly in cold regions as for the same amount of heating or cooling nanofluids require smaller heating systems compared to conventional fluids used in such applications which can save space, power and related cost as well.

2.4.2. Solar Energy Technologies

Nanofluids offer great potentials in solar energy technologies-based applications because of their enhanced absorption, scattering and optical properties as well as other thermophysical properties such as heat capacity. In addition to many other beneficial features nanofluids offer several key advantages in solar power plants including passing through pumps and plumbing of minisystems without any adverse effects, directly absorption solar energy, high absorption in the solar range and low emittance in the infrared, enhanced heat convection and radiation transfer, and enhancing absorption efficiency by changing the size and concentration of nanoparticles [65]. However, comparatively very fewer investigations have been made on the application of nanofluids in this highly demanding sector. In recent years, numbers of research investigations have been made to assess the applicability and performance of nanofluids in solar energy technologies particularly in various solar collectors [65-75]. Most of these studies demonstrated that nanofluids can considerably improve the efficiency of solar collectors and key findings from some of the representative studies are followed here.

Taylor et al. [65] revealed that the power tower solar collectors can be benefited from the potential efficiency improvements due to using a nanofluid as working fluid and their experiments on a laboratory-scale nanofluid dish receiver indicated that up to 10% increase in efficiency relative to a conventional fluid is possible. By comparing the energy and revenue generated in a conventional solar thermal plant to a nanofluid-based one, they found that a 100 MWe capacity solar thermal power tower operating in a solar resource could generate about $3.5 million more per year by incorporating a nanofluid receiver [65].

Experimental study of Lenert and Wang [72] on concentrated solar power application demonstrated that the efficiency increases more than 35% with nanofluid. Yousefi et al. [73, 74] reported about 28.3% increase in efficiency of flat-plate solar collectors using nanofluids and the enhanced efficiency further increase with increasing the concentration of MWCNT used in their study.

Typical solar thermal-energy storage facilities such as solar thermal power plants require the storage medium to have high heat capacity and thermal conductivity. Thus applying nanofluids with their higher thermal conductivity as well as heat capacity (compared to their base fluids) can remarkably enhance the capability and performance of solar thermal power plants [68]

Above mentioned exciting findings clearly indicate that like many other applications these new fluids can revolutionize the solar energy technologies for harvesting solar energy for numerous applications. More details on the research development and findings on the usability and performance of nanofluids in solar collectors can be found in a recent review study by Javadi et al. [76].

2.5. New Nanofluids

A novel class of nanofluids which is named as "IoNanofluids" has recently been devised by our group [77]. IoNanofluids are defined by the dispersion of nanoparticles in ionic liquids (ILs) only and they also exhibit high thermal conductivity, high heat capacity, non-volatile, designable, and green solvent characteristics [77-80]. These new nanofluids have emerged as possible alternatives to current engineering fluids for advanced heat transfer applications, namely in small volume heat exchangers, cooling technologies and other chemical engineering and green energy-based applications [81-82]. Other attractive features of IoNanofluids are that they are designable and fine-tunable through their base ionic liquids for desired properties and tasks. These fluids can also be used for the development of new pigments for solar energy-based applications [83].

Ionic liquids have high volumetric heat capacity and good thermal conductivity and the combination of nanomaterials with these liquids creates great expectation considering the enhancement of the thermal properties. The use of nanoparticles as enhancing agents of the fluids properties enables the association of little quantities of different nanoparticles with different ionic liquids, thus obtaining flexible and designable (on a molecular level) substances that can be conceived according to the properties needed for a certain application. Results from our research showed that IoNanofluids containing MWCNT exhibit enhanced thermal conductivity and specific heat capacity compared to their base ionic liquids [77-80]. The increase in thermal conductivity of IoNanofluids was attributed to the interactions between the ionic liquid and the MWCNT resulting change in the structure of the liquid phase and to the electrical and geometric properties of the IL-nanomaterial interphase, found to be much thicker than predicted by existing theories [79, 80]. These properties further increase with the concentration of nanoparticles and fluid temperature in some extent.

Very recently Ferreira et al. [84] reported thermal and transport properties of quaternary phosphonium ionic liquids and their dispersed IoNanofluids containing MWCNT. They demonstrated that thermal conductivity and heat capacity of their IoNanofluids are higher than those of their base ILs and these properties increase with increasing the concentration of MWCNT. Interestingly the viscosity of IoNanofluids was found significantly lower as compared to the base ionic liquids. This is particularly fascinating as having low viscosity and high thermal conductivity IoNanofluids can be better alternative in numerous heat transfer applications compared to base ILs. They also found that like ionic liquids the studied IoNanofluids are also thermally stable.

The possible scientific, technical and economic success of IoNanofluids open new markets, as new products for engineering fluids, heat insulators, catalysts, etc., can be envisaged. In addition, the replacement of environmentally aggressive chemicals, foams, solid composites are in daily order, imposed by regulations or social responsibility. In recent reviews the authors have shown the exciting properties of IoNanofluids and highlighted their potential benefits and applications in different areas [81, 82, 85].

3. CHALLENGES OF NANOFLUIDS RESEARCH

The major challenges of nanofluids research include achieving long term stability of nanofluids, reliability and accuracy of experimental systems used to measure various properties of nanofluids, consistency and resolving uncertainties in measured data and findings, lack of fundamental understanding of heat transfer mechanisms, model development considering underlying mechanisms in macro to nanoscale, and practical application based research.

Although some innovative techniques were bought forward for the preparation of nanofluids, the main challenges of having long-term stable nanofluids without adding any dispersant mostly remain unsolved as besides sonications researchers commonly used various types and quantities of surfactants to achieve short-term stability of their nanofluids. Long term stability of nanofluids is of high importance particularly for their practical applications and more systematic studies are needed on the stability issue of synthesized nanofluids.

Researchers working on nanofluids have shown tremendous attention to the thermal properties, particularly thermal conductivity of nanofluids rather than application based investigations such as convective and boiling heat transfer features, microchannel flow, droplet spreading, and so on. Although applications of nanofluids as advanced cooling media appear to be promising, the advancement toward concrete understanding on the observed properties and features of nanofluids as well as their development for commercial applications remain challenging mainly due to the lack of agreement among the data from different research groups, lack of understanding of the mechanisms, and unsystematic measurements and sample preparation. Hence, proper sample preparation and repeatable and more systematic experimental studies on measuring any properties of nanofluids are worthwhile.

Another key issue is to identifying the actual heat transfer mechanisms and to develop models that can explain the observed thermal conductivity of nanofluids. As mentioned before despite numerous research efforts devoted there remain controversies regarding the heat transfer mechanisms and there are unidentified multi-scale factors and thus no widely accepted model is yet exist. The main challenge is to develop model for the thermal conductivity of nanofluids by taking into account all potential macro to molecular scale activities of both phases.

In the point of view of thermodynamics (Gibbs formulation) we are dealing with a suspension or emulsion with different degrees of aggregation of the nanoparticles and therefore the phase is not homogeneous especially when surfactants are used to stabilize the systems and to avoid microphase separation and stratification. Therefore, any theory to be developed to interpret the behavior of these fluids (i.e., nanoscale systems) is faced to apply the thermodynamics of multiphase (at least biphasic) heterogeneous systems.

This approach has not been followed by the current research and its solution is one of the most significant challenges that nanofluids research faces within the next years, trying to derive the heat transfer and thermodynamic properties of these nanosystems from constitutive equations and by involving statistical thermodynamics of "solvent" and nanomaterials shape (particles, spheres, rods, films, etc.). Hypothesis as property additivity, found applicable in some cases to properties like density and heat capacity, have to be tested at molecular level. However, transport properties pose a much bigger challenge, as in liquid state theory of molecular fluids, as the actual knowledge of structure and interactions for liquid densities is far from being resolved.

Two interesting fields proved that the nanofluids based on ionic liquids (termed as IoNanofluids) and IL based natural nanomaterials (IoBiofluids) are very challenging systems from the structure and applications view points. However the research so far developed in these areas proved that the preparation of these nanofluids, the characterization of the nanomaterials including the particle distribution analysis, and the study of the interface nanofluid/nanomaterial by imaging techniques like TEM are very important issues, necessary for the minimum understanding of their behavior and properties. In order to overcome those enormous challenges in research and development of nanofluids, besides careful characterization and preparation of sample nanofluids, extensive systematic and rigorous experimental together with analytical studies on various aspects of nanofluids are to be performed.

CONCLUSION AND OUTLOOK

Since coning the concept of nanofluids and despite the actual controversies about the mechanisms of heat transfer in nanofluids, there is no doubt that the quantity and quality of the research and applications of these novel fluids has increased almost exponentially. Although significant progress has been made, variability in the heat transfer characteristics still present that may be the results of the preparation of the nanofluids being quite delicate, and that the thermodynamic stability of these systems in long term periods is not achieved. In fact, a nanofluid does not necessarily mean a simple mixture of solid nanoparticles and a liquid in the thermodynamic definition, and the techniques used by different researchers are sometimes ill-defined.

Since dispersions of nanoparticles in nanofluids are not fully homogenous and there are always different degrees of aggregation or clustering of the nanoparticles it is very challenging to develop model for such suspensions of inhomogeneous and unknown structured agglomeration of dispersed nanoparticles. The research so far developed in this area proved that the preparation of these nanofluids, the characterization of the nanomaterials including the particle distribution analysis, and the study of the interface nanofluid/nanomaterial by imaging techniques like TEM are very important issues, necessary for the minimum understanding of their behavior and properties. If these nanofluids are stable, they can be proposed as heat transfer agents of the 21st century for cooling devices or systems that respond more efficiently to the challenges of great heat loads, higher power engines and brighter optical devices, increased transportation, micromechanics, instrumentation, HVAC and medical applications.

According to some researchers if agglomerations of nanoparticles yield positive impact on the thermal conductivity of nanofluids, it is unclear why and how ultrasonic disruption (sonication) and/or addition of dispersants (surfactants) are necessary since both of these techniques actually have negative impact on agglomeration of dispersed nanoparticles in base fluids. Regarding another important mechanisms, it is understandable that molecules of the host fluid can easily absorbed at the surface of nanoparticles and form interfacial layer of unknown dimensions (particularly thickness) and properties. The thermophysical properties of such interfacial layer (nanolayer) will be different from those of the host fluid and nanoparticles and in general the magnitude of thermal conductivity of nanolayer should be between those of base fluids and nanoparticles. Thus if nanolayer fact is considered and any additional liquids/solid dispersants are used, nanofluids are basically multiphasic complex mixtures. Nevertheless, the biggest challenge to incorporate the presence of nanolayer is to determine the size and properties of nanolayer which need molecular level understanding of structures of nanolayer and its formation in relation to surface chemistry and other inter molecular level activities. Therefore both nanolayer and nanoparticle agglomeration mechanisms still need to be studied comprehensively in order to better understanding and to incorporate them in thermal conductivity modeling for nanofluids. Computer studies, using molecular dynamics based numerical studies can give excellent insights about the structure, the preferential paths for heat conduction, the Newtonian or non-Newtonian, in essence, the dynamics of momentum, heat and mass transfer behaviors of the nanofluids.

Having pose the fundamental challenges in this field, the success of many technological applications of nanofluids requires a pragmatic approach, whereby the experiment, theoretical development and correlation can help to construct a methodology for property estimation for process design, namely in heat transfer and storage devices.

When dealing with heat transfer applications, the value of the heat exchange surface and the value of the heat transfer coefficients are crucial for an optimal technological design of most devices influencing directly its cost (construction and operational). Therefore, the applications of these systems in the industry have to be rationalized, if we want that the heat transfer fluids currently in use are to be replaced by environmentally friendly heat transfer fluids which are more efficient and more durable.

Two new classes of nanofluids named as IoNanofluids and IoBiofluids show great promises as well as new challenges. Preliminary research findings showed that these new nanofluids could be better alternative than the base ionic liquids in cooling applications.

Nevertheless progress in developing nanofluids for practical applications is mainly impeded due to inconsistencies in results and lack of fundamental understanding of the thermophysical properties of these novel fluids. Thus, besides resolving all inconsistencies and controversies as well as overcoming the challenges in research in this emerging field, nanofluids must meet the following conditions: homogeneous dispersion, stable over a great period of time (not producing phase separation even at a micro scale), and be free of additives, such as surfactants or salts. In addition, one of envisaged applications involves their use as heat transfer fluids and a control of viscosity is crucial in order to create fluids with sensible heat transfer coefficients in dynamic regime (good fluidity).

REFERENCES

[1] Choi, S. U. S. In *Developments and applications of non-Newtonian flows*; Siginer, D. A.; Wang, H. P.; Eds.; ASME Publishing: New York,USA,1995; FED-Vol. 231/MD-Vol. 66, pp 99-105.

[2] Yang, Y. M.; Maa, J. R. *Int. J. Heat Mass Transfer* 1984, 27, 145-147.

[3] Masuda, H; Ebata, A.; Teramae, K.; Hishinuma, N. *Netsu Bussei* 1993, 4, 227-233.

[4] Hashin, Z.; Shtrikman, S. *J. Appl. Phys.* 1962, 33, 3125–3131.

[5] Prasher, R.; Phelan, P.; Bhattacharya, P. *Nano Lett.* 2006, 6, 1529-1534.

[6] Keblinski, P.; Prasher, R.; Eapen, J. *J. Nanopart. Res.* 2008, 10, 1089-1097.

[7] Murshed, S. M. S. *J Nanopart Res.* 2009, 11, 511–512.

[8] Murshed, S. M. S.; Leong, K. C.; Yang, C. *Appl. Therm. Eng.* 2008, 28, 2109-2125.

[9] Chandrasekar, M.; Suresh S. *Heat Transf. Eng.* 2009, 30, 1136-1150.

[10] Lee, J. H.; Lee, S. H.; Choi, C. J.; Jang, S. P.; Choi, S. U. S. *Int. J. Micro-Nano Scale Transport* 2010, 1, 269-322.

[11] Saidur, R.; Leong, K. Y.; Mohammad, H. A. *Ren. Sust. En. Rev.* 2011, 15, 1646-1668.

[12] Murshed, S. M. S.; Nieto de Castro, C. A. In *Green Solvents I: Properties and Applications in Chemistry*, Ali M.; Inamuddin; Eds.; Springer, London, UK 2012, Ch. 14, pp.397-415.

[13] Buongiorno et al. *J. Appl. Phys.* 2009, 106, 094312 (14 pp).

[14] Maxwell, J. C., *A treatise on electricity and magnetism*, Clarendon Press: Oxford, UK. 1891.

[15] Nan, C. W.; Birringer, R.; Clarke, D. R.; Gleiter H. *J. Appl. Phys.* 1997, 81, 6692-6699.

[16] Lee, W.H.; Rhee, C. K. et al. *Nanoscale Res Lett.* 2011, 6, 258 (11 pp).

[17] Venerus, D. C.; Buongiorno, J. et al. *Appl. Rheol.* 2010, 20, 44582 (7 pp).

[18] Einstein, A. *Investigations on the Theory of the Brownian Movement*, Dover Publications, Inc.: New York, USA, 1956.

[19] Das, S. K.; Choi, S. U. S.; Patel, H. E. *Heat Transf. Eng.* 2006, 27, 3-19.

[20] Murshed, S. M. S.; Nieto de Castro, C. A.; Lourenço, M. J. V.; Lopes, M. L. M.; Santos, F. J. V. *Ren. Sust. En. Rev.* 2011, 15, 2342-2354.

[21] Murshed, S. M. S. *Heat Transf. Eng.* 2012, 33, 722-731.

[22] Wang, L.; Fan, J. *Nanoscale Res. Lett.* 2010, 5, 1241-1252.

[23] Yu, W.; Xie, H.; Chen, L.; Li, Y. *Colloid. Surf. A,* 2010, 355, 109-113.

[24] Feng, X.; Ma, H.; Huang S. et al. *J. Phys. Chem. B,* 2006, 110, 12311-12317.

[25] Boleininger, J.; Kurz, A.; Reuss, V.; Sönnichsen, C. *Phys. Chem. Chem. Phys.* 2006, 8, 3824-3827.

[26] Wei, X.; Wang, L. Particuology 2010, 8, 262-271.

[27] Zhu, H. T.; Zhang, C. Y.; Tang, Y. M.; Wang, J. X. *J. Phys. Chem. C*, 2007, 111, 1646-1650.

[28] Yu, W.; Xie, H. *J. Nanomat.* 2012, 2012, 435873.

[29] Zhang, X.; Gu, H.; Fujii, M. *Exp. Therm. Fluid Sci.* 2007, 31, 593–599.

[30] Choi, S. U. S.; Zhang, Z. G.; Yu, W.; Lockwood, F. E.; Grulke, E. A. *Appl. Phys. Lett.* 2001, 79, 2252-2254.

[31] Amrollahi, A.; Hamidi, A. A.; Rashidi, A. M. *Nanotechnology* 2008, 19, 315701.

[32] Xie, H.; Lee, H.; Youn, W.; Choi, M. *J. Appl. Phys.* 2003, 94, 4967-4971.

[33] Eastman, J. A.; Choi, S. U. S.; Li, S.; Thompson, L. J. In *Proc. Symp. Nanophase Nanocomposite Materials II*, Boston, USA, 1997.

[34] Krishnamurthy, S.; Bhattacharya, P.; Phelan, P. E.; Prasher, R. S. *Nano Lett.*2006, 6, 419–423.

[35] Wang, X.; Xu, X.; Choi; S. U. S. *J. Thermophys. Heat Transf.* 1999, 13, 474-480.

[36] Xie, H.; Wang, J.; Xi, T.; Liu, Y.; Ai, F.; Wu, Q. *J Appl. Phys.* 2002, 91, 4568-4572.

[37] Wang B. X.; Zhou, L. P.; Peng, X. F. *Int. J. Heat. Mass Transf.* 2003, 46, 2665-2672.

[38] Kumar, D. H.; Patel, H. E.; Kumar, V. R. R.; Sundararajan, T.; Pradeep, T.; Das, S. K. *Phys. Rev. Lett.* 2004, 93, 4301–4304.

[39] Murshed, S. M. S.; Leong, K. C.; Yang, C. *Int. J. Therm. Sci.* 2005, 44, 367-373.

[40] Murshed, S. M. S.; Leong, K. C.; Yang, C. *Int. J. Therm. Sci.*2008, 47, 560-568.

[41] Wen, D.; Ding, Y. *Int. J. Heat Mass Transf.* 2004, 47, 5181-5188.

[42] Jung, J. Y.; Oh, H. S., Kwak, H. Y. In *Proceedings of the IMECE*, ASME Heat Transfer Division, Chicago, USA, 2006.

[43] Pak, B. C.; Cho, Y. I. *Exp. Heat Transf.* 1998, 11, 151-170.

[44] Xuan, Y.; Li, Q. *J. Heat Transf.* 2003, 125, 151-155.

[45] Kulkarni, D. P.; Namburu, P. K.; Bargar, H. E.; Das, D. K. *Heat Transf. Eng.* 2008, 29,1027-1035.

[46] Murshed, S. M. S.; Leong, K. C.; Yang, C.; Nguyen, N. T. *Int. J. Nanosci.* 2008, 7, 325-331.

[47] Buongiorno J. *J. Heat Transf.* 2006, 128, 240-250.

[48] Murshed, S. M. S.; Nieto de Castro, C. A. *Rec. Pat. Nanotechnol.* 2013, 7, 216-223.

[49] Kim, H.; Kim, J.; Kim, M. *Nuclear Eng. Technol.* 2006, 39, 61-68.

[50] Kim, H. D.; Kim, J.; Kim, M. *Int. J. Multiphase Flow* 2007, 33, 691–706.

[51] Park, K. J.; Jung, D.; Shim, S. E. *Int. J. Multiphase Flow* 2009, 35, 525-532.

[52] Jung, J. Y., Kim, E. S.; Kang, Y. T. *Int. J. Heat Mass Transf.* 2012, 55 1941-1946.

[53] Kole, M.; Dey, T. K. *Appl. Therm. Eng.* 2012, 37,112-119.

[54] Bruggeman, D. A. G. *Ann. Phys.* 1935, 24,636-679.

[55] Hamilton, R. L.; Crosser, O. K. *Ind. Eng. Chem. Fund.* 1962, 1, 182–191.

[56] Keblinski, P.; Phillpot, S. R.; Choi, S. U. S.; Eastman, J. A. *Int. J. Heat Mass Transf.* 2002, 45, 855-863.

[57] Eastman, J. A.; Phillpot, S. R.; Choi, S. U. S.; Keblinski, P. *Ann. Rev. Mater. Res.* 2004, 34, 219-246.

[58] Taylor, R. A. et al. *J. Appl. Phys.* 2013, 113, 011301.

[59] Wen, D.; Lin, G.; Vafaei, S.; Zhang, K. *Particuology* 2009, 7, 141-150.

[60] Wong, K. V.; De Leon, O. *Adv. Mech. Eng.* 2010, 2010,519659.

[61] Roberts, N. A.; Walker, D. G. *Appl. Therm. Eng.* 2010, 30, 2499-2504.

[62] Rafati, M.; Hamidi, A. A.; Niaser M. S. *Appl. Therm. Eng.* 2012, 45-46, 09-14.

[63] Yousefi, T.; Mousavi, S. A.; Farahbakhsh, B.; Saghir, M. Z. *Microelec. Reliab.* 2013, 53, 1954-1961.

[64] Kulkarni, D. P.; Das, D. K.; Vajjha R. S. *Appl. Energy* 2009, 86, 2566-2573

[65] Taylor, R. A., Phelan, P. E.; Otanicar, T. P., Walker, C. A., Nguyen, M.; Trimble, S.; Prasher, R. *J. Ren. Sust. Energy* 2011, 3, 023104.

[66] Tyagi, H.; Phelan, P. E.; Prasher, R. *J. Solar Ener. Eng.* 2009, 131, 041004.

[67] Otanicar, T.; Phelan, P. E.; Prasher, R. S.; Rosengarten, G.; Taylor, R. A. *J. Ren. Sust. Energy* 2010, 2, 033102.

[68] Shin, D.; Banerjee, D. *Int. J. Heat Mass Transf.* 2011, 54, 1064-1070.

[69] Taylor, R.A.; Phelan, P.E.; Otanicar, T.P.; Adrian, R.; Prasher, R. S. *Nanoscale Res. Lett.* 2011, 6, 225.

[70] Sani, E.; Mercatelli, L.; Barison, S., Pagura, C.; Agresti, F.; Colla, L.; Sansoni, P. *Sol. Energ. Mat. Sol. Cell.* 2011, 95, 2994–3000.

[71] Lu, L.; Liu, Z.-H.; Xiao, H.-S. *Sol. Energy* 2011, 85, 379–387.

[72] Lenert, A.; Wang, E.N. *Sol. Energy* 2012, 86,253–265.

[73] Yousefi, T.; Veysi, F.; Shojaeizadeh, E.; Zinadini, S. *Ren. Energy* 2012, 39, 293–298.

[74] Yousefi, T.; Veysi, F.; Shojaeizadeh, E.; Zinadini, *Exp. Therm. Fluid. Sci.* 2012, 39, 207–212.

[75] Saidur, R.; Meng, T.C.; Said, Z.; Hasanuzzaman, M.; Kamyar, A. *Int. J. Heat Mass Transf.* 2012, 55, 5899-5907.

[76] Javadi, F. S.; Saidur, R.; Kamalisarvestani, M. *Ren. Sust. Ener. Rev.*, 2013, 28, 232–245.

[77] Nieto de Castro, C. A.; Lourenço, M. J. V.; Ribeiro, A. P. C.; Langa, E.; Vieira, S. I. C. *J. Chem. Eng. Data* 2010, 55, 653-661.

[78] Nieto de Castro, C. A.; Murshed, S. M. S.; Lourenço, M. J. V.; Santos, F. J. V.; Matos Lopes, M. L.; França, J. M. P. *Int. J. Therm. Sci.* 2012, 62, 34-39.

[79] Ribeiro, A. P. C.; Vieira, S. I. C.; Goodrich, P.; Hardacre, C.; Lourenço, M. J. V.; Nieto de Castro, C. A. *J. Nanofluid.* 2013, 2, 55-62.

[80] França, J. M. P.; Vieira, S. I. C.; Lourenço, M. J. V.; Murshed, S. M. S.; Nieto de Castro C. A. *J. Chem. Eng. Data* 2013, 58, 467-476.

[81] Nieto de Castro, C. A.; Murshed, S. M. S.; Lourenço, M. J. V.; Santos, F. J. V.; Matos Lopes, M. L.; França, J. M. P. In *Green Solvents I: Properties and Applications in Chemistry*, Ali M.; Inamuddin; Eds.; Springer, London, UK 2012, Ch. 8, pp.233-249.

[82] Murshed, S. M. S.; Nieto de Castro C. A.; Lourenço, M. J. V.; Lopes, M. L. M.; Santos, F. J. V. *J. Phys.: Conf. Ser.* 2012, 395, 012117.

[83] Vieira, S. I. C.; Lourenço, M. J. V.; Alves, J. M., Nieto de Castro, C. A. *J. Nanofluid.* 2012, 1,148-154.

[84] Ferreira, G. M.; Simões, P. N.; Ferreira, A. F.; Fonseca, M. A.; Oliveira, M. S. A.; Trino, A. S. M.; *J. Chem. Thermodyn.* 2013, 64, 80-92.

[85] Nieto de Castro, C. A.; Ribeiro, A. P. C.; Vieira, S. I. C.; Lourenço, M. J. V.; Santos, F. J. V.; Murshed S. M. S.; Goodrich, P.; Hardacre, C. In *Ionic Liquids- New Aspects for the Future*, Kadokawa, J.; Ed.; Intech, Rijeka, Croatia, 2013 Ch. 7, pp.165-193.

INDEX

#

21st century, 261, 273

A

acetic acid, 66
acid, 9, 23, 25, 78, 82, 99, 137
acidic, 158
acidity, 6
actuation, 104
actuators, 54
additives, viii, 3, 4, 6, 77, 79, 90, 105, 197, 201, 274
adhesion, 201
adsorption, 113, 118, 133, 134, 139, 143, 144, 145, 233
advancement(s), 29, 30, 243, 264, 272
adverse effects, 270
aerosols, 13
aerospace, 16, 54
aggregation, viii, 5, 8, 10, 37, 38, 40, 42, 45, 77, 78, 82, 83, 91, 92, 95, 98, 99, 101, 243, 262, 272, 273
Al2O3 particles, 196, 202, 203
alcohols, 237
algorithm, 248
alters, 119, 144, 245
aluminium, 12, 16, 155, 156, 157, 158, 180, 181, 182, 183, 184, 186, 188, 189, 205, 206
aluminum oxide, 197
aluminum surface, 209
amino, 5, 24
ammonia, 104
ammonium, 3, 56, 78
amplitude, 12, 103
anatase, 137
anisotropy, 92
Applications, v, 27, 48, 73, 101, 130, 191, 235, 237, 239, 241, 242, 269, 275, 277

arithmetic, 38
assessment, 7
assets, 17
asymmetry, 248
atmosphere, 199
atmospheric pressure, 198, 199, 201, 202, 203, 205, 219
atoms, 33, 34, 264
Au nanoparticles, 8

B

barriers, 10, 17
baths, 11, 12
batteries, 54
behaviors, 66, 67, 72, 115, 128, 274
beneficial effect, 32
benefits, 72, 216, 261
benzene, 56, 91, 195
biopolymer, 5
birefringence, 240
bismuth, 16, 140
Boltzmann constant, 35, 85, 144
bonding, 32, 33, 222
bonds, 241
boric acid, 203
bottom-up, 4, 264
bounds, 37, 87, 88, 262
Brazil, 26
Brownian motion, 10, 32, 39, 46, 62, 139, 157, 160, 219, 233, 269
bulk materials, 82

C

calcium, 16, 18, 205
calcium carbonate, 18
cancer, 234, 237, 241, 269

cancerous cells, 241
candidates, 78, 241
capillary, vii, viii, 133, 134, 135, 140, 141, 144, 148, 151, 216, 217, 220, 221, 226, 230, 234
capital intensive, 17
carbon, viii, 1, 16, 18, 26, 30, 53, 54, 55, 56, 61, 63, 71, 72, 83, 91, 99, 100, 110, 137, 138, 155, 197, 198, 240, 241
carbon nanotubes, v, viii, 1, 18, 20, 21, 22, 25, 26, 30, 53, 54, 55, 56, 61, 71, 72, 83, 91, 155, 197, 198
carboxylic acid, 78
carefulness, 265
catalyst, 17
cation, 78, 104
cationic surfactants, 9
C-C, 241
cellulose, 16
cerium, 16
challenges, viii, 54, 55, 261, 262, 263, 272, 273, 274
chemical(s), 3, 4, 5, 6, 9, 11, 12, 13, 15, 17, 18, 28, 29, 30, 54, 56, 71, 134, 135, 156, 216, 241, 251, 264, 271, 272
chemical industry, 29
chemical interaction, 11
chemical properties, 17
chemical reactions, 11
chemical vapor deposition, 56, 71, 264
CHF, 193, 194, 195, 196, 197, 198, 199, 200, 201, 202, 203, 204, 211, 267
Chicago, 276
China, 18, 19, 20, 23, 24, 25, 27, 254
chitosan, 56, 57, 61, 66, 67, 68, 70
circulation, 242
classes, 9, 274
cleaning, 134
clustering, 5, 32, 40, 62, 83, 101, 113, 137, 160, 269, 273
clusters, 5, 56, 89, 101, 118, 133, 140, 147, 151, 258
CMC, 196
C-N, 241
CO2, 2
coatings, 17
cobalt, 16, 80, 92
collisions, 10, 12, 33, 157, 196, 245
color, 241, 242
commercial, 1, 3, 14, 15, 16, 17, 77, 78, 242, 243, 251, 264, 265, 270, 272
communication, 156
community(s), 16, 54, 134
compatibility, 30, 94
competitiveness, 18
complex numbers, 244

complexity, 56, 247
composites, 38, 272
composition, 9, 13, 17, 18, 134, 136
compression, 10, 11, 12
computing, 112, 156
concentration, 12, 44, 115, 125, 205
condensation, 3, 4, 205
conditioning, 133, 156, 269
conductance, 103
conduction, 33, 38, 40, 56, 83, 87, 101, 103, 105, 166, 245, 262, 269, 274
conference, 55, 265
configuration, 87, 239, 243
congress, 75
construction, 274
consumption, 4, 78, 104
contamination, 2, 3, 9, 10, 11, 13, 18, 134, 135
controversial, viii, 2, 110, 266
controversies, 62, 262, 269, 272, 273, 274
Convective heat transfer, 267
convention, 239
COOH, 20, 21
cooling, vii, 12, 29, 30, 31, 43, 46, 47, 48, 54, 77, 78, 101, 102, 104, 105, 133, 156, 166, 205, 216, 263, 269, 270, 271, 272, 273, 274
cooling process, 270
coordination, 6
copper, 1, 4, 16, 36, 54, 70, 137, 138, 156, 166, 167, 169, 190, 197, 198, 199, 202, 203, 205, 206, 207, 208, 209, 264
copyright, 82, 88, 89, 90, 93, 96, 97, 102, 103, 104, 135, 206, 207, 210
correlation(s), 30, 85, 127, 128, 129, 162, 163, 174, 176, 177, 178, 179, 197, 200, 202, 208, 209, 210, 220, 221, 267, 274
correlation coefficient, 128
cosmetic(s), 16, 17, 18
cost, 3, 4, 5, 17, 18, 116, 156, 240, 241, 250, 251, 270, 274
covering, 45, 69, 206
CPI, 18
creep, 243
critical analysis, vii
Croatia, 277
crystal structure, 33
crystalline, 6, 9, 39
crystals, 34
cure, 11
CVD, 71, 264
cycles, 10, 11, 12, 88, 90
cycling, 88, 251

D

damages, 11
damping, 77, 101
decay, 88, 92, 127
defects, 33
defence, 16
deformation, 103
degradation, 11, 201
deposition, 46, 47, 133, 147, 198, 199, 201, 202, 205, 211, 264, 268
depth, 11, 215, 216, 217, 218, 220, 221, 222, 224, 225, 226, 227, 233, 234
destruction, 98, 194
detachment, 211, 224, 225
detection, 78, 242
deviation, 38, 80, 161, 174, 175, 219, 251
dielectric constant, 245, 246
differential scanning, 70
differential scanning calorimetry, 70
diffusion, 243, 267
diffusivity, 2, 53, 70, 217, 266
diodes, 54
dipoles, 244
discomfort, 11
discontinuity, 39, 79
disorder, 104
dispersion, 3, 4, 7, 9, 10, 11, 12, 13, 14, 18, 32, 33, 36, 37, 45, 47, 55, 56, 60, 61, 70, 83, 91, 98, 99, 100, 112, 113, 118, 121, 157, 158, 201, 218, 225, 271, 274
dispersity, 137
displacement, 143
distilled water, 6, 66, 112, 157, 158, 159, 161, 181, 183, 184, 186, 188
distribution, 2, 6, 7, 13, 14, 145, 167, 229, 246, 248, 251, 264, 273
DNA, 216, 240, 242
doping, 238
drawing, 45
Droplet formation, 225
drug delivery, 234, 242, 269
drying, 4
DSC, 70
durability, 2, 17, 137
Dynamic, 140, 146, 149, 150, 151, 239
Dynamic contact angle, 149, 150, 151
dynamic control, 252
dynamic viscosity, 31, 163, 197, 210, 218

E

East Asia, 18
editors, viii
electric field, 242
electrical properties, 65, 110
electricity, 54, 275
electrolyte, 199
electromagnetic, 238, 240, 245
electromagnetic fields, 238, 240
electron(s), 35, 36, 245, 246
electron microscopy, 246
electronic systems, 30, 156
elongation, 220
emission, 201, 247, 248
employment, 16
emulsions, 28
encapsulation, 216
energy, vii, 1, 2, 7, 10, 12, 13, 14, 16, 32, 35, 54, 60, 65, 72, 78, 85, 104, 134, 135, 136, 137, 139, 140, 144, 145, 146, 147, 169, 171, 172, 181, 201, 211, 219, 238, 240, 241, 247, 248, 263, 269, 270, 271
energy density, 12
energy input, 13, 169, 241
energy transfer, 211
engineering, vii, viii, 2, 16, 17, 18, 28, 65, 133, 234, 237, 240, 249, 251, 252, 269, 271, 272
England, 25
entropy, 34
environment(s), 17, 18, 67, 78, 241
environmental contamination, 2
environmental standards, 5
equilibrium, 85, 101, 133, 134, 135, 136, 140, 143, 144, 146, 147, 151
Equilibrium contact angle, 137, 141, 143
equipment, 2, 11, 12, 13, 14, 110, 156, 164, 189, 219
e-readers, 242
erosion, 10, 13
ester, 89, 92, 94, 200
ethanol, 6, 8, 140
ethylene, 1, 4, 5, 30, 32, 41, 56, 60, 61, 66, 67, 70, 71, 78, 110, 112, 125, 126, 147, 195, 200, 201, 262
ethylene glycol, 1, 4, 30, 32, 41, 56, 60, 61, 66, 67, 70, 71, 78, 110, 112, 125, 126, 147, 195, 200, 201, 262
EU, 18
Europe, 16, 18
European Commission, 18, 28
evaporation, 3, 5, 54, 137, 138, 140, 205
evolution, viii, 134, 147, 261
exercise, 110, 249, 262, 263
exposure, 11, 89, 90, 91, 92, 94, 241, 251

extinction, 243, 244, 247, 253
extraction, 133
extrusion, 224

F

fabrication, 221, 237, 239, 246, 250, 251, 264
fatty acids, 78, 79
ferrite, 90, 92, 95
ferromagnetic, 85, 104, 240
field theory, 79
filament, 228
fillers, 54, 61
film thickness, 144
films, 273
filters, 104, 237, 238, 239, 240, 243, 247, 249
filtration, 17
financial, 151
financial support, 151
first generation, 249
flexibility, 85
flotation, 8
flow field, 150, 151
fluctuations, 85, 87, 172
fluorescence, 218
foams, 9, 272
food, 16
force, 4, 10, 11, 18, 62, 90, 113, 134, 137, 140, 146,
    217, 224, 240
formation, viii, 4, 6, 12, 13, 63, 86, 87, 91, 101, 198,
    201, 215, 216, 217, 218, 220, 221, 222, 223, 224,
    225, 226, 227, 228, 229, 230, 231, 232, 233, 234,
    274
formula, 111, 141
fouling, 267
France, 28, 48, 77, 130, 234, 254
free energy, 134, 136, 144
friction, 136, 144, 148, 150, 151, 156, 170, 172, 173,
    174, 175, 177, 178, 180, 186, 187, 188, 189
fullerene, 16, 26
functionalization, 18, 56, 57, 66, 251

G

gallium, 16
geometrical parameters, 225
geometry, 3, 13, 72, 146, 216, 217, 218, 220, 229,
    230, 231, 234, 243, 244
Georgia, 29, 51
Germany, 18, 19, 22, 23, 24, 25, 49, 51, 75, 121,
    223, 255, 257
glucose, 4

glycerin, 83, 99, 100
glycerol, 66
glycol, 5, 9, 30, 56, 61, 66, 70, 71, 83, 99, 100, 129,
    199
gold nanoparticles, 251
google, 244
grain boundaries, 33
grain size, 26
graph, 162, 183
graphite, 1, 15, 16, 18
gravitational force, 137
gravity, 10, 31, 140
growth, 8, 16, 17, 28, 54, 55, 204, 211, 224, 242,
    251, 252, 262, 265
guidelines, 11

H

hardness, 17
harvesting, 238, 247, 250, 263, 271
health, 11, 17
hearing loss, 11
heat capacity, 69, 270, 271, 273
heat loss, 205, 241
heat removal, 30, 104, 151, 238
Heat transfer, 104, 168, 176, 178
height, 145, 195, 197
heptane, 80, 83, 90, 93, 94
heterogeneous systems, 273
hexane, 83, 90, 93, 94, 95
history, 5, 67
host, 56, 241, 243, 264, 268, 274
hot spots, 46
House, 49, 257, 258
HRTEM, 5
human, 18
human health, 18
Hunter, 27, 76, 152
hybrid, 59, 60, 83, 97
hydrazine, 6
hydrogen, 32, 33
hydrogen bonds, 32
hydrophobicity, 17, 150
hydroxide, 78
hydroxyl, 9
hydroxyl groups, 9
hyperthermia, 241, 242, 247, 248
hypothesis, 33, 40, 243, 247
hysteresis, 88, 89, 134, 135

I

ID, 21, 22, 214
ideal, 17, 241, 243, 251
identification, vii, 1
identity, vii, 1, 4
image(s), 6, 14, 15, 54, 86, 87, 89, 91, 98, 148, 198, 202, 223, 224, 225, 228, 229, 231
immersion, 12, 30, 42
improvements, 270
impurities, 14, 15, 33, 90
India, 19, 20, 22, 23, 24, 25, 77
indium, 5, 204
industry(s), 5, 6, 16, 17, 18, 29, 54, 243, 263, 270, 274
inertia, 172
infrastructure, 17
inhomogeneity, 134, 135
insulation, 134, 166, 173
insulators, 33, 240, 272
integration, 216
integrity, 2
interface, 7, 9, 35, 38, 39, 62, 79, 134, 135, 136, 139, 144, 219, 222, 227, 234, 242, 269, 273
interfacial layer, 63, 65, 274
interphase, 271
intrinsic viscosity, 45, 69, 112
investment, 17
ions, 7, 78
Iran, 19, 24, 25, 193
iron, 16, 78, 80, 83, 88, 92, 94, 104
irradiation, 4, 6, 12, 264
isolation, 30, 216
issues, 4, 6, 261, 264, 273

J

Japan, 23
Jordan, 257

K

kerosene, 78, 80, 82, 84, 85, 86, 87, 88, 89, 90, 93, 94, 95, 98, 99
kinetics, 7
Korea, 153

L

laminar, 70, 156, 157, 173, 174, 175, 176, 177, 178, 181, 182, 183, 184, 187, 188, 189, 190, 267

laser ablation, 264
lasers, 104, 237
lattices, 33, 63
layering, 45, 136
lead, 11, 12, 14, 16, 31, 35, 46, 113, 115, 167, 205, 208
legislation, vii, 1, 3
lens, 242
life cycle, 17
life sciences, 16
lifetime, 251
light, 36, 83, 94, 104, 234, 237, 238, 241, 242, 243, 244, 245, 246, 248, 249, 251, 257
light scattering, 246
linear dependence, 110
liquid interfaces, 226, 243
liquid phase, 2, 4, 7, 18, 63, 271
liquids, 1, 2, 3, 5, 30, 32, 36, 42, 56, 133, 134, 151, 216, 217, 218, 237, 238, 241, 242, 271, 273, 274
lithium, 54, 80, 88
low temperatures, 32, 248
lubricants, 78
Luo, 51, 235, 256
lying, 160, 163

M

macromolecules, 134
magnesium, 16
magnet, 79, 103
magnetic field, viii, 77, 78, 79, 84, 85, 86, 87, 88, 89, 90, 91, 92, 94, 101, 102, 105, 202, 240
magnetic moment, 84, 85
Magnetic nanofluids, 77, 78
magnetic particles, 77, 84, 85, 91, 92
magnetism, 275
magnetization, 86, 89, 90
magnitude, 30, 32, 33, 42, 44, 47, 54, 65, 119, 121, 155, 217, 232, 238, 274
majority, 240, 242, 252
man, 107
management, 17, 29, 30, 54, 69, 72, 104, 156, 216, 263, 269
manipulation, vii, viii, 1, 215, 216, 217, 218, 221, 225, 233, 234
manufacturing, 18, 54, 110, 156, 216
mass, 5, 17, 35, 70, 121, 157, 167, 169, 171, 172, 190, 195, 197, 200, 274
materials, 3, 5, 9, 10, 16, 17, 18, 35, 38, 54, 67, 69, 78, 92, 94, 112, 128, 208, 222, 238, 239, 243, 245, 249
materials science, 54, 249
matrix(es), 11, 37, 245

matter, 240
Maxwell equations, 244
measurement(s), 4, 6, 7, 8, 9, 33, 36, 88, 95, 98, 101, 103, 117, 119, 135, 137, 138, 140, 157, 159, 161, 163, 164, 165, 166, 168, 169, 171, 172, 173, 177, 178, 182, 189, 198, 216, 219, 223, 248, 261, 262, 263, 265, 272
mechanical properties, 17, 18
mechanisms, 10, 62, 268
media, 9, 67, 161, 244, 248, 272
medical, 2, 16, 54, 110, 216, 241, 269, 274
medicine, 18, 78
melting, 5, 17, 34
melting temperature, 34
MEMS, 54, 78, 101, 269, 270
metal oxides, 1, 121
metals, 1, 5, 36, 54, 245
meter, 159, 162, 166, 167, 169, 171
methodology, vii, 1, 2, 6, 15, 177, 264, 274
methyl cellulose, 196
Miami, 254
Micro-electromechanical systems (MEMS), 270
microelectronics, 18, 29, 46, 47, 54, 110, 216
micrometer, 156
microscope, 167, 223
microscopy, 15, 83, 86, 89, 91, 98, 167
microstructure, 6
migration, 197, 203
military, 16
miniature, 78, 104, 105
miniaturization, 17
Ministry of Education, 151
Minneapolis, 191
mixing, 10, 238
model system, 78, 104
models, 32, 36, 37, 38, 39, 40, 44, 45, 46, 53, 62, 63, 64, 67, 69, 72, 80, 109, 111, 112, 113, 115, 116, 117, 118, 127, 128, 129, 136, 151, 262, 263, 266, 268, 272
modifications, 245
mold, 221
molecular dynamics, 40, 274
molecules, 6, 7, 9, 10, 98, 136, 150, 219, 242, 264, 274
momentum, 274
morphology, 2, 15, 137, 198, 250
Moses, 29
MR, 80, 88, 89, 92, 103, 151
multiphase materials, 262
multiwalled carbon nanotubes, 60

N

nanoelectronics, 54
nanofabrication, 252
nanofibers, 1, 3, 16
nanohorns, 241
nanomaterials, 1, 2, 3, 4, 6, 7, 11, 14, 15, 16, 17, 18, 26, 28, 54, 55, 56, 63, 79, 121, 238, 251, 271, 273
nanometer(s), 54, 77, 146, 194, 216, 251
Nanoparticles, v, 18, 19, 24, 109, 130, 131, 150, 151, 194, 197, 198, 199, 200, 201, 202, 205, 225
nanorods, 1, 3, 62, 63, 244, 246
nanostructures, 4, 264
nanosystems, 273
nanotechnology(s), 18, 28, 237
nanotube, 16, 53, 54, 63, 71, 72, 91, 117, 129
nanowires, 1, 33, 34
NEMS, 78, 101
neural network, 209, 210
New South Wales, 237
Newtonian fluids, 46, 196
next generation, 72, 110, 156
NHS, 57, 59
nickel, 89
nonionic surfactants, 9
nucleation, 204, 205, 206, 211

O

OH, 20, 21, 22
oleic acid, 19, 78, 82, 91, 114, 137
one dimension, 33
operating range, 181, 189
operations, 13
optical fiber, 78
optical properties, 18, 237, 238, 239, 243, 244, 245, 246, 247, 248, 249, 250, 251, 252, 270
Optics, 27, 238, 255, 257, 258
optimal performance, 240
optimization, 14, 56, 210, 252
optimization method, 210
optoelectronics, 250
ores, 8
organic solvents, 238
overlap, 86
oxide nanoparticles, viii, 6, 9, 104, 109, 110, 112, 115, 119, 128
oxygen, 221
oxygen plasma, 221

P

parallel, 37, 38, 63, 84, 85, 87, 88, 89, 91, 92, 101
particle morphology, 7
patents, 17, 268
pathways, 40
permeability, 17, 262
permission, 82, 88, 89, 90, 93, 96, 97, 102, 103, 104, 135, 206, 207, 210
permit, 43
petroleum, 134
pH, 8, 9, 17, 91, 112, 113, 114, 118, 157, 158, 162
pharmaceutical, 17
phase boundaries, 10
phase transitions, 104
phonons, 33
phosphate, 197
photolithography, 221
photons, 240, 241
photothermal, 242
physical chemistry, 134
physical interaction, 2
physical properties, 61, 134, 211
physics, 133, 212, 213
pipeline, 250
Planck constant, 144
plants, 194, 271
platinum, 6, 16, 221
PMMA, 221
polar, 7, 9, 32
polarization, 64
policy, 5
polydimethylsiloxane, 81, 221, 222
polydispersity, 217, 246
polymer(s), 5, 10, 16, 18, 19, 25, 54, 251
polymerase, 216
polymerase chain reaction, 216
polymer-based composites, 54
polymerization, 251
polyvinyl alcohol, 3, 197
pool boiling, 193, 194, 204
porosity, 43
Portugal, viii, 1, 26, 53, 73, 109, 129, 215, 261
potential benefits, 269, 272
power generation, 156, 194
power plants, 270, 271
precipitation, 7, 91
preparation, vii, viii, 1, 2, 3, 4, 6, 53, 56, 78, 110, 117, 119, 263, 264, 272, 273
pressure gauge, 205
prevention, 2
probability, 10, 63, 246, 252
probability density function, 63, 246, 252

probe, 10, 12, 13, 104
producers, 2
product performance, 17
production costs, 17
propagation, 87, 245, 247
proportionality, 141, 164
propylene, 199
protection, 11, 17, 18, 203
protective role, 8
protein crystallization, 216
prototype, 241, 252
pumps, 223, 238, 239, 270
pure water, 71, 195, 196, 197, 198, 199, 200, 202, 203
purity, 4, 15, 18, 26, 264
PVA, 3
PVP, 6, 8, 18, 19, 25, 56, 57, 251

Q

quantum dot(s), 16, 241
quaternary ammonium, 78

R

radial distance, 144
radiation, 238, 243, 244, 247, 248, 252, 256, 257, 258, 270
radius, 39, 45, 46, 64, 86, 140, 145, 146, 229, 244, 246, 251, 252
reactants, 6
reaction rate, 4, 143
reactivity, 2
reading, 173
reagents, 3
recovery, 222
red shift, 251
refractive index, 242, 244, 246, 252
regression, 109, 127, 129, 210
regulations, 17, 272
relaxation, 39, 246, 253
reliability, 177, 272
relief, 164
repulsion, 8, 10, 78, 112, 133
requirements, 2, 4
researchers, vii, 2, 15, 45, 54, 56, 57, 62, 67, 72, 109, 110, 111, 116, 126, 129, 133, 134, 155, 182, 194, 201, 211, 216, 217, 224, 240, 241, 252, 262, 264, 265, 266, 269, 272, 273, 274
reserves, 78
residues, 10, 13
resistance, 33, 37, 39, 169, 205, 223

resolution, 122, 242
resources, 2, 17
response, 92
retinol, 16
revenue, 270
rheology, 66, 67
Rheology, 51, 68, 89, 102, 122, 153
risk(s), 10, 11, 12, 17, 18
rods, 273
Romanus, 254
room temperature, 5, 34, 41, 57, 60, 83, 96, 98, 115, 116, 117, 121, 127, 219, 221, 226, 227, 228, 232, 266
roughness, 34, 134, 135, 177, 193, 194, 195, 203, 204, 205, 206, 207, 208, 209, 210, 268
routes, 17, 264
Royal Society, 49
rubber, 166
Russia, 22, 23

S

safety, 11, 17, 18
salts, 5, 241, 274
saturation, 86, 88, 89, 92, 137, 205
savings, 2
scatter, 159, 242
scattering, 33, 34, 35, 39, 110, 237, 238, 240, 242, 243, 244, 245, 246, 247, 248, 253, 257, 270
schema, 78, 88, 89, 104
scholarship, 151
science, vii, 2, 16, 134, 249, 270
sediment, 7, 8
sedimentation, 4, 7, 8, 77, 91, 99, 100, 113, 156, 193, 195, 198, 199, 205, 206, 226
segregation, 91
selectivity, 241
selenium, 16
self-assembly, 246
semiconductor(s), 16, 18, 33, 54, 240
sensing, 238
sensors, 18, 54, 78, 104, 221, 237, 240
SES, 22
shape, 4, 6, 12, 13, 18, 54, 62, 63, 64, 72, 110, 113, 116, 134, 136, 140, 156, 216, 218, 221, 224, 227, 228, 229, 234, 238, 240, 242, 244, 250, 263, 268, 273
shear, 5, 7, 10, 42, 43, 44, 45, 46, 65, 66, 67, 68, 69, 72, 101, 102, 109, 110, 112, 113, 122, 124, 125, 128, 141, 145, 147, 148, 164, 190, 217, 218, 220, 224, 227, 263
shear rates, 5, 45, 65, 67, 69, 110, 112, 113, 125, 141, 147, 148

shock, 10, 12
shock waves, 10
showing, 7, 102, 229, 262
side effects, 13, 14
signals, 169
signs, 186
silane, 23, 24
silica, 9, 16, 18, 26, 109, 134, 199, 202, 204, 250, 251
silicon, 16, 18, 26, 33, 34, 56, 140
Silicone oil, 68, 129
silver, 1, 3, 5, 6, 8, 16, 18, 26, 167, 240, 245, 264
simulation(s), 17, 33, 39, 40, 65, 245, 248
Singapore, 133, 151
SiO_2, 9, 19, 23, 24, 90, 96, 97, 111, 114, 117, 121, 122, 124, 125, 126, 128, 199, 204
Size effect, 34, 36
social responsibility, 272
society, 17
sodium, 3, 4, 6, 8, 56, 91, 98, 195, 197, 200, 204, 217
sodium dodecyl sulfate (SDS), 56, 204, 217
soft lithography, 221
software, 148
solar collectors, 240, 241, 270, 271
solar system, 241
solid phase, 2, 81
solid surfaces, 134
solubility, 6, 9, 13, 17
solution, 5, 6, 7, 9, 10, 78, 98, 137, 138, 144, 147, 148, 150, 151, 162, 194, 196, 198, 217, 243, 244, 248, 264, 273
solvation, 94
solvents, 9, 17, 55
South Africa, 155
South Korea, 19, 25
Spain, 22
specific gravity, 17
specific heat, 31, 35, 53, 55, 56, 70, 159, 170, 189, 208, 263, 266, 271
Specific heat capacity, 69
specific surface, 198
spectroscopy, 5
spin, 16, 17, 133, 134
stability, vii, 1, 2, 4, 6, 7, 8, 9, 10, 47, 55, 56, 61, 78, 90, 91, 98, 99, 100, 116, 121, 137, 140, 251, 264, 272, 273
stabilization, 56, 57, 78, 79, 82, 251
standard deviation, 246
standardization, 13
state(s), vii, viii, 2, 8, 10, 29, 56, 59, 104, 134, 172, 237, 240, 247, 264, 265, 267, 273
steel, 12, 167, 195, 197, 203, 205

Index

287

storage, 4, 7, 62, 69, 166, 241, 271, 274
stratification, 272
stress, 15, 65, 122, 124, 144, 145, 164, 181, 190, 220, 224, 226
stretching, 10, 11, 12
structural defects, 240
structure, 5, 6, 9, 10, 85, 87, 90, 150, 271, 273, 274
structure formation, 85, 87, 90
structuring, 134
substrate(s), 140, 151, 238
sulfate, 3, 4
Sun, 19, 25, 257
superparamagnetic, 84
supplier(s), vii, 1, 2, 14, 15, 18, 26, 137, 218, 251
surface area, 13, 156, 169, 189, 211
surface chemistry, 18, 62, 269, 274
surface energy, 18, 147
surface modification, 9
surface properties, 62
surface tension, 13, 31, 134, 137, 139, 140, 143, 198, 200, 208
surface treatment, 23
surfactant(s), 3, 4, 6, 8, 9, 56, 57, 59, 60, 65, 66, 68, 72, 78, 81, 87, 91, 97, 98, 110, 114, 118, 119, 134, 137, 195, 197, 198, 199, 200, 204, 217, 218, 219, 251, 272, 274
suspensions, 1, 10, 11, 37, 38, 40, 45, 54, 56, 62, 63, 66, 67, 69, 77, 78, 80, 85, 92, 110, 111, 112, 113, 157, 162, 194, 198, 201, 238, 240, 241, 251, 261, 262, 264, 273
Switzerland, 219
synthesis, 3, 4, 5, 6, 8, 9, 17, 216, 237, 251, 264

T

Taiwan, 218
tanks, 241
target, 15
techniques, 3, 4, 10, 42, 56, 57, 112, 117, 118, 119, 135, 194, 216, 217, 222, 237, 239, 245, 248, 263, 264, 265, 272, 273, 274
technology(s), vii, 4, 5, 16, 17, 30, 133, 134, 156, 216, 234, 237, 238, 239, 242, 250, 259, 269, 270, 271
TEM, 6, 54, 87, 101, 137, 138, 273
temperature dependence, vii, 29, 33, 42, 97, 98, 119, 129, 215, 218, 220, 221, 233
tension(s), 3, 137, 139, 140, 143, 147, 151, 215, 217, 218, 219, 220, 221, 224, 225, 226, 227, 229, 230, 232, 233
testing, 2, 17, 18, 170, 171, 190, 263
textbook, 178, 181, 182
Thailand, 193, 211

Thermal conductivity, 41, 42, 61, 80, 84, 88, 89, 91, 94, 95, 98, 162
thermal energy, 35, 85
thermal expansion, 32
thermal properties, viii, 3, 42, 53, 55, 56, 78, 90, 110, 237, 238, 262, 263, 265, 269, 271, 272
thermal resistance, 9, 39, 40, 63, 65, 79, 190, 205, 206
thermodynamic properties, 273
thermodynamics, 272, 273
thin films, 33, 34, 36
thinning, 42, 43, 46, 65, 66, 67, 68, 72, 109, 110, 128, 141, 147, 263, 267
time periods, 137
tin, 204, 205
titanate, 16, 117
titania, 9, 34, 41, 109
titanium, 1, 16, 18, 26, 156, 218
T-junction, viii, 215, 216, 217, 218, 220, 221, 222, 223, 224, 225, 226, 227, 228, 231, 233, 234
toluene, 80
top-down, 4, 264
toxicity, vii, 1, 3, 11
tracks, 95
trade, 1
training, 11
transducer, 12, 166, 171
transfer performance, viii, 31, 55, 70, 72, 115, 206, 216, 263, 267
transformation, 6, 8, 264
transition temperature, 218, 234
transmission, 194
transparency, 221
transport, vii, 33, 38, 39, 47, 62, 77, 79, 101, 102, 238, 248, 269, 271, 273
transportation, 4, 54, 110, 156, 216, 269, 270, 274
treatment, 4, 9, 17, 57, 221, 241, 269
trial, 14
tumor, 234, 242, 247, 248
tumor growth, 242
tungsten, 16
turbulence, 181
Turkey, 18, 19, 22, 23, 24, 25, 193

U

UK, 73, 76, 130, 235, 275, 277
ultrasound, 12, 28
underlying mechanisms, 53, 109, 129, 140, 267, 272
uniform, 37, 56, 90, 91, 95, 102, 169, 175, 178, 223, 246
United States (USA), 5, 18, 19, 20, 21, 22, 23, 24, 25, 28, 54, 73, 76, 121, 129, 130, 133, 137, 152,

155, 191, 216, 221, 223, 234, 254, 255, 256, 257, 258, 261, 275, 276
UV, 5, 6, 7, 241
universal gas constant, 34
universities, 16, 17
UV light, 241

V

vacuum, 3, 5, 22, 244
validation, 177, 264
valve, 95, 164, 167
vapor, 3, 4, 31, 134, 137, 194, 205, 240, 251, 264
variables, 32, 42, 208, 209
variations, 84, 101, 134, 150, 246
varieties, 264
vector, 253
velocity, 8, 13, 33, 35, 46, 145, 146, 147, 148, 178, 180, 182, 183, 184, 185, 188, 220, 226, 229, 246, 252
ventilation, 133
vessels, 12, 13
vibration, 61, 196
Viscosity, v, 42, 45, 65, 69, 96, 109, 110, 111, 115, 116, 117, 118, 125, 129, 138, 141, 142, 162, 163, 164
volatility, 43

W

wall temperature, 166, 169, 175, 194, 198

Washington, 49, 255
waste, 11
waste disposal, 11
water vapor, 208
wavelengths, 242
websites, 14
wettability, 9, 46, 133, 196, 197, 198, 202, 204, 211
wetting, vii, viii, 134, 135, 136, 137, 140
wires, 169, 202
workers, 11, 54, 262
worldwide, vii, 54, 216, 262

X

xanthan gum, 5
XRD, 6, 98

Y

yield, 17, 31, 111, 274

Z

zinc, 16, 18, 26, 203, 241
zinc oxide, 16, 26, 203, 241
zirconia, 202
ZnO, 5, 9, 24, 25, 147, 198, 199, 204, 262